Lyme Borreliosis

NATO ASI Series

Advanced Science Institutes Series

A series presenting the results of activities sponsored by the NATO Science Committee, which aims at the dissemination of advanced scientific and technological knowledge, with a view to strengthening links between scientific communities.

The series is published by an international board of publishers in conjunction with the NATO Scientific Affairs Division

A	Life Sciences	Plenum Publishing Corporation
B	Physics	New York and London
C	Mathematical and Physical Sciences	Kluwer Academic Publishers
D	Behavioral and Social Sciences	Dordrecht, Boston, and London
E	Applied Sciences	
F	Computer and Systems Sciences	Springer-Verlag
G	Ecological Sciences	Berlin, Heidelberg, New York, London,
H	Cell Biology	Paris, Tokyo, Hong Kong, and Barcelona
I	Global Environmental Change	

Recent Volumes in this Series

Volume 256 —Advances in the Biomechanics of the Hand and Wrist
edited by F. Schuind, K. N. An, W. P. Cooney III, and
M. Garcia-Elias

Volume 257 —Vascular Endothelium: Physiological Basis of Clinical Problems II
edited by John D. Catravas, Allan D. Callow, and Una S. Ryan

Volume 258 —A Multidisciplinary Approach to Myelin Diseases II
edited by S. Salvati

Volume 259 —Experimental and Theoretical Advances in Biological Pattern
Formation
edited by Hans G. Othmer, Philip K. Maini, and James D. Murray

Volume 260 —Lyme Borreliosis
edited by John S. Axford and David H. E. Rees

Volume 261 —New Generation Vaccines: The Role of Basic Immunology
edited by Gregory Gregoriadis, Brenda McCormack,
Anthony C. Allison, and George Poste

Volume 262 —Radiolabeled Blood Elements: Recent Advances in Techniques and
Applications
edited by J. Martin-Comin, M. L. Thakur, C. Piera, M. Roca, and
F. Lameña

Series A: Life Sciences

Lyme Borreliosis

Edited by

John S. Axford and David H. E. Rees

St. George's Hospital Medical School
University of London
London, United Kingdom

Springer Science+Business Media, LLC

Proceedings of a NATO Advanced Research Workshop on
Lyme Borreliosis,
held May 19–20, 1993,
in London, United Kingdom

NATO-PCO-DATA BASE

The electronic index to the NATO ASI Series provides full bibliographical references (with keywords and/or abstracts) to more than 30,000 contributions from international scientists published in all sections of the NATO ASI Series. Access to the NATO-PCO-DATA BASE is possible in two ways:

—via online FILE 128 (NATO-PCO-DATA BASE) hosted by ESRIN, Via Galileo Galilei, I-00044 Frascati, Italy

—via CD-ROM "NATO Science and Technology Disk" with user-friendly retrieval software in English, French, and German (©WTV GmbH and DATAWARE Technologies, Inc. 1989). The CD-ROM also contains the AGARD Aerospace Database.

The CD-ROM can be ordered through any member of the Board of Publishers or through NATO-PCO, Overijse, Belgium.

Library of Congress Cataloging-in-Publication Data

Lyme borreliosis / edited by John S. Axford and David H.E. Rees.
        p.   cm. -- (NATO ASI series. Series A, Life sciences ; v. 260)
    "Published in cooperation with NATO Scientific Affairs Division."
    "Proceedings of a NATO Advanced Research Workshop on Lyme Borreliosis, held May 19-20, 1993 in London, United Kingdom"--T.p. verso.
    Includes bibliographical references and index.
    ISBN 978-1-4613-6024-7   ISBN 978-1-4615-2415-1 (eBook)
    DOI 10.1007/978-1-4615-2415-1
    1. Lyme disease--Congresses.   I. Axford, John S.   II. Rees, David H. E.   III. North Atlantic Treaty Organization. Scientific Affairs Division.   IV. NATO Advanced Research Workshop on Lyme Borreliosis (1993 : London, England) V. Series.
    [DNLM: 1. Lyme Disease--congresses.   2. Borrelia burgdorferi--congresses.   WC 406 L9853 1994]
    RC155.5.L93 1994
    616.8'2--dc20
    DNLM/DLC
    for Library of Congress   94-2223
                                                              CIP

Additional material to this book can be downloaded from http://extra.springer.com.

ISBN 978-1-4613-6024-7

©1994 Springer Science+Business Media New York
Originally published by Plenum Press, New York in 1994
Softcover reprint of the hardcover 1st edition 1994

PREFACE

Lyme Borreliosis is a worldwide infectious disease causing a multisystem illness with considerable morbidity, particularly in North America and Europe. The causative agent is the spirochaete *Borrelia burgdorferi*, which is usually transmitted by the ixodid tick from animal reservoirs. This book is formed by contributions from the Second European Symposium on Lyme Borreliosis, held at St George's Hospital Medical School, London in 1993, which reviewed the current state of knowledge of the condition with regard to clinical manifestations, diagnosis, treatment, ecology, epidemiology, biology and immunopathogenesis.

In this book, important data is reviewed concerning the clinical manifestations of Lyme Borreliosis. It seems that strain variation of the spirochaete is the main cause of regional differences seen in the clinical presentation of patients. One striking example of this, is the relatively high incidence of Lyme arthritis in the USA and apparent rarity of this manifestion in some areas of Europe. These important studies open the way for exciting new research that focuses on the immunological and molecular mechanisms that result in disease. A full insight into the ecology of *Borrelia burgdorferi* is essential to a balanced understanding of the disease and a number of excellent reviews on this subject are included. Significant advances with regard to the biology of *Borrelia burgdorferi* and the immunopathogenic mechanisms that result in disease have been made, enabling the role of the B and T lymphocytes in disease to be established and the development of sophisticated phenotyping methods, improved diagnostic tests and effective vaccines.

The meeting brought together most European and several American experts in the field and was a resounding success. This book summarises the proceedings of the Second European Symposium on Lyme Borreliosis and will give the reader up-to-the-minute knowledge on this important and fascinating condition.

JOHN S. AXFORD
DAVID H. E. REES

CONTENTS

The Henry Fuller Lecture 1
   Pathogenesis, Diagnosis and Treatment of Lyme Arthritis
   Professor Allen C. Steere

CLINICAL MANIFESTATIONS AND TREATMENT

Clinical Manifestations of Lyme Borreliosis in an Italian Endemic Region 7
   G. Bianchi, L. Buffrini, P. Monteforte, V. Garzia, M. C. Grignolo,
   G. L. Mancardi, A. Parodi, F. Crovato and G. Rovetta

Lyme Borreliosis in Children 13
   H. J. Christen, F. Hanefield, H. E. Eiffert and R. Thomssen

Lyme Carditis 21
   J. Evans, L. Rosenfeld and R. Schoen

A Recombinant OspA and OspB Based Lyme Disease Vaccine 25
   E. Fikrig, F. S. Kantor, S. W. Barthold and R. A. Flavell

Acrodermatitis Chronica Atrophicans and Lyme Arthritis : Study of Czech
   Lyme Borreliosis Patients Having both Syndromes 33
   J. Hercogova, M. Valesova, M. Tomankova and D. Frösslova

Aspects of Lyme Carditis 39
   M. R. van der Linde and H. J. G. M. Crijns

Borrelia Burgdorferi Shown by PCR from Skin Biopsy Specimen After a Fly
   Bite 45
   J. Oksi, I. Helander, H. Aho, M. Marjamäki and M. K. Viljanen

Lyme Arthritis : Experience from Somerset, England 49
   T. G. Palferman

No Evidence to Implicate *Borrelia Burgdorferi* in the Pathogenesis of Dilated
Cardiomyopathy in the UK 55
D. H. E. Rees, P. Keeling, J. McKenna and J. S. Axford

Evidence for Lyme Disease in Urban Park Workers : A Potential New Health
Hazard for City Inhabitants 61
D. H. E. Rees and J.S. Axford

Management of a Deer Tick Bite 69
S. K. Sood, M. B. Salzman, L. Carmody, L. G. Rubin and J. Piesman

ECOLOGY & EPIDEMIOLOGY

Lyme Borreliosis in Australia 75
R. D. Barry, B. J. Hudson, D. R. Shafren and M. C. Wills

Geographic Diversity of Lyme Borreliosis 83
G. Bianchi and G. Rovetta

Parallelism Between Lyme and Tick-Borne Encephalitis (TBE) Sero-
Epidemiology Following Occupational Exposure in South Germany 89
J. Clement, H. Leirs, P. McKenna, V. Armour, D. Ward, J. Groen,
A. Osterhaus and C. Kunz

Investigation of Genetic Changes Associated with Attenuation of *Borrelia
Burgdorferi* by *In Vitro* Cultivation 95
A. R. Emilianus, K. J. Cann, D. J. M. Wright and L. C. Archard

Phytoecological Mapping of *Ixodes Ricinus* as an Approach to the Distribution
of Lyme Borreliosis in France 105
B. Gilot, C. Guiguen, B. Degeilh, B. Doche, J. Pichot and J. C. Beaucournu

The Ecology of Lyme Borreliosis in Sweden 113
T. G. T. Jaenson, S. Bergström, H. A. Mejlon, L. Noppa, B. Olsen and
L. Talleklint

Lyme Disease in Scotland - Results of a Serological Study in Sheep 117
G. B. B. Mitchell and I. W. Smith

The Ecology of Lyme Borreliosis in the UK 125
P. Nuttal, S. Randolph, D. Carey, N. Craine, A. Livesley and L. Gern

The relative Contributions of Transovarial and Transstadial Transmission to the
Maintenance of Tick-Borne Diseases 131
S. E. Randolph

Epidemiology of Lyme Disease in Italy 135
    G. Trevisan, J. Simeoni, P. Conci, A. M. Bassot, C. Nobile, G. Stinco,
    G. Bianchi and G. Rovetta

Role of Host Density in the Ecology of Lyme Disease 139
    T. E. Awerbuch and A. Spielman

Borrelia Burgdorferi Studies in Man and Ticks in Scotland 147
    S. M. Curtin & T. H. Pennington

BIOLOGY AND IMMUNOPATHOGENESIS

Expression of Public Idiotypes in Patients with Lyme Arthritis 155
    J. S. Axford, R. A. Watts, A. A. Long, D. A. Isenberg and A. C. Steere

Cellular Immune Reactions to *Borrelia Burgdorferi* - The T-cell-Macrophage
    Axis 169
    G. R. Burmester, T. Häupl and M. Rittig

OspB Sequence Variation to *Borrelia Burgdorferi* along the Coast of Maine 175
    D. A. Caporale, T. D. Kocher, R. P. Smith, P. W. Rand and E. H. Lacombe

Multiple Amino Acid Sequence Alignment of the Major Outer Surface Proteins
    OspA and OspB of Various *Borrelia Burgdorferi* Strains 181
    W. Fellinger, M. Reindl, G. Stöffler and B. Redl

Lyme Disease in an Experimental Cat Model 187
    M. D. Gibson, C. R. Young, M. T. Omran, K. Palma, J. F. Edwards and
    J. A. Rawlings

Borrelia Burgdorferi Infection in Mice : Aspects of Inflammation and Immune
    Responses 201
    N. Honarvar, E. Böggemeyer, C. Galanos, M. Modolell, D. Vestweber,
    R. Wallich, M. D. Kramer, U. E. Schaible and M. M. Simon

Chemotaxonomy of *Borrelia* 211
    M. A. Livesley and P. A. Nuttall

Phenotypic and Genotypic Analysis of Chinese *Borrelia Burgdorferi* from
    Various Sources 217
    L. Muqing, W. Jianhui and Z. Zhefu

Lyme Disease in an Experimental Mouse Model 227
    S. Reddy, M. D. Gibson, J. Rawlings, G. Stoica and C. R. Young

Sectional Uptake and Cytosolic Processing of *Borrelia Burgdorferi* by Human
    Phagocytes 241
  M. G. Rittig, M. Kressel, T. Haupl and G. R. Burmester

Physical and Genetic Maps of the *Borrelia Burgdorferi* Sensu Lato
    Chromosomes 249
  I. Saint Girons, I. G. Old, C. Ojaimi, J. MacDougall and B. E. Davidson

Repeated DNA Sequences on Circular and Linear Plasmids of *Borrelia*
    Burgdorferi Sensu Lato 253
  W. R. Zückert, E. Filipuzzi-Jenny, J. Meister-Turner,
  M. Stalhammar-Carlemalm and J. Meyer

Biochemical and Immunological Analysis of a Polymorphic Low-Molecular-
    Weight Lipoprotein of *Borrelia Burgdorferi* 261
  R. Wallich, C. Helmes, U. E. Schaible, M. D. Kramer and M. M. Simon

DIAGNOSIS

Detection of Lyme Disease Spirochaete DNA in Clinical Samples 269
  K. J. Cann, M. L. Wilson, C. Akintunde, L. Archard and D. J. M. Wright

Clinical and Serological Study of Lyme Borreliosis in a Population of
    Neurological Patients 279
  E. Capello, G. Bianchi, P. Monteforte, L. Buffrini, A. Schenone, S. Ratto,
  N. Dagnino, G. Rovetta and G. L. Mancardi

Pitfalls in the Laboratory Diagnosis of Lyme Borreliosis 285
  S. J. Cutler and D. J. M. Wright

Serodiagnosis of Lyme Disease in the UK 291
  E. Guy, I. Ferguson, R. Sorouri-Zanjani and S. O'Connell

PCR-Based Detection of CSF *Borrelia Burgdorferi* as a Predictor of Treatment
    Response in CNS Lyme Borreliosis 295
  J. Halperin, T. Keller and M. Whitman

Diagnostic Detection of *Borrelia Burgdorferi* DNA by the Polymerase Chain
    Reaction (PCR) 303
  A-M. Lebech

Improved Detection of Immunoglobulin M in Sera of Erythema Migrans
    Patients by Western Blotting with a Local *Borrelia Burgdorferi* Skin
    Isolate 307
  S. Rijpkema, H. Kuiper, M. Molkenboer, A. van Dam and J. Schellekens

Differentiation of *Borrelia burgdorferi* Isolates from Ticks and Humans by
Different Monoclonal Antibodies in Immunofluorescence 315
A. Schönberg and C. Loser

The use of PCR in the Direct Detection of *Borrelia Burgdorferi* from *Ixodes
Dammini* 321
R. Sun, S. L. Barmat, S. H. McQuilkin, L. S. Risley and R. Diaco

Borrelia Burgdorferi Detected in the Blood, Synovium and Skin of Patients
with Lyme Arthritis 327
M. Valešová, J. Hercogová and D. Hulínská

Index 331

PATHOGENESIS, DIAGNOSIS AND TREATMENT OF LYME ARTHRITIS

Professor Allen C Steere

Chief, Rheumatology/Immunology, New England Medical Center
Tufts University School of Medicine, Boston, MA 02111, USA

ABSTRACT

Lyme arthritis was recognised as a new nosological entity in 1975 because of geographic clustering of the disease in children in Lyme, Connecticut, who were thought to have juvenile rheumatoid arthritis (1). Joint disease in these children was characterised by brief, recurrent attacks of asymmetric swelling and pain in a few large joints, especially the knee. It then became apparent that Lyme arthritis was part of a multisystem illness that usually began with a characteristic annular skin lesion, erythema migrans (2). This skin lesion had been described previously in Europe, but it had not been associated there with arthritis.

It is now known that joint involvement may be a feature of Lyme borreliosis in both the United States and Europe, but it seems to be a less common manifestation of the illness in Europe. This is probably owing to strain variations in the spirochaete. Subgroup 1 strains of Borrelia burgdorferi seem to be particularly arthritogenic. To date, only this strain has been isolated in the United States, whereas all three subgroups of the spirochaete have been found in Europe (3).

NATURAL HISTORY

During the late 1970s, before the role of antibiotic treatment was known in Lyme disease, we studied prospectively 55 patients with erythema migrans and the initial manifestation of their illness to determine the natural history of the subsequent arthritis (4). Of the 55 untreated patients, 11 (20%) had no later manifestations of Lyme disease. 10 of the patients (18%) subsequently had intermittent episodes of joint, periarticular, or musculoskeletal pain, but they never developed objective joint abnormalities. After a period of weeks to months 28 patients (51%) had one episode or multiple intermittent attacks of frank arthritis.

Most patients developed large knee effusions, but other large joints , the temporomandibular joints, or periarticular sites were sometimes affected. Attacks of joint swelling or periarticular pain were often brief, lasting only weeks. In most cases only one or two joints or periarticular sites were affected at a time. The remaining 6 patients (11%) developed chronic synovitis later in the illness; of these, 2 (4%) had erosions, and 1 (2%) had permanent joint disability.

The number of patients who continued to have attacks of arthritis decreased by about 10-20% each year (4). Attacks of knee swelling sometimes became longer during the second or third year of illness, however, lasting months rather than weeks. It was during this period that a small percentage of patients developed chronic arthritis, which we have defined as one year or more of continuous joint inflammation. Brief episodes of joint pain sometimes occurred after frank arthritis had disappeared (5). The total cumulative duration of arthritis was significantly shorter in young children than in adults.

PATHOGENESIS

In affected patients B burgdorferi probably spreads to joints during the first days or weeks of the illness. Such patients often appear systemically ill and may have vague, migratory musculoskeletal pain in bursae, tendons, muscle, bone or joints. The specific immune response to the spirochaete is usually minimal (6,7). Early responses are often restricted to the 41 kilodalton flagella antigen, the 21 kilodalton outer surface protein C (OspC), or the 58 and 66 kilodalton heat shock proteins of the spirochaete. The specific IgM response, which usually peaks between the third and sixth week of infection, is often associated with polyclonal activation of B cells, including raised total serum IgM levels (8), and the presence of cryoprecipitates, circulating immune complexes (9), and occasionally, rheumatoid factor (10), antinuclear antibodies, or anti-cardiolipin antibodies (11). In most instances, these non-specific immune phenomena are not thought to cause symptoms.

Months later the disease seems to localise to one or a few joints, which become markedly inflammed and swollen. By this time patients usually have strong cellular and humoral immune responses to 12 or more spirochaetal polypeptides (6,7). In the United States the final point in expansion of the antibody response is often the recognition of outer surface proteins A and B (OspA and OspB) near the beginning of prolonged episodes of arthritis, a median of 18 months after disease onset (12). In preliminary analysis, expression of IgA antibodies bearing idiotype 16/6 was significantly increased during episodes of arthritis, but not before that time (13).

As in rheumatoid arthritis, the synovial lesion in Lyme arthritis shows synovial cell hyperplasia, vascular proliferation, and a heavy infiltration of mononuclear cells (14) - a histological picture that is suggestive of a delayed hypersensitivity immune response. In a study by Yssel et al all 18 T cell clones studied from patients with Lyme arthritis produced a T helper type 1-like cytokine hypersensitivity response which seems to be important in the control of intracellular pathogens. Preliminary evidence suggests that B burgdorferi spriochaetes may survive in fibroblasts and in several other eukaryotic cell types (16), but the mechanism of their survival in synovial tissue is not yet known. In several instances B burgdorferi has been cultured from synovial fluid (17) or seen in synovial tissue (14), but it has been difficult to demonstrate the spirochaete within the joint by these methods.

It is hoped that the polymerase chain reaction may provide a sensitive method to show the presence of the spirochaete within the joint.

B burgdorferi is a potent inducer of interleukin 1 (IL-1) (18), a cytokine with pro-inflammatory effects on synoviocytes and chondrocytes that may lead to destruction of cartilage and bone. Because episodic attacks are characteristic of Lyme arthritis, we studied the clinical course of Lyme arthritis in 83 patients in relation to the synovial fluid concentrations of IL-1ß and its natural receptor antagonist (IL-1ra) (19). Patients with high concentrations of IL-1ra and low concentrations of IL-1ß had rapid resolution of attacks of arthritis, whereas patients with the reverse pattern of cytokine concentrations had long intervals to recovery. In contrast, concentrations of tumour necrosis factor, a cytokine that shares many biological features with IL-1, showed no correlation with the course of arthritis. Thus the balance between IL-1ß and IL-1ra would appear to be important in recovery from episodes of arthritis.

The fact that only a small percentage of patients develop chronic Lyme arthritis suggests that host factors may determine the severity and duration of the arthritis. To learn whether genetically determined variations in the host immune response might account for such outcomes we determined the immunogenic profiles of 130 patients with various manifestations of Lyme disease (20). Of the 80 patients with arthritis, 57% of those with chronic arthritis had the HLA-DR4 specificity and 43% had HLA-DR2; altogether, 89% had HLA-DR2 or HLA-DR4, or both, as compared with 27% of those with only short attacks of arthritis (relative risk 22; $p=0.00006$). Furthermore, the presence of HLA-DR4 in patients with arthritis was associated with lack of response to antibiotic treatment ($p=0.01$). In these 80 patients we subsequently showed that only the combination of the HLA-DR4 specificity and antibody reactivity with outer surface proteins A and B (OspA and OspB) of the spirochaete was associated with this prolonged resistance to treatment (12). In inbred strains of mice an association has also been noted between chronic arthritis and certain H-2 haplotypes, the equivalent of the human HLA-DR allele (21). These observations suggest that certain Class II major histocompatibility genes determine a host immune response that results in chronic arthritis and lack of response to antibiotic treatment.

It is not yet clear whether the situation is the same in Europe. Other subgroups of B burgdorferi more commonly cause infection there; antibody responses to the OspA and OspB proteins are unusual, and the occurrence of arthritis, particularly chronic arthritis, seems to be rare.

DIAGNOSIS

The diagnosis of Lyme disease is based upon recognition of a characteristic clinical picture, a history of exposure in an endemic area, and after the first weeks of infection a raised antibody response to B burgdorferi. Patients with Lyme arthritis usually have recurrent, brief attacks of oligoarticular arthritis in large joints accompanied by markedly raised antibody responses to B burgdorferi; western blots frequently show reactivity with 12 or more spirochaetal polypeptides (7). In Europe, however, where infection may occur with any one of the three subgroups of B burgdorferi, the antibody response may be more restricted. On both continents serological testing for Lyme disease is not standardised and different laboratories may get different results.

In addition, patients with past infection often remain seropositive for years; and some patients, particularly in Europe, may have asymptomatic infection. If these patients develop another illness, especially if it is another type of arthritis, the positive serological test for Lyme disease may cause diagnostic confusion.

In adults Lyme arthritis is most like Reiter's syndrome or reactive arthritis, which are types of arthritis associated with HLA-B27. Unlike Reiter's syndrome, we have not seen chronic enthesopathy, sacroiliitis, or urethritis in Lyme arthritis. In children Lyme arthritis is most similar to pauciarticular juvenile rheumatoid arthritis. Lyme arthritis can usually be distinguished from these entities by the brevity of initial attacks and by the markedly increased antibody response to B burgdorferi.

A small percentage of patients develop fibromyalgia or chronic fatigue soon after Lyme disease, suggesting that B burgdorferi is one of the infectious agents that may trigger these disabling syndromes (22). Even when triggered by Lyme disease, these syndromes, in our experience, do not respond to antibiotic treatment. Chronic fatigue or fibromyalgia of unknown cause is all too often misdiagnosed as Lyme disease and treated inappropriately with prolonged courses of antibiotics (23).

TREATMENT

Lyme arthritis can usually be treated successfully with standard oral or intravenous antibiotic regimens, including doxycycline 100mg twice a day for 30 days; amoxycillin and probenecid 500mg of each four times a day for 30 days'; or intravenous ceftriaxone 2g once a day for two weeks (24). The response may be slow, however, and several months are sometimes required for the complete resolution of arthritis. Regardless of the antibiotic used, the route of administration, or the number of courses given, a small subset of patients with Lyme arthritis does not respond to antibiotic treatment. These patients have an increased frequency of the HLA-DR4 specificity and antibody reactivity to outer surface proteins A and B (OspA and OspB) of the spirochaete (12). We treat this group of patients with non-steroidal anti-inflammatory agents or intra-articular steroids, or if necessary, with arthroscopic synovectomy (25).

REFERENCES

(1) Steere AC, Malawista SE, Snydman DR et al. Lyme arthritis an epidemic of oligoarticular arthritis in children and adults in three Connecticut communities. Arthritis Rheum (1977); 20 : 7-17.

(2) Steere AC, Malawista SE, Hardin JA, Ruddy S, Askenase PE, Andiman WA. Erythema chronicum migrans and Lyme arthritis : the enlarging clinical spectrum. Ann Intern Med (1977); 86 : 685-98

(3) Baranton G, Postic D, Saint Girons I, et al. Delineation of Borrelia burgdorferi sensu stricto, Borrelia garinii sp. nov., and group VS461 associated with Lyme borreliosis. Int J Sysy Bacteriol (1992); 42 : 378-83.

(4) Steere AC, Schoen RT, Taylor E. The clinical evolution of Lyme arthritis. Ann Intern Med (1987); 107 : 725-31.

(5) Szer IS, Taylor E, Steere AC. The long-term course of children with Lyme arthritis. N Engl J Med (1991); 325 : 159-63.

(6) Yoshinari NH, Reinhardt BN, Steere AC. T cell responses to polypeptide fractions of Borrelia burgdorferi in patients with Lyme arthritis. Arthritis Rheum (1991); 34 : 707-13.

(7)   Dressler F, Whalen JA, Reinhardt BN, Steere AC. Western blotting in the serodiagnosis of Lyme disease. J Infect Dise (1993); 167 : 392-400.

(8)   Steere AC, Hardin JA, Ruddy S, Mummaw JG, Malawista SE. Lyme arthritis : correlation of serum and cryoglobulin IgM with activity and serum IgG with remission. Arthritis Rheum (1979); 22 : 471-83.

(9)   Hardin JA, Steere AC, Malawista SE. Immune complexes and the evolution of Lyme arthritis : dissemination and localization of abnormal Clq binding activity. N Engl J Med (1979); 301 : 1358-63.

(10)   Kujala GA, Steere AC, Davis JS IV. IgM rheumatoid factor in Lyme disease : correlation with disease activity, total serum IgM, and IgM antibody to Borrelia burgdorferi. J Rheumatol (1987); 14 : 772-6.

(11)   Mackworth-Young CG, Harris EN, Steere AC et al. Anti-cardiolipin antibodies in Lyme disease. Arthritis Rheum (1988); 31 : 1052-6.

(12)   Kalish RA, Leong JM, Steere AC. Delay in the immune response to outer-surface proteins (Osp) A and B of B burgdorferi : correlation with arthritis and treatment failure in susceptible patients with Lyme disease (Abstract). Arthritis Rheum (1991); 34 : S43.

(13)   Axford JS, Watts RA, Long AA, Isenberg DA, Steere AC. Expression of public idiotypes in patients with Lyme arthritis. Ann Rheum Dise (1993); 52: 199-205.

(14)   Steere AC, Duray PH, Butcher EC. Spirochaetal antigens and lymphoid cell surface markers in Lyme synovitis : comparison with rheumatoid synovium and tonsillar lymphoid tissue. Arthritis Rheum (1988); 31 : 487-95.

(15)   Yssel H, Shanafelt M-C, Soderberg C, Schneider PV, Anzola J, Peltz G. Borrelia burgdorferi activates a T helper type 1-like T cell subset in Lyme arthritis. J Exp Med (1991); 174 : 593-601.

(16)   Georgilis K, Peacocke M, Klempner MS. Fibroblasts protect the Lyme disease spirochaete, Borrelia burgdorferi, from ceftriaxone in vitro. J Infect Dis (1992); 166 : 440-4.

(17)   Schmidi J, Hunziker T, Moesli P, Schaad UB. Cultivation of Borrelia burgdorferi from joint fluid three months after treatment of facial palsy due to Lyme borreliosis. J Infect Dis (1988); 158 : 905-6.

(18)   Miller LC, Isa S, Vannier E, Georgilis K, Steere AC, Dinarella CA. Live Borrelia burgdorferi preferentially activate IL-1ß gene expression and protein synthesis over the interleukin-1 receptor antagonist. J Clin Invest (1992); 90 : 906-12.

(19)   Miller LC, Lynch EA, Isa S, Logan JW, Dinarello CA, Steere AC. Balance of synovial fluid IL-1ß receptor antagonist and recovery from Lyme arthritis. Lancet (1992); 341 : 146-8.

(20)   Steere AC, Dwyer E, Winchester R. Association of chronic Lyme arthritis with HLA-DR4 and HLA-DR2 alleles. N Engl J Med (1990); 323 : 219-23.

(21)   Schaible UE, Kramer MD, Wallich R, Tram T, Simon MM. Experimental Borrelia burgdorferi infection in inbred mouse strains : antibody response and association of H-2 genes with resistance and susceptibility to development of arthritis. Eur J Immunol (1991); 21 : 2397-405.

(22)   Dinerman H, Steere AC. Lyme disease associated with fibromyalgia. Ann Intern Med (1992); 117 : 281-5.

(23)   Steere AC, Taylor E, McHugh GL, Logigian EL. The overdiagnosis of Lyme disease. JAMA In Press

(24)   Steere AC. Lyme Disease. N Engl J Med (1989); 321 : 586-96.

(25)   Schoen RT, Aversa JM, Rahn DW, Steere AC. Treatment of refractory chronic Lyme arthritis with arthroscopic synovectomy. Arthritis Rheum (1991); 34 : 1056-60.

CLINICAL MANIFESTATIONS OF LYME BORRELIOSIS

IN AN ITALIAN ENDEMIC REGION

Gerolamo Bianchi, Laura Buffrini, Patrizia Monteforte, Vincenzo Garzia,
Maria Clara Grignolo, Gian Luigi Mancardi *, Aurora Parodi **,
Franco Crovato ***, Guido Rovetta

Italian Group for the Study of Lyme Borreliosis (IGSLB)
DIMI - Division of Rheumatology, Institute E.Bruzzone-Rheumatological
Centre,* Institute of Clinical Neurology ,** Institute of Dermatology, University
of Genova,*** Division of Dermatology, U.S.L. XVIII, Chiavari, Italy

INTRODUCTION

Lyme borreliosis (LB) is a multi-system disease caused by the tick - trasmitted spirochete *Borrelia burgdorferi* (Bb). The disease is characterized by early and late phases and by clinical manifestations that may involve the skin, heart, joints, eyes, central and peripheral nervous system.

LB is endemic in Italy in two areas: Liguria, located in the north-west of the country, where the disease was first described (1), and Friuli-Venezia-Giulia located in the north-east of Italy.

In 1990 Italian researchers involved with LB founded the Italian Group for the Study of Lyme Borreliosis (IGSLB) with the aim to collect clinical and epidemiological data related to LB (2). This study concerns to LB patients resident in Liguria, one of the Italian endemic areas.

PATIENTS AND METHODS

During the last eight years 1422 patients complaining of symptoms compatible with a Bb infection were referred to us by dermatologists, neurologists, and general practitioners, or they were seen at our clinic for the evaluation of musculo-skeletal symptoms. The reasons for referral were cutaneous in 45% of patients, neurologic in 15%, articular in 38%, and cardiac in 2%. We think that the fairly homogeneous distribution of the cause of consultation should have avoided significant selection bias in our casistic.

The set of criteria (table 1) we used to diagnose LB was: residence in an endemic area or a documented tick bite, the presence of antibodies against Bb in the serum, CSF and synovial fluid, and the presence of at least two of the clinical manifestations of Bb infection: systemic, articular, neurological and cardiac, with exclusion of other recognizable diseases (3).

Alternatively, residence in an endemic area or a documented tick bite, and one of the typical cutaneous lesions: Erythema migrans (EM), lymphocytoma cutis (LC), and acrodermatitis chronica atrophycans (ACA) led us to make diagnosis of LBp 12X cutis (LC), and acrodermatitis.

Table 1. Diagnostic Criteria

1 Endemic area
2 Tick bite
3 Antibody titer (IgG) against Bb ≥ 256
4 Two manifestations of LB: systemic,articular,neurological,cardiac
5 EM, LC, ACA

A : 1 or 2 plus 3 and 4
B : 1 or 2 plus 5
* esclusion of other recognizable diseases.

Antibodies against Bb were detected by means of indirect immunofluorescence (IFA), and later by ELISA. Specimens were pre-absorbed with Treponema phagedenis to avoid cross-reactions (4).

RESULTS

Lyme borreliosis was confirmed in fourteen percent of the patients, one-hundred and twenty-one were women and seventy-four were men. Mean age was forty-six years, with a range between three and eighty-five years. In thirty-one patients, sixteen percent, we did not detect antibodies against Bb. Twenty-one of seronegative patients had EM, eight had LC and two had ACA.

As regards the clinical involvement observed forty-five percent of patients had dermatologic manifestations, thirty-nine percent articular, thirty-one neurologic, and three

Table 2. Clinical involvement

	n	%
Dermatologic	89	45
Articular	76	39
Neurologic	63	31
Cardiac	4	3

percent cardiac manifestations of the disease (table 2). We also observed one female suffering from an uveitis.

EM was the most frequently observed cutaneous manifestation in our patients: it was present in sixty-two people, corresponding to thirty-one percent of the one-hundred and ninety-five patients. ACA was reported by only two patients (1%), and lymphocytoma by eighteen (9%). Lichen sclerosus et atrophicans was present in eight patients (5%).

In the majority of these patients, dermatologic involvement was the only clinical manifestation of the disease (80%). EM was associated to other organ system manifestations in twenty percent of dermatologic patients.

The prevalence of arthritis, defined as joint pain and swelling observed by a physician, in this series of patients was seventy-six out of one-hundred and ninety-five, corresponding to thirty-nine percent.

The pattern of articular involvement in our patients confirms that arthritis of LB is more often a mono- or oligo-arthritis. Knees were the most frequently involved joints, accounting for forty-seven percent of arthritis, followed by wrists with twenty-five percent. Chronic arthritis was present in ten out of seventy-six patients, the thirteen percent. Arthralgias were reported by one-hundred and twenty-seven out of one-hundred and ninety-five of our patients, sixty-five percent, and were most frequently poly-articular (table 3).

Table 3. Articular manifestations

	n	%
Arthralgias	127/195	65
Mono-arthritis	27	35
Oligo-arthritis	33	46
Poly-arthritis	16	19
Total	76	100

Arthritis constituted the only clinical manifestation in fifty-one patients with articular involvement. However, arthritis was accompanied by other clinical manifestations in twenty-five patients, and in these cases it was most frequently associated to neurologic involvement, especially cranial neuritis and late neurologic disease (table 4).

Neurologic manifestations occurred in sixty-three patients. Cranial neuritis was the most frequently observed abnormality (52%), accounting for half of the neurologic patients. Radiculoneuritis was present in eleven cases (18%), while nine patients (14%) complained of classical meningopolyneuritis (Bannwarth syndrome). Central nervous system involvement was present in ten (16%) of the patients.

Again, the majority of patients (70%) had neurologic involvement as the only manifestation of LB. As mentioned before, neurological symptoms were frequently accompanied by arthritis (25%).

Cardiac abnormalities were observed in four cases (3%), all these patients had atrioventricular conduction disturbances. To date we have not yet observed patients with myocarditis.

Table 4. Clinical manifestations in patients with articular involvement

	n	%
Only arthritis	51	67
Arthritis and :		
neurologic	11	14
EM	8	10
EM and neurologic	2	3
EM and cardiac	1	2
EM, neurologic and cardiac	3	4
Total	76	100

If we consider the number of clinical manifestations observed in our casistic, it is clear that the majority of patients (166/195, 85%) presented only one manifestation of the disease, and of these cases seventy-one had only skin involvement, fifty-one arhtritis, and forty-four neurologic manifestations.

Twenty-three patients were affected by two manifestations, and these were most frequently represented by articular and neurological disease.

Patients with three and four clinical manifestations of LB represented, in our casistic, a minority, thus suggesting a less serious clinical pattern of Bb infection in Italian patients.

CONCLUSIONS

In our patients, as previously reported (5), the prevalence of dermatologic involvement is lower than in the USA but the same as in Europe. The prevalence of arthritis is half-way between USA and Europe. The prevalence of neurologic manifestations is higher than in the USA and is not different from other European countries.

Our results suggest that LB in Liguria may be less serious than in other areas. This could be due to the presence of different strains of Bb, which could have less pathogenic capacity. Furthermore, this difference could be due to variations in HLA-related host response. However, it must be taken into account that ascertainement bias might play a role.

Bias may be due both to selection of patients by different specialists and to low sensitivity and specificity of currently used diagnostic criteria. Consideration must also be given to the fact that Liguria is an endemic region for Rickettsia conorii and general pratictioners habitually treat subject bitten by ticks with tetracycline. This could also explain the differences in the clinical pattern of LB in our region.

REFERENCES

1. Crovato F., Nazzari G., Fumarola D., Rovetta G., Cimmino M., Bianchi G. Lyme disease in Italy: first reported case.Ann. Rheum. Dis. 1985; 44:570.
2. Rovetta G., Trevisan G., Cinco M., Crovato F., Parodi A., .Mancardi G.L., Fumarola D., Bianchi G. Italian group for the Lyme Borreliosis. Epidemiology of Lyme Borrelisis in Italy. Proceedings and

abstracts of the Fifth International Conference on Lyme Borreliosis, Washington, D.C., May 30-June 2, 1992. Bethesda,Md: Fed. Am. Soc. Exp. Biol. 1992; A340.

3. Bianchi G., Rovetta G., Monteforte P et al. Articular involvement in european patients with Lyme disease. A report of 32 italian patients. Br. J. Rheum. 1990; 29:178-80

4. Bianchi G., Rovetta G., Buffrini L. et al. Lyme arthritis: evaluation of antibodies against Borrelia burgdorferi in IFA and ELISA methods. Hungarian Rheumatology 1991; 32: 89.

5. Monteforte P., Rovetta G., Buffrini L., Garzia V., Parodi A., Mancardi G.L., Crovato F., Bianchi G. Clinical manifestations of Lyme Borreliosis in Italy. Proceedings and abstracts of the Fifth International Conference on Lyme Borreliosis, Washington, D.C., May 30-June 2, 1992. Bethesda,Md: Fed. Am. Soc. Exp. Biol. 1992; A1.

LYME BORRELIOSIS IN CHILDREN

Hans-Jürgen Christen[1], Folker Hanefeld[1], Helmut Eiffert[2]
and Reiner Thomssen[2]

[1] Department of Pediatrics
[2] Department of Medical Microbilogy
University Hospital
D-3400 Goettingen, F.R.G.

Lyme borreliosis is a common infectious disorder in children. The daily life and play routines in particular make them more likely than adults to be bitten by ticks and thus more likely to be infected by Borrelia burgdorferi. Lyme borreliosis as a clinical entity was first recognized in children suffering from arthritis (1). In the meantime, research in this field has mainly been focused on the disease in adults. Epidemiological data about the true incidence of Lyme borreliosis in childhood are still sparce. The whole clinical spectrum of Lyme borreliosis described in adults is also found in children, but age-specific differences in the relative frequency of certain manifestations, in the course of the disease as well as in diagnostic pecularities are evident (2-9).

PREVALENCE OF ANTIBODIES AGAINST BORRELIA BURGDORFERI

In a serosurvey the prevalence of antibodies against Borrelia burgdorferi in children from Lower Saxony/Germany was investigated. Blood samples of 574 healthy children were collected over a two-years-period. The infection rate measured by IgG antibodies was 2.6%. There was a clear increase beyond the age of 5 years (Table 1). IgG-positive cases in the first year of life reflect maternal antibody transfer. IgM antibodies were found in 1 to 3% of the cases.

In the same way the serum of delivering women and the cord blood samples was tested for specific IgM and IgG antibodies against Borrelia burgdorferi. The study included two obstetrical departments, one with a predominantely rural catchment area (n = 517) and another with an urban clientele exclusively (n = 2131). The infection rate was the same in both groups, resulting in 0.8% for specific IgM antibodies and 7.0% for specific IgG antibodies. In conclusion, infection rate in childhood is clearly lower than in adults. This means that in the individual case a positive test result is of higher diagnostic value in children than in adults.

Lyme Borreliosis, Edited by J.S. Axford and
D.H.E. Rees, Plenum Press, New York, 1994

Table 1. Age-specific prevalence of IgG and IgM antibodies against Borrelia burgdorferi in healthy children from Lower Saxony/Germany

Age	Children tested n	IgG-positive (serum) %	IgM-positive (serum) %
< 1 y.	68	2.9	0.0
1 - 2 y.	162	1.2	0.6
3 - 4 y.	112	0.9	0.0
5 - 7 y.	122	4.9	2.5
> 7 y.	110	3.6	1.8
Total	574	2.6	0.7

PROSPECTIVE STUDY ON THE EPIDEMIOLOGY AND CLINICAL MANIFE-STATIONS OF LYME BORRELIOSIS IN CHILDHOOD

A prospective hospital-based multicentre study on Lyme borreliosis in children was started in 1986 including seven pediatric departments which supply a geographically well-defined region in the Northeast of Germany with a population of 350,000 children (10). For certain questions pediatric cases from other hospitals in West Germany were included who were examined by the same serological methods in our laboratory. Special emphasis was laid on neurological disorders. Lyme borreliosis was diagnosed on the basis of the detection of specific antibodies against Borrelia burgdorferi using an immunoglobulin M (IgM) capture assay developed by Wassmann er al. (11) and an IgG enzyme-linked immunosorbent assay (ELISA) according to Eiffert et al. (12). Diagnosis of neuroborreliosis was established by detection of specific IgM antibodies in CSF.

Neuroborreliosis in childhood

The study examined 208 children with Lyme borreliosis (Table 2), of whom 169 had Lyme neuroborreliosis which was diagnosed by the detection of specific IgM-antibodies in the cerebrospinal fluid.

Facial palsy and aseptic meningitis accounted for nearly 90% of all cases with neuroborreliosis (Figure 1). Meningoradiculoneuritis with peripheral nerve involvement (Bannwarth's syndrome) was diagnosed only in 3.6% of the children (n = 6), although this is the most frequent symptom of Lyme neuroborreliosis in adult patients. Unusual neurological syndromes in which borrelial infection was diagnosed included isolated abducens nerve palsy, oculomotor palsy, acute ataxia, chronic headache and three cases with the typical features of Gullian-Barré syndrome. In conclusion, there are some striking differences in the clinical spectrum between children and adults with facial palsy and aseptic meningitis as predominant features of neuroborreliosis in childhood.

Table 2. Prospective multicenter study on Lyme borreliosis in childhood

Disorders	Total n	thereof: Lyme borreliosis n
Neurologic disorders	**547**	**169**
Acute facial palsy[1]	218	93
Aseptic meningitis[2]	283	46
Recurrent headache	7	7
Meningoencephalitis	7	6
Bannwarth's syndrome[3]	6	6
Guillain-Barré's syndrome	5	3
Abducens nerve palsy	4	2
Isolated ptosis	1	1
Acute ataxia	2	1
Pseudotumor cerebri	2	1
Multiple sclerosis	9	--
Other neurologic symptoms	3	3
Dermatologic disorders	**30**	**30**
Erythema migrans	24	24
Borrelial lymphocytoma	5	5
Morphea	1	1
Acute arthritis	**15**	**9**
Gonarthritis	10	7
Oligoarthritis	2	2
Polyarthritis	3	--
Total	**592**	**208**

[1]Cases in which the facial palsy was the first and dominant symptom, and in which there were no clinical signs of peripheral radiculoneuritis

[2]The concept of "aseptic meningitis" describes those cases of meningitis with a supposedly infectious genesis, but for which there are neither microscopic findings nor bacteriological cultures which prove it

[3]Cases with clinical signs of peripheral radiculoneuritis

It was of special interest to investigate the relative frequency of Lyme borreliosis in these disorders in childhood. For this epidemiological approach only cases from Lower Saxony were studied.

The yearly incidence of Lyme neuroborreliosis in Lower Saxony was 5.8/100,000 children ages 1 to 13. The manifestation index was 0.16, compared with the presence of antibodies against Borrelia burgdorferi for children in the same age group and region. This means one case of Lyme neuroborreliosis per 620 infected children.

In a three-years-period (1987 to 1989) 74 consecutive cases with acute peripheral facial palsy were studied for Lyme borreliosis. Lyme borreliosis proved to be the most frequently verifiable cause of acute peripheral facial palsy in children. In one third of the cases (32.9%) borrelial infection was diagnosed by detection of specific IgM antibodies in CSF. In contrast to facial palsy of unknown etiology cases with Lyme borreliosis occurred between May and November only. In this period Lyme borreliosis accounted for nearly every second case of facial palsy in children (25 of 53 cases from May to November). Bilateral facial palsy was, without exception, found to be caused by Lyme borreliosis; thus it can be considered a specific neurological sign of this infection.

In the three-years-period (1987 to 1989) 240 consecutive cases with aseptic meningitis were studied (Figure 2). In 11.7% of these children Borrelia infection was diagnosed. Lyme borreliosis was the third most common cause of aseptic meningitis in childhood. It accounted for one third of the cases with known etiology and was as freqent as mumps. Aseptic meningitis caused by Borrelia infection appeared solely from June to November, significantly different from cases with aseptic meningitis from other causes (p < 0.05; Fisher's exact test).

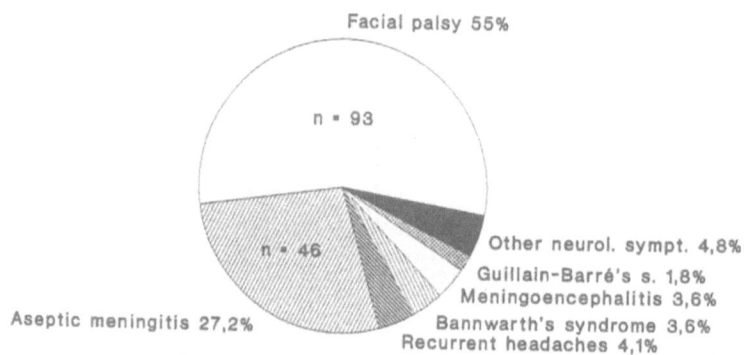

Figure 1. Clinical spectrum of Lyme neuroborreliosis in children (n = 169)

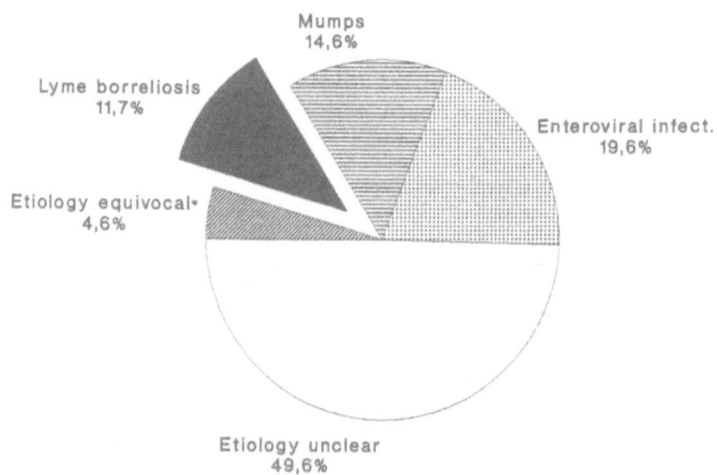

Figure 2. Etiology of aseptic meningitis in childhood (Lower Saxony, 1987-1989, n = 240); *etiology equivocal, because cross-reactions or concomitant infections could not be ruled out

Neither the patient's history nor clinical findings were helpful for the diagnosis of neuroborreliosis. A preceding tick bite and/or erythema migrans were recalled only in one third of the cases presenting with aseptic meningitis (17 out of 46 cases) and half of the cases suffering from facial palsy (54 out of 93 cases). In facial palsy the tick bite or erythema migrans was mostly located in the head-neck region, usually on the same side as subsequent facial palsy suggesting a direct invasion via the affected nerve by Borrelia burgdorferi. In contrast, adults experience most bites on their lower extremities. This may be an explanation for the fact that Bannwarth's syndrome, the typical neurological disorder in adults, occurs only rarely in pediatric patients.

CSF findings are diagnostic in cases with facial palsy. With two exceptions all children with Lyme borreliosis revealed an inflammatory CSF syndrome (Table 3). This was true irrespective whether clinical signs of meningitis were present or not. Only one fourth of these children (25.8%) presented with signs of meningeal inflammation. In facial palsy of unknown etiology CSF findings were usually normal confirming the validity of the IgM test in use.

Table 3. CSF findings for acute peripheral facial palsy compared according to cause of palsy

	Cell count (/ ul) Min. - Max. (Median) (Cases)	Protein content (mg%) Min. - Max. (Median) (Cases)	CSF/blood glucose ratio Min. - Max. (Median) (Cases)
Lyme borreliosis	1 - 960* (128) (n = 93)	11 - 339* (49) (n = 84)	0,37 - 0,87 (0,59) (n = 67)
Other cause	0 - 77* (1) (n = 120)	10 - 61* (25) (n = 95)	0,32 - 1,00 (0,59) (n = 86)

*$p < 0,05$; Wilcoxon-Mann-Whitney test

On the contrary, in aseptic meningitis CSF findings did not allow a distinction according to etiology. There were significant differences in the distributions of CSF cell count and protein content according to the cause of the meningitis, but in the individual case of aseptic meningitis this allowed no differentiation to be made (Table 4). Thus in this disorder diagnosis of Lyme borreliosis could only be established by means of antibody testing in CSF.

The IgM capture ELISA made possible the simple and reliable detection of intrathecal antibody synthesis by calculating the difference between the extinction values measured under saturation conditions in CSF and serum (method described by 13). Intrathecal IgM antibody synthesis could be demonstrated in three quaters of the cases (72.6%). In these cases the difference between the extinction values of CSF and serum was greater than 0.1 with a relative accumulation between the difference values of 1 and 2.

Table 4. CSF findings for aseptic meningitis compared according to cause of infection

	Cell count (/ ul) Min. - Max. (Median) (Cases)	Protein content (mg%) Min. - Max. (Median) (Cases)	CSF/blood glucose ratio Min. - Max. (Median) (Cases)
Borrelia meningitis	13 - 3456* (258) (n = 46)	26 - 210* (63) (n = 41)	0,33 - 1,00 (0,57) (n = 41)
Other cause of aseptic meningitis	6 - 5808* (145) (n = 224)	7 - 260* (36) (n = 206)	0,34 - 0,98 (0,60) (n = 202)

*$p < 0,05$; Wilcoxon-Mann-Whitney test

The following pecularities for antibody testing in neuroborreliosis in childhood are worth mentioning:

1. Demonstration of IgM antibodies in CSF proved to be the deciding diagnostic criterion for acute neuroborreliosis. In contrast to adults, testing for IgG antibodies in children is of minor diagnostic value. IgG antibodies were only found in serum in a quarter of the cases (23.8%) and in CSF in a few cases (5.4%) at the onset of the disease.

2. Seronegative antibody tests do not exclude neuroborreliosis: In 10.1% of the cases the CSF was positive for IgM antibodies but the serum was negative when tested first. This underlines the importance of CSF tests for antibodies against Borrelia burgdorferi in order to prevent false-negative diagnoses.

3. Antibody results are not helpful in evaluating the course of the disease. IgM antibodies in CSF are still detectable after several months in spite of normal results for cell count and protein content. On follow-up examination after an intervall of more than three months no less than half of the children still showed IgM antibodies in serum (59.1%) and one-third in CSF (39.1%). For this reason, a repeated lumbar puncture after antibiotic therapy seems not neccessary if the patient is free of symptoms.

Dermatological manifestations

Because only inpatients were included in the study, the clinical pictures examined (Table 2) cannot be considered representative for Lyme borreliosis in children. Erythema migrans is by far the most common symptom of Borrelia infection in childhood. Twenty-four children were investigated with a well-defined erythema migrans, i.e. with a history of a tick bite at the site of the erythema migrans. Only 11 out of 24 children revealed elevated IgM antibody titers and only one of them were positive for IgG antibodies. This discrepancy could be due to the time interval between the tick bite and manifestation of the eryhtema migrans indicating the length of the infection. The median timespan for the antibody-positive group was 19 days, much longer than in the antibody-negative group (14 days). In conclusion, testing for specific antibodies is of little value in differential diagnosis of erythema migrans. Its characteristic appearance usually allows a clinical diagnosis.

The same is true for Borrelial lymphocytoma, another characteristic skin lesion of Lyme borreliosis which mainly affects children and adolescents. Most common sites are the ears, the nipples and the scrotum. In our series five children presented with this symptom located at the ears in all cases. IgM antibodies against Borrelia burgdorferi were found only in two out of five children and none of them revealed elevated IgG antibody titres.

Acrodermatitis chronica atrophicans was not observed in our series. In the recent literature there have been only few pediatric case reports on this disorder which mainly affects adults (14).

Lyme arthritis

Lyme arthritis consists of brief, self-limited attacks with joint effusion most commonly involving the knee. Without a history of a preceding tick-bite or an erythema migrans, these are the cases of 'classical' Lyme arthritis, diagnosis may be difficult in the individual case. Because of the high infection rate false-positive results have to be considered. In our series Lyme arthritis was diagnosed in nine children, seven of whom suffered from gonarthritis. Only three of these nine children presented with a history of a tick bite and/or erythema migrans. With only one exception, all children revealed elevated IgM antibody titres, and four elevated IgG antibody titres. In two children with arthritis, the diagnosis could be confirmed by detection of IgM antibodies in joint fluid; one child only showed IgM antibodies against Borrelia burgdorferi in the joint fluid and not in the serum samples.

Therapy and prognosis

Lyme borreliosis in children proved to be an illness with an acute course and good prognosis. All children suffering from neuroborreliosis received intravenous antibiotic therapy, most with high doses (500,000 IU/ body weight/d) of penicillin G (86.7%). Treatment lasted 10 to 14 days almost without exception (92.1%). All children showed a rapid recovery without any neurological sequelae. Although the rate of spontaneous remission may be very high, it is not possible to predict on a patient-by-patient basis just when this will occur. Therefore consistent antibiotic therapy is indicated in every case.

SUMMARY

Children are more likely than adults to be bitten by ticks and thus more likely to be infected by Borrelia burgdorferi. In a serosurvey the infection rate measured by IgG antibodies was 2.6%. In a prospective hospital-based multicentre study 208 children with Lyme borreliosis were examined of whom 169 had Lyme neuroborreliosis which was diagnosed by detection of specific IgM-antibodies in the CSF using an IgM capture ELISA. The yearly incidence of Lyme neuroborreliosis was 5.8 cases/100,000 children ages 1 to 13. The manifestation index was 0.16, or one case of Lyme neuroborreliosis per 620 infected children. Facial palsy and aseptic meningitis accounted for nearly 90% of all cases with neuroborreliosis indicating striking differences in the clinical spectrum between children and adults. Lyme borreliosis proved to be the most frequently verifiable cause of acute peripheral facial palsy in children, causing every second case of this disorder in summer and autumn. Lyme borreliosis was the third most frequent cause of aseptic meningitis in childhood. Inflammatory changes of the cerebrospinal fluid along with the presence of specific antibodies are the conditio sine qua non for diagnosis of Lyme neuroborreliosis. High dose intravenous penicillin G proved to be highly effective in pediatric Lyme neuroborreliosis.

REFERENCES

1. Steere AC, Malawista SE, Snydman DR, Shope RE, Andiman WA, Ross MR, Steele FM. Lyme arthritis: an epidemic of oligoarticular arthritis in children and adults in three Connecticut communities. Arthritis Rheum 1977;20:7-17

2. Huppertz HI. Childhood Lyme borreliosis in Europe. Eur J Pediatr 1990;149:814-21

3. Jörbeck HJ, Gustafsson PM, Lind HC, Stiernstedt G. Tick-borne Borrelia-meningitis in children. An outbreak in the Kalmar area during the summer of 1984. Acta Paediatr Scand 1987;76:228-33

4. Kristoferitsch W. Neuropathien bei Lyme-Borreliose. Wien New York: Springer; 1989

5. Pfister HW, Einhäupl KM, Wilske B, Preac-Mursic V. Bannwarth's syndrome and the enlarged neurological spectrum of arthropod-borne borreliosis. Zentralbl Bakteriol Mikrobiol Hyg A 1986;263:343-7

6. Stiernstedt G. Tick-borne Borrelia infection in Sweden. Scand J Infect Dis 1985;45 [Suppl]:1-70

7. Stiernstedt G, Sköldenberg B, Gårde A, Kolmodin G, Jörbeck H, Svenungsson B, Carlström A. Clinical manifestations of Borrelia infections of the nervous system. Zentralbl Bakteriol Mikrobiol Hyg A 1986;263:289-96

8. Hansen K, Lebech AM. The clinical and epidemiological profile of Lyme neuroborreliosis in Denmark 1985-1990. Brain 1992;115:399-423

9. Weber K, Burgdorfer W (eds). Aspects of Lyme borreliosis. Berlin Heidelberg: Springer; 1993

10. Christen HJ, Hanefeld F, Eiffert H, Thomssen R. Epidemiology and clinical manifestations of Lyme borreliosis in childhood. Acta Paediatr 1993;82 [Suppl 386]:1-75

11. Wassmann K, Borg-von Zepelin M, Zimmermann O, Stadler M, Eiffert H, Thomssen R. Determination of immunoglobulin M antibody against Borrelia burgdorferi to differentiate between acute and past infections. In: Stanek G, Kristoferitsch W, Pletschette M, Barbour AG, Flamm H, eds. Lyme Borreliosis II. Zentralbl Bakteriol Mikrobiol Hyg 1989;Suppl 18:281-9

12. Eiffert H, Lotter H, Thomssen R. Use of peroxidase-labelled antigen for the detection of antibodies to Borrelia burgdorferi in human and animal sera. Scand J Infect Dis 1991;23:79-87

13. Hansen K, Lebech AM. Lyme neuroborreliosis: a new sensitive diagnostic assay for intrathecal synthesis of Borrelia burgdorferi-specific immunoglobulin G, A, and M. Ann Neurol 1991;30:197-205

14. Nadal D, Gundelfinger R, Flueler U, Boltshauser E. Acrodermatitis chronica atrophicans. Arch Dis Child 1988;63:72-4

LYME CARDITIS

Janine Evans, Lynda Rosenfeld, Robert T. Schoen

Sections of Rheumatology and Cardiology
Yale University School of Medicine
New Haven, Connecticut   06510

INTRODUCTION

Lyme carditis complicates infection with *Borrelia burgdorferi* in 8% of untreated individuals with erythema migrans (EM)[1]. In patients treated for early Lyme disease carditis occurs much less frequently[2]. Carditis typically develops within 3 to 5 weeks after the appearance of EM however, heart block can be the sole presenting symptom of Lyme disease[3].

Atrioventricular (AV) conduction disturbances, myocarditis, and pericarditis are the most common manifestions of Lyme carditis.   Chronic dilated cardiomyopathy and other conduction disturbances have also been described. Lyme carditis typically resolves spontaneously without sequelae although some patients require temporary pacemakers.

We have recently evaluated 7 patients who presented with Lyme carditis during the summer of 1992, the largest group presenting in one year since the original description from this institution in 1980.   We report our experience and review the literature.

EPIDEMIOLOGY

The majority of cases occur in young males during the summer months[4]. Six of the 7 patients in our series were male. The age range varied between 12-52 years (mean 28.5). In contrast with other reported cases only 2 patients recalled having a tickbite and 3 had a rash consistent with EM. Prior to the development of heart block, most patients had nonspecific prodromal symptoms including fatigue, headache, fever, and arthralgais, which in retrospect were attributed to Lyme disease. One patient developed Lyme meningitis. Lightheadedness, palpitations and shortness of breath were the most common presenting cardiac complaints. One individual was completely assymptomatic and was diagnosed during a routine physical examination.

Lyme Borreliosis, Edited by J.S. Axford and
D.H.E. Rees, Plenum Press, New York, 1994

CONDUCTION ABNORMALITIES

All of our patients had evidence of high degree AV block. 6 patients had periods of complete heart block and the 7th had a Wenckeback rhythm. AV block is the most common abnormality associated with Lyme disease. In their original series, Steere et al found fluctuating degrees of AV block in 18 of 20 patients[4]. Patients with high degree AV block developed symptoms suggestive of cardiac disease in contrast to patients with first degree heart block alone who typically were assymptomatic. The degree of AV block fluctuated rapidly, progressing from first degree to complete heart block and back in minutes. Escape rates are variable and frequently patients require insertion of a temporary pacemaker.

Electrophysiology studies revealed that the AV node is most commonly involved site[5]. Blockage may also occur in the HIS bundles and combined proximal and distal conduction disturbances have also been reported. Ventricular dysrhythmias have also been reported but are uncommon. In the majority of cases AV block is self limited and resolves even without antibiotic therapy. Resolution of heart block typically progresses from complete heart block to Wenckebach and finally first degree heart block before a return to normal sinus rhythm.

MYOCARDITIS

Only 1 patient had EKG changes suggestive of myocardial involvement. 3 patients had abnormal echocardiograms although most of the abnormalities were interpreted as rate related and not a reflection of underlying myocarditis. Steere et al found myocarditis in 13 of 20 patients[4]. Evidence for myocarditis included T wave flattening or inversion, intraventricular conduction defects, ventricular premature contractions, and a small pericardial effusion. Recent studies of endomyocardial biopsy specimens have demonstrated the presence of infiltrates of lymphocytes and monocytes. Endocardial regions frequently have dense areas of plasma cells and lymphocytes[6]. Occasionally focal infiltrates around muscle fibers and blood vessels are found but myocyte necrosis is unusual. Spirochetes compatible with *B. burgdorferi* have been demonstrated near lymphoid cells and in the endocardium.

OTHER CARDIAC LESIONS

All of our patients had normal echocardiograms at 6-9 month follow up. Chronic cardiomyopathy associated with Lyme disease has been reported in Europe. Klein et al performed serologic studies in 54 consecutive patients with chronic heart failure[7]. 32.7% of patients had anti-*B. burgdorferi* antibodies in their serum when tested by ELISA assay. Spirochetes were demonstrated on endomyocardial biopsy in one of the patients. The response to treatment with antibiotics is variable [8,9]. There have been no cases of chronic cardiomyopathy from Lyme disease reported in the United States. Strain variation may account for this difference.

SEROLOGIC TESTING

All of our patients had serologic evidence of *B. burgdorferi* infection. One patient initially had negative Lyme disease serologies however he seroconverted on repeat testing one week later. Almost all reported cases in the literature were seropositive for *B. burgdorferi* and typically had elevations in the IgM component. Serologic confirmation distinguishes Lyme carditis from other infections with cardiac manifestations such as Coxsacki virus A and B, Echovirus 6 and 8, hepatitis B, Yersinia enterocolitica, Rocky Mountain spotted fever, and acute rheumatic fever.

TREATMENT AND PROGNOSIS

All of our patients were treated with antibiotics (6 with ceftriaxone, one with doxycycline) with reolution of complete heart block within 48 hours. All had variable degrees of first and second degree heart block for up to two weeks during treatment. Six patients initally required monitoring in an intensive care unit and one required a temporary pacemaker. All patients were in normal sinus rhythm at one month follow up. Three patients had residual symptoms at 1 month followup; 2 complained of fatigue and arthraligas and 1 noted exertional dyspnea. One patient had minor musculoskeletal symptoms at 6-9 month follow up. The majority of cases reported in the literature had complete resolution of carditis even in the absence of antibiotic treatment [4]. There have been two reported deaths from Lyme carditis. The first individual had coexisting infections with *B. burgdorferi* and *Babesiosia microti*[10]. The second death occured in a young male who had positive Lyme disease serologies and died following an appendectomy [11]. He had histologic changes consistent with Lyme carditis although no spirochetes were noted on Warthin Starry stain.

SUMMARY

Lyme carditis remains a significant complication in Lyme disease. It may present quite dramatically especially in indiviuals who lack definitive symptoms of Lyme disease. Complete resolution occurs in the majority of cases although many patients require intensive care unit monitoring and some need insertion of a temporary pacemaker. We found no evidence of long term cardiac sequelae after antibiotic treatment.

REFERENCES

1. A.C. Steere, T.F. Broderick, S.E. Malawista, Erthema chronicum migrans and Lyme arthritis: Epidemiologic evidence of a tick vector, *Am J Epidemiol.* 108: 312-321 (1978).
2. A. Rubin, C. Servera, P. Nikitin, A. McAllister, G.P. Wormser, R. B. Nadelman, Prospective evaluation of heart block complicating early Lyme disease, *PACE.* 15: 252-255 (1992).
3. F. Baylac-Domengetroy, C. Vieyres, R. Barraine, Complete heart block as the sole presentation of Lyme disease, *Arch Int Med.* 151:1240 (1991).
4. A.C. Steere, W.P. Balsford, M. Weinberg, J. Alexander, H.J. Berger, S. Wolfson, S.E. Malawista, Lyme carditis: cardiac abnormalities of Lyme disease. *Ann Int Med.* 93:8-16 (1980).
5. M.R. Van der Linde, Lyme carditis: clnical characteristics of 105 cases, *Scan J Infect Dis.* 77:81-84 (1991).
6. M.R. Van der Linde, H.J.G.M. Crijins, J. de Koning, J.A.A. Hoogkamp-Korstanje, J.J. de Graaf, D.A. Piers, A. Van der Galien, K.I. Lee, Range of atrioventricular conduction disturbances in Lyme borreliosis: a report of four cases and review of other published reports, *Br. Heart J.* 63: 162-168 (1990).
7. J.Klein , G. Stanek, R. Bittner, R. Horvat, C. Holzinger, D. Glogar, Lyme borreliosis as a cause of myocarditis and heart muscle disease, *Eur Heart J.* 12 Supp: 73-75 (1991).
8. R. Gasser, J Dusleag, E. Reisinger, R. Stauber, B. Feigl, S. Pongratz, W. Klein, C. Furian, K. Pierer, Reversal by ceftriaxone of dilated cardiomyopathy Borrelia burgdorferi infection, *Lancet.* 339: 1174-1175 (1992).
9. J. Bergler-Klein, d. Glogar, G. Stanek, Clinical outcome of *Borrelia burgdorferi* related dilated cardiomyopathy after antibiotic treatment, *Lancet.* 340: 317-318 (1992).
10. L.C. Marcus, A.C. Steere, P.H. Duray, A.E. Anderson, E.B. Mahoney, Fatal pancarditis in a

patient with coexistent Lyme disease and *Babesiosis, Ann Int Med.* 103: 374-376 (1985).

11. N.R.B. Cary, D.J.M. Wright, S.J. Currer, L.M. Shapiro, A.A. Grace, Fatal Lyme carditis and endodermal heterotropia of the atrioventricular node, *Postgrad Med J.* 66: 134-136 (1990).

A RECOMBINANT OspA AND OspB BASED LYME DISEASE VACCINE

Erol Fikrig,[1] Fred S. Kantor,[2] Stephen W. Barthold,[3] and
Richard A. Flavell[4]

[1]Section of Rheumatology
[2]Section of Allergy and Clinical Immunology
[3]Section of Comparative Medicine
[4]Section of Immunobiology, and Howard Hughes Medical Institute
Yale University School of Medicine
P.O. Box 3333
310 Cedar Street
New Haven, CT 0617

INTRODUCTION

Lyme borreliosis is a tick borne infection caused by the spirochete, Borrelia burgdorferi(1). Disease is often marked by a skin rash (erythema migrans), and can progress through several stages to cause arthritis, carditis or neurologic symptoms(2). The initial tick bite and skin rash can be difficult to identify, therefore, an effective vaccine should prove useful in endemic areas.

An understanding of protective immunity to B. burgdorferi infection has been aided by animal models of Lyme disease. A model of Lyme borreliosis in the immunocompetent C3H mouse mimics features of human Lyme disease. Infected mice develop arthritis and carditis, with an intermittent disease course and persistent infection for up to one year after inoculation(3-6). Fourteen days after infection the animals exhibit acute arthritis and carditis. Over several months, joint and heart disease resolves, with intermittent episodes of exacerbation. In contrast, severe combined immunodeficient (SCID) mice develop persistent and progressive arthritis and carditis(7,8). Other models of Lyme disease include hamsters, rabbits, rats and gerbils, in which infection can be established, but disease manifestations are mild unless animals are infected as neonates or immunosuppressed as adults(9-14).

Lyme Borreliosis, Edited by J.S. Axford and
D.H.E. Rees, Plenum Press, New York, 1994

We showed that passive immunization of C3H mice with anti-B. burgdorferi sera protected mice from infection(15). Fourteen days after infection with 10^4 B. burgdorferi strain N40, administered via intradermal inoculation, the animals were sacrificed, tissues were cultured for spirochetes, and joints and hearts were examined microscopically for inflammatory disease. These studies indicated that humoral immunity was sufficient for protection.

Outer surface proteins A and B (OspA, OspB) are two B. burgdorferi surface proteins, and potential vaccine candidates. Recombinant OspA and OspB, from Borrelia strain N40 have been cloned, expressed and purified as fusion proteins with glutathione transferase(15-17). Mice were vaccinated with OspA in complete Freund's adjuvant; vaccination was followed by two weekly booster injections in incomplete adjuvant. Two weeks after the final booster injection, organs were cultured for B. burgdorferi and examined for histopathology (joints and hearts). Mice immunized with either recombinant OspA or OspB developed a strong Immunoglobulin G antibody response to the antigens. Mice vaccinated with OspA are protected from infection when challenged intradermally with an inoculum of 10^4 B. burgdorferi N40, and mice vaccinated with OspB are protected from intradermal infection when challenged with an inoculum of 10^2 B. burgdorferi N40, showing that both OspA and OspB are vaccine candidates(17).

When C3H mice are immunized passively with polyclonal or monoclonal antibodies to OspA or OspB, to determine if protection could be shown with antibody to these proteins, mice are protected from infection(15)(17). Similar conclusions have been reached by Schaible and his colleagues, who have shown that the passive administration of monoclonal OspA antibodies protect SCID mice from B. burgdorferi infection(18-20).

B. burgdorferi can persist in host tissue for a long time, and both hamsters and mice can remain chronically infected, suggesting that B. burgdorferi has developed mechanisms to evade host defenses, possibly analogous to Borrelia hermsii, which has been shown to vary the expression of its surface antigens(21). Time-course studies determine, however, that OspA mediated protective immunity is long lasting. When vaccinated mice were challenged with B. burgdorferi up to four months after immunization, or sacrificed six months after challenge infection, protection was seen to be complete(22).

The breadth of protection afforded by vaccination with the OspA or B antigens has not been fully explored. Initial studies in hamsters showed that passive immunization of animals with B. burgdorferi antisera did not protect the animals from infection with heterologous strains of B. burgdorferi(10). Studies to develop a taxonomic classification system of Borrelia strains are progressing. DNA- and RNA-hybridization studies identify at least two distinct groups of B. burgdorferi, with strains B31 and 20047 representing these groups(23). Wilske and colleagues have shown that heterogeneity of the Osp proteins is present, as evidenced by variable reactivity of different B. burgdorferi strains with monoclonal antibodies to OspA and OspB(24). Furthermore, spirochetes with mutations in the Osps have been described. Sadzienne and colleagues have isolated B. burgdorferi in vitro that lack either OspA or OspB, or both(25). Rosa and colleagues have also shown that homologous recombination between OspA and OspB can occur, resulting in hybrid surface proteins(26). The degree to which strain variability or mutations influences protective immunity has not been elucidated.

We have analyzed variability in OspB to determine how B. burgdorferi may escape immune destruction(17). We have demonstrated that B. burgdorferi N40 which evade vaccination immunity to OspB have a truncated form of OspB, due to a TAA stop codon at nucleotide 577. These spirochetes have been identified as ΔN40. In contrast, B. burgdorferi N40 that express full-length OspB are unable to infect mice immunized with OspB. Mapping of the OspB antibody response shows that epitopes in the C terminus of OspB are surface-exposed and bind protective monoclonal and polyclonal antibodies. This suggests that the C terminus of OspB is important for eliciting a protective immune response to OspB. Truncation or modification of outer surface proteins that do not bind protective antibody may be a means by which Bb evades host defenses.

Cross-protection studies in C3H mice have investigated the breadth of protection eliciting OspA immunization. In these experiments mice vaccinated with OspA or OspB were challenged with several selected strains of B. burgdorferi(15)(27). These studies are somewhat limited, for many strains must first be passaged through C3H mice to become reproducibly infectious. This may influence the interpretation of results, in that the B. burgdorferi tested may not be representative of natural populations. We performed experiments to determine if vaccination with OspA protected mice from infection with five selected B. burgdorferi strains. Strain 297 represented an isolate from a patient with neuroborreliosis. Strain B31 and CD16 represented isolates from the northeast and midwestern USA. Strain 25015 has an unusual pattern for OspA on SDS-polyacrylamide gel electrophoresis, migrating at 33 kDa rather than 31-kDa(28). Cross-protection extended to all strains except Borrelia strain 25015, indicating a broad specificity of protection(15)(27). To determine the reason that OspA cross protection did not extend to strain 25015, we cloned and sequenced OspA-25015. The nucleotide sequences showed 40 amino acid differences between the two strains(27). Most of the differences are located in the C terminus of the protein, suggesting that epitopes that bind protective antibody are present in this region, analogous to our results with OspB. Indeed, monoclonal antibodies, prepared against B. burgdorferi N40, that protect mice against infection with B. burgdorferi N40 do not react with OspA-25015 by immunoblot analysis.

We mapped the epitopes on OspA-N40 bound by patients' sera and monoclonal and polyclonal antibodies to determine the location or nature of protective epitopes on OspA(29). Protective monoclonal antibodies (mAbs) bound to conformationally dependent epitopes in the carboxyl terminus of the OspA protein. Similarly the expanded OspA response in sera from patients with chronic Lyme disease, mice vaccinated with OspA or mice chronically infected with B. burgdorferi bind to several OspA epitopes including the regions bound by protective antibody.

Sera from selected patients with different stages of Lyme disease were used to passively immunize mice against B. burgdorferi challenge to determine if human antibodies could protect the animals from infection. Sera from 2 patients with late Lyme disease which contained strong antibody reactivity to proteins in B. burgdorferi lysates, including antibodies to the OspA and B, partially protected mice from infection following challenge with a small inoculum (10^2) of B. burgdorferi. Mice immunized with sera from either of these 2 patients developed significantly fewer infections with B. burgdorferi (patient 1:5% infected mice, patient 2:25%) relative to control mice (90%). In contrast, sera from 2 patients with early or late Lyme disease that lacked antibodies reactive to OspA and OspB did not confer protection. The protective immunity in our studies appeared to be related, at least in part, to the presence of a strong humoral response to OspA and OspB. These results suggest that during prolonged infection, some patients develop an immune response that may be partially protective against B. burgdorferi reinfection. Therefore, although most patients do not mount a strong humoral response to OspA or OspB during natural infection, vaccination with OspA or OspB may elicit protective immunity.

While immunization with OspA provides protection against infection with B. burgdorferi we wished to test whether vaccination could also be therapeutic. C3H mice were actively immunized with OspA at different intervals after infection with B. burgdorferi to determine the effect of postexposure vaccination on the course of murine Lyme borreliosis. When vaccination was commenced in the early stages (5 to 14 days) of infection, active immunization with OspA partially cleared spirochetes from the bloodstream but did not eliminate them from other tissues or alter the course of joint or heart disease. Commencement of vaccination at 60 days after infection (at which time joint or heart disease is resolving), however, reduced both the number of infected mice and individual joints with arthritis, a result suggesting an acceleration of the resolution phase of the disease. Postexposure immunization with OspA may therefore partially alter the course of murine Lyme arthritis.

Finally, we have evaluated an OspA based vaccine against B. burgdorferi infection transmitted by tick bite. Several groups have indicated that vector-mediated transmission of disease may be influenced by anti-inflammatory and immunosuppressive substances within tick salivary glands(30-33). We allowed Ixodes dammini nymphal ticks, infected with B. burgdorferi N40, to feed on OspA immunized mice(34). The OspA immunized mice were protected from infection. Moreover, spirochetes were eliminated from ticks that fed on OspA immunized mice, but were present in ticks that had fed upon control animals, as determined by immunofluorescence, dark field microscopy and culture. Engorged ticks were also stored until their molt to adults. Even at this time point, the ticks remained free of spirochete infection, indicating that B. burgdorferi destruction was complete. This suggests that the efficacy of an Osp vaccine is based on two mechanisms, the first causes antibody-mediated destruction of B. burgdorferi in the host, and the second eliminates spirochetes in the tick prior to disease transmission.

More research is needed to develop an effective and safe vaccine against Lyme disease. Clinical studies with purified preparations of OspA or OspB in adjuvant that is acceptable for human use, will determine if OspA or OspB are safe and effective in humans.

Table 1. Protection of mice from B. burgdorferi infection.[a]

Immunization	Challenge (No. of Spirochetes)	Protection from infection
Active		
OspA-N40	10^4 B. burgdorferi N40	Yes
	B. burgdorferi N40 (T)	Yes
	10^4 B. burgdorferi 25015	No
OspA-25015	10^4 B. burgdorferi 25015	Yes
	10^4 B. burgdorferi N40	No
OspB-N40	10^2 B. burgdorferi N40	Yes
	10^2 B. burgdorferi ΔN40	No
Passive		
Borrelia antisera	10^4 B. burgdorferi N40	Yes
C terminal OspA mAb (IgG3)	10^4 B. burgdorferi N40	Yes
C terminal OspA mAb (IgG1)	10^4 B. burgdorferi N40	No
N terminal OspA mAb (IgG2a)	10^4 B. burgdorferi N40	No
C terminal OspB mAb (IgG2a)	10^2 B. burgdorferi N40	Yes
N terminal OspB mAb (IgG2a)	10^2 B. burgdorferi N40	No

[a] Mice were actively immunized with B. burgdorferi antigen, or passively immunized with Borrelia antisera or antibodies, and were then challenged with B. burgdorferi (by intradermal syringe or tickbite (indicated by T)), and assayed for protection by culture and histopathology. B. burgdorferi DN40 has a truncated OspB.

ACKNOWLEDGEMENTS

We thank Nancy Marcantonio, Kathleen Deponte, Deborah Beck, Lucy Kim, Gordon Terwilliger, Manchuan Chen and Hong Tao for technical assistance. This work was supported by NIH grant P01-AI-30548, CDC grant U5-CCU-106581, the Mathers Foundation, a Lederle Young Investigator Award in Vaccine Development from the Infectious Disease Society of America and a Deans Young Faculty Award. Richard A. Flavell is an Investigator of the Howard Hughes Medical Institute.

REFERENCES

1. Steere, A.C., R.L. Grodzicki, A.N. Kornblatt, J.E. Craft, A.G. Barbour, W. Burgdorfer, G.P. Schmid, E. Johnson and S.E. Malawista. 1983. The spirochetal etiology of Lyme disease. *N. Engl. J. Med.* 308:733.

2. Steere, A.C. 1989. Lyme disease. *N. Engl. J. Med.* 321:586.

3. Barthold, S.W., D.S. Beck, G.M. Hansen, G.A. Terwilliger and K.D. Moody. 1990. Lyme borreliosis in selected strains and ages of laboratory mice. *J. Infect. Dis.* 162:133.

4. Barthold, S.W. 1991. Infectivity of Borrelia burgdorferi Relative to Route of Inoculation and Genotype in Laboratory Mice. *J. Infect. Dis.* 163:419.

5. Barthold, S.W., D.H. Persing, A.L. Armstrong and R.A. Peeples. 1991. Kinetics of Borrelia burgdorferi: Dissemination and evolution of disease following intradermal innoculation of mice. *Am. J. Pathol.* 139:263.

6. Barthold, S.W., M.S. DeSouza, J.L. Janotka, A.L. Smith and D.H. Persing. 1993. Chronic Lyme borreliosis in the laboratory mouse. *Am. J. Pathol.* :(in press).

7. Schaible, U., M. Kramer, C. Museteanu, H. Zimmer, H. Mossman and M.M. Simon. 1989. The severe combine immunodeficient mouse: a laboratory model of Lyme arthritis and carditis. *J. Exp. Med.* 170:240.

8. Barthold, S.W., C.L. Sidman and C. Sidman. 1992. Lyme borreliosis in genetically susceptible severe combined immunodeficient mice. *Am. J. Trop. Med. Hyg* 47:605.

9. Barthold, S.W., K.D. Moody, G.A. Terwilliger, P.H. Duray, R.O. Jacoby and A.C. Steere. 1988. Experimental Lyme arthritis in rats infected with Borrelia burgdorferi. *J Inf Dis* 157:842.

10. Johnson, R.C., C. Kodner, M. Russell and P.H. Duray. 1988. Experimental infection of hamsters with Borrelia burgdorferi. *Ann NY Acad Sci* 539:259.

11. Moody, K.D., S.W. Barthold, G.A. Terwilliger], D.S. Beck, G.M. Hansen and R.O. Jacoby. 1990. Experimental chronic Lyme borreliosis in Lewis rats. *Am J Trop Med Hyg* 42:65.

12. Schmitz, J.L., R.F. Schell, A. Hejke, D.M. England, S.M. Callister and R.F. Schell. 1989. Induction of Lyme arthritis in LSH Hamsters. *Am J Pathol* 134:1113.

13. Kornblatt, A.N., A.C. Steere and D.G. Brownstein. 1984. Experimental Lyme disease in rabbits: spirochetes found in erythema migrans and blood. *Infect Immun* 46:220.

14. Preac-Mursic, V. et al. 1990. *Infection* 18:332.

15. Fikrig, E., S.W. Barthold, F.S. Kantor and R.A. Flavell. 1990. Protection of mice against the Lyme disease agent by immunizing with recombinant OspA. *Science* 250:553.

16. Fikrig, E., S.W. Barthold, N. Marcantonio, K. Deponte, F.S. Kantor and R.A. Flavell. 1992. Roles of OspA, OspB and flagellin in protective immunity to Lyme borreliosis in laboratory mice. *Infect. Immun.* 59:553.

17. Fikrig, E., H. Tao, F.S. Kantor, S.W. Barthold and R.A. Flavell. 1993. Borrelia burgdorferi lacking the carboxyl terminus of OspB evade immune destruction. *Proc. Natl. Acad. SCi. U.S.A* :(In press).

18. Schaible, U.E., M.D. Kramer, K. Eichmann, M. Modolell, C. Museteanu and M.M. Simon. 1990. Monoclonal antibodies specific for the outer surface protein A (OspA) of Borrelia burgdorferi prevent Lyme borreliosis in severe combined immunodeficiency (scid) mice. *Proc Natl Acad Sci USA* 87:3768.

19. Simon, M.M., U.E. Schaible, M.D. Kramer, C. Eckerskorn, C. Museteanu, H.K. Muller-Hermelink and R. Wallich. 1991. Recombinant outer surface protein A from Borrelia burgdorferi induces antibodies protective against spirochetal infection in mice. *J. Infect. Dis* 164:123.

20. Simon, M.M., U.E. Schaible, R. Wallich and M.D. Kramer. 1991. A mouse model for Borrelia burgdorferi infection: approach to a vaccine against Lyme disease. *Immunol Today* 12:11.

21. Barbour, A.G. 1989. Antigenic variation in relapsing fever Borrelia species: Genetic aspects. *In* Mobile DNA. D. E. Berg and M. M. Howe, American Society for Microbiology. Washington D. C. 9th, 783.

22. Fikrig, E., S.W. Barthold, F.S. Kantor and R.A. Flavell. 1992. Long term protection of mice from Lyme disease by immunizing with recombinant OspA. *Infect. Immun.* 60:773.

23. Postic, D., C. Edlinger, C. Richaud, F. GRimont, Y. Dufresne, P. Perolat, G. Baranton and P.A.D. Grimont. 1990. Two Genomic Species in Borrelia burgdorferi. *Pasteur Institute: Research Microbiology* 141:465.

24. Wilske, B., V. Preac-Mursic, G. Schierz, R. Kuhbeck, A.G. Barbour and M. Kramer. 1988. Antigenic variability of Borrelia burgdorferi. *Ann NY Acad Sci* 539:126.

25. Sadziene, A., P. Rosa, D. Hogan and A. Barbour. 1992. Antibody resistant mutants of Borrelia burgdorferi: In vitro selection and characterization. *J Exp Med* 176:799.

26. Rosa, P., D. Hogan and T. Schwan. 1992. Recombination between major surface protein genes of Borrelia burgdorferi. *Mol. Microbiol.* 6:3031.

27. Fikrig, E., S.W. Barthold, D.H. Persing, X. Sen, F.S. Kantor and R.A. Flavell. 1992. Borrelia burgdorferi strain 25015: Characterization of OspA and vaccination against infection. *J. Immunol.* 148:2256.

28. Anderson, J., S. Barthold and L.A. Magnarelli. 1991. Infectious but nonpathogenic isolate of Borrelia burgdorferi. *J. Clin. Microbiol.* 28:2693.

29. Sears, J., E. Fikrig, T. Nakagawa, K. Deponte, N. Marcantonio, F. Kantor and R.A. Flavell. 1991. Molecular mapping of OspA-mediated protection against Borrelia burgdorferi, the Lyme disease agent. *J. Immunol.* 147:1995.

30. Ribeiro, J.M.C., G.T. Makoul, D.R. Robinson and A. Spielman. 1985. Antihemostatic, antinflammatory, and immunosuppressive properties of the saliva of a tick, Ixodes dammini. *J. Exp. Med.* 161:332.

31. Ribeiro, J., T. Mather, J. Piesman and A. Spielman. 1987. Dissemination and salivary delivery of Lyme disease spirochetes in vector ticks (Acari: Ixodidae) . *J Med Entomol* 24:201.

32. Ribeiro, J.M.C., J.J. Weis and S.R.Telford, III. 1990. Saliva of the tick Ixodes dammini inhibits neutrophil function. *Exp. Parasitol* 70:382.

33. Titus, R.G. and J.M.C. Ribeiro. 1988. Salivary gland lysates from the sand fly Lutzomyia longipalpis enhance Leishmania infectivity. *Science* 239:1306.

34. Fikrig, E.F., S.R. Telford, S.W. Barthold, F.S. Kantor, A. Spielman and R.A. Flavell. 1992. Elimination of Borrelia burgdorferi from vector ticks feeding on OspA-immunized mice. *Proc Natl Acad Sci U.S.A.* 89:5418.

ACRODERMATITIS CHRONICA ATROPHICANS AND LYME ARTHRITIS: STUDY OF CZECH LYME BORRELIOSIS PATIENTS HAVING BOTH SYNDROMES

Jana Hercogova,[1] Marika Valesova,[2] Marta Tomankova[1] and Dagmar Frösslova[1]

[1] Dermatology Department, 2nd Medical School
[2] 1st Internal Medicine, 3rd Medical School
Charles University, Prague, Czech Republic

INTRODUCTION

When Alan Steere first described Lyme disease in 1977, the primary syndrome that he described was a new form of arthritis that could be associated with the occurrence of a tick bite and skin rash (Steere et al. 1977). Since then our knowledge of this new disease has grown such that a large spectrum of differing syndromes have now been shown to be associated with the disease. Indeed one of the most notable characteristics of the disease is its multi-systemic, multi-syndromic nature. In general the most frequent syndromes can broadly broken down into three major categories, namely, those affecting the skin (acute disease: erythema migrans, Borrelia lymphocytoma; chronic disease: acrodermatitis chronica atrophicans: ACA), those affecting the joints (Lyme arthritis: LA) and those affecting the nervous system (neuroborreliosis). In many instances the three categories have been studied independently of each other with little attention being paid to the multi-syndromic nature of the disease, whereby patients often suffer from syndromes falling into two or more of these broad syndrome categories.

The disorder ACA was first described in 1883 (Buchwald 1883) and is thus the earliest documented syndrome which is now associated with Lyme borreliosis. ACA usually begins with the gradual development of an inconspicuous skin inflammation which is most frequently localized over the joints of the extremities, especially the foot, knee, elbow, wrist joints and along the extensors. As ACA does not affect real acral parts, the term arthrodermatitis was used by Oppenheim (1931) but without any connection to joint involvement. However, as early as 1921, Jessner described a patient with ACA, arthralgias and laryngitis and a year later reported that bone atrophy of the extremities could be observed for eight further patients with ACA by means of X-ray examination (Jessner 1921, 1922). In more recent years it has been possible to isolate *Borrelia burgdorferi* from ACA lesion (Asbrink and Hovmark 1985), from synovial fluid (Snydman et al. 1986, Schmidli et al. 1988) and synovium (Valesova et al. 1989) thus confirming the diagnosis. Since this time the number of reported incidence of patients having both disorders has risen considerably. It was thus of particular interest to us to study coincidence of these two disorders among Czech patients suffering from Lyme borreliosis.

MATERIAL AND METHODS

Between January 1987 and April 1993, 72 patients were treated for ACA (54 females, 18 males, aged between 22 - 80 years).

According to Asbrink and Hovmark (1987) ACA was diagnosed on the basis of the following criteria: a) clinical picture, b) histopathological findings (mostly bandlike lymphocytic infiltrate with a moderate to rich admixture of plasma cells and dilatation of blood vessels), c) the presence of increased levels of IgG antibodies to *Borrelia burgdorferi* by ELISA method using a sonicated preparation of *Borrelia burgdorferi* as the antigen and/or observance of spirochetes by electron microscopy or successful *Borrelia burgdorferi* isolation from the skin biopsy specimen (Hulinska et al. 1989 and Hercogova et al. 1993). The serological examination, cultivation and electron microscopy were provided in the Reference Laboratory for Lyme Borreliosis, National Health Institute, Prague headed by Dr. D. Hulinska.

The diagnosis of LA was established in patients met the following criteria: a) clinical symptoms of arthritis, b) there was the presence of increased levels of IgG antibodies to *Borrelia burgdorferi*, c) they recalled having had a tick bite or erythema migrans or visited the area endemic for Lyme borreliosis, if other rheumatic diseases could be excluded (e.g. if they were negative for rheumatic factor, antinuclear antibodies etc.). LA was defined as acute if the arthritis only lasted for a few days or a period of less than six months, and as chronic arthritis if the attacks lasted for more than six months.

RESULTS

The inflammatory phase of ACA was observed in 68 patients, the atrophy phase of ACA was present in three patients and fibrous nodules near the elbows occured in two patients being the only manifestation of Lyme borreliosis in one of them. 47 patients had ACA lesions on one extremity (mostly on the lower extremity, n = 36). In the remaining 25 patients more than one area was involved including the trunk (n = 3) and the face (n = 2).

ACA was the only manifestation of Lyme borreliosis observed in 37 patients while in 35 patients (48.6 %) extracutaneous complications developed. Table 1.

Intermittent episodes of arthralgia or migratory musculosceletal pain was present in 26 of 72 patients (36.1 %). LA was diagnosed in nine of 72 patients with ACA (12.5 %) and was classified as acute in six patients and as chronic arthritis in the remaining three patients. Of the nine patients with both ACA and LA, six were females and three were males, aged between 31 - 75 years.

ACA lesions preceeded LA in seven patients, in one patient ACA developed 6 months after the onset of arthritis and in the remaining patient both disorders were noticed concurrently. Two patients had a history of erythema migrans (4 and 6 months before ACA developed), three patients were repeatedly bitten by a tick and all but two patients visited an area where Lyme borreliosis was known to be endemic. Five patients had ACA lesions on more than one extremity and in two patients cutaneous involvement of the trunk was also present. The inflammatory phase of ACA was found in eight cases, in one patient the atrophy phase of ACA was present while in another patient fibrous nodules were observed near the elbows.

Table 1. Extracutaneous manifestations in 72 patients with acrodermatitis chronica atrophicans.

Fatique	12
Intermittent arthralgia	22
Lyme arthritis	9
Musculoskeletal pain	4
Paraesthesias	5
Cephalea	3
Depressions	2
Loss of concentration	3
Vertigo	3
Diplopia	1
Carpal tunnel syndrome	3
Myositis	2
Cranial neuritis (VII. cranial nerve)	1
Peripheral neuropathy	7
Myelopolyradiculoneuritis	1
Encephaloradiculoneuritis	1
Encephalomyelopolyneuritis	1
Encephalopathy	2
Regional lymphadenopathy	2

Table 2. Joints involved in nine patients with acrodermatitis chronica atrophicans and Lyme arthritis.

Sternoclavicular	1
Shoulder	3
Elbow	2
Radiocarpal	2
Metacarpophalangeal	5
Finger proximal interphalangeal	2
Hip	1
Knee	2
Talocrural	2

Joint involvement at the time of ACA diagnosis included: monoarthritis (n = 5), oligo-arthritis (n = 2), polyarthritis (n = 2), asymmetrical in all but one patients. The arthritic changes were localized underlying ACA lesions in seven patients.

ACA and LA were the only manifestations of Lyme borreliosis observed in two patients, but the remaining seven patients complained of fatique (n = 3), paraesthesias (n = 3), one patient suffered from diplopia, cephalea and vertigo. Carpal tunnel syndrome were diagnosed by electromyographic studies in three patients and encephalopolyneuritis was observed in two patients based on the CSF examination, electroencephalography and electromyography.

After antibiotic therapy with parenteral penicillin (n = 3), cefotaxime (n = 3), ceftria-xone (n = 1), tetracycline (n = 1) or with oral doxycycline (n = 1) ACA lesions disappeared within 8 months in six patients, improved or did not get worse in three patients including one of these was with atrophy phase of ACA. LA resolved in five patients within 8 months but in three cases recurrent attacks of arthritis occured after therapy. These attacks of arthritis required further antibiotics (parenteral penicillin, doxycycline, cefotaxime or ceftri-axone) and affected either the joints previously involved (n = 2) or the other joints (n = 1). All the affected joints are shown in the Table 2. In one patient, treated with ceftriaxone, the biopsy shown that myositis of musculus gastrocnemius developed during therapy. At the end of the follow up period (1.5 - 5 years), in seven patients tested the high levels of IgG antibodies to *Borrelia burgdorferi* were still detected.

DISCUSSION

Both ACA and LA disorders are chronic manifestations as they develop weeks to years after the inicial tick bite. Chronic arthritis occurs in approximately 60 % of untreated cases of Lyme borreliosis in the U.S.A. (Steere 1989), but is less frequent in the Europe. In a study of Swedish ACA patients (Hovmark, Asbrink and Olsson 1986) involvement of joints and/or bones underlying the cutaneous lesions were found by means of X-ray examination in 16 of 50 patients. Weber and Neubert (1986) reported joint involvement in 4 of 21 patients with ACA, Aberer and Klade (1991) described a Sudeck-like atrophy in two cases of ACA and one patient with arthritis of the knee. Thus the finding that nine of 72 ACA patients suffered from LA in this study is comparable to those of other studies.

Typical LA is asymmerical oligoarticular arthritis, primarily found in the large jonits, especially the knee (Steere 1989). In our study, gonitis was observed in two patients being bilateral in one of them. The involvement of the other large joints, e.g. shoulder, elbow, wrist, hip, ankle, occured in all patients, but was accompanied with polyarthritis of metacarpo-phalangeal and/or proximal interphalangeal joints in five patients. Symmetrical joint involvement was seen in two patients, i both of them the arthritis affected the joints not underlying the cutaneous lesions of ACA.

Concomitant with ACA, neurological manifestations have been also described in the literature in approximately 30 - 50 % of cases (Kristoferitsch 1989). In our set of patients the neurological involvement was present in 15 of 72 cases (20.8 %), but in four of the nine patients with combined ACA and LA. The antibiotic therapy of the patients with both ACA and LA resolved or improved inflammatory ACA lesions but did not prevent recurrent attacks of arthritis in three patients and development of myositis in one another patient. Thus therapy of LA has yet to be solved.

CONCLUSIONS

1. The frequency of LA in Czech ACA patients is relatively low (in nine of 72 patients, e.g. 12.5 %).

2. Although both ACA and LA occur during the chronic stage of Lyme borreliosis, the sequence in which they occur is variable.

3. In our study antibiotic therapy, including 3rd generation cephalosporins, did not prevent the further complications (recurrent attacks of arthrtitis and myositis).

REFERENCES

Aberer, E., and Klade, H., 1991, Cutaneous manifestations of Lyme borreliosis, *Infection* 19:284.

Asbrink, E., and Hovmark, A., 1985, Successful cultivation of spirochetes from skin lesions of patients with erythema chronicum migrans Afzelius and acrodermatitis chronica atrophicans, *Acta Pathol Microbiol Immunol Scand* B 93:161.

Asbrink, E., and Hovmark, A., 1987, Cutaneous manifestations in Ixodes-borne Borrelia spirochetosis, *Intern J Dermatol* 26:215.

Buchwald, A., 1883, Ein Fall von diffuser idiopatischer Hautatrophie, *Vierteljahresschr Dermatol* 15:553.

Hercogova, J., Tomankova, M., Plch, J., Jirous, J., Hulinska, D., Frösslova, D., and Bartak, P., 1993, Borrelia burgdorferi isolates from erythema migrans, *Archives Dermatol* - in press.

Hulinska, D., Jirous, J., Valesova, M., and Hercogova, J., 1989, Ultrastructure of Borrelia burgdorferi in tissues of patients with Lyme disease, *J Bas Microbiol* 29:73.

Jessner, M., 1921, Zur Kenntnis der Acrodermatitis chronica atrophicans, *Arch Dermatol (Berlin)* 134:478.

Jessner, M., 1922, Weiter Beitrag zur Kenntnis der Acrodermatitis chronica atrophicans,*Arch Dermatol (Berlin)* 139:294.

Kristoferitsch, W., 1989, " Neuropathien bei Lyme-Borreliose," Springer-Verlag, Wien.

Oppenheim, M., 1931, Atrophien, *in:*"Handbuch der Haut- und Geschlechtskrankheiten," Springer-Verlag, Berlin.

Schmidli, J., Hunziker, T., Moesli, P., and Schaad, U.B., 1988, Cultivation of Borrelia burgdorferi from joint fluid three months after treatment of facial palsy due to Lyme borreliosis, *J Infect Dis* 158:905.

Snydman, D.R., Schenkein, D.P., Berardi, V.P., Lastavica, C.C., and Pariser, K.M., 1986, Borrelia burgdorferi in joint fluid in chronic Lyme arthritis, *Ann Intern Med* 104:798.

Steere, A.C., Malawista, S.E., Snydman, D.R., Shope, R.E., Andiman, W.A., Ross, M.R., and Steele, F.M., 1977, Lyme arthritis: an epidemic of oligoarticular arthritis in children and adults in three Connecticut communities, *Arthritis Rheum* 20:7.

Steere, A.C., 1989, Lyme disease, *N Engl J Med* 321:586.

Valesova, M., Trnavsky, K., Hulinska, D., Alusik, S., Janousek, J., and Jirous, J., 1989, Case report: Detection of Borrelia in the synovial tissue of a patient with Lyme borreliosis, detected by electron microscopy, *J Rheumatol* 16:1502

Weber, K., and Neubert, U., 1986, Clinical features of early erythema migrans disease and related disorders, *Zbl Bakt Hyg* A263:209.

ASPECTS OF LYME CARDITIS

Marcel R. van der Linde[1], Harry J.G.M. Crijns[2]

[1]Department of Cardiology
Nij Smellinghe Hospital
9202 NN Drachten
The Netherlands
[2]Department of Cardiology
Thoraxcenter
University Hospital of Groningen
9713 EZ Groningen
The Netherlands

INTRODUCTION

The cardiac manifestations of Lyme borreliosis (Lyme carditis) were first characterized by Steere et al[1] in 1980. In this article, atrioventricular block, perimyocarditis and mild heart failure were described as the features of Lyme carditis. The first specific articles on Lyme carditis in Europe were published in 1984[2,3], although already in 1973 a French patient was described with typical features of Lyme carditis (erythema migrans, arthralgias, Wenckebach atrioventricular (AV) block and perimyocarditis), but the diagnosis could only be confirmed retrospectively in 1985, when the immunofluorescence assay became available for B. burgdorferi and the preserved serum was found positive in this patient[4]. Other authors, however, had previously related cardiac complaints and abnormalities to other manifestations of Lyme borreliosis[5,6]. In recent years, the number of publications on Lyme carditis has increased rapidly, extending the pattern of Lyme carditis with other cardiac manifestations and showing the diversity of the possible expressions of this disease[7-13].

Nevertheless, it took several years before this manifestation of Lyme borreliosis was generally known as a possible cause of cardiac problems in everyday clinical practice. The several International Conferences on Lyme borreliosis and the many local symposia contributed significantly to the awareness of Lyme carditis as a treatable cause of heart disease.

In this article, the clinical features, diagnostic tools and therapy of Lyme carditis are discussed, including recent data on the subject.

Lyme Borreliosis, Edited by J.S. Axford and
D.H.E. Rees, Plenum Press, New York, 1994

INCIDENCE AND CLINICAL FEATURES

Incidence

In the North American literature, the incidence of Lyme carditis was estimated as occuring in 8% of patients with Lyme borreliosis[14]. In a recent prospective study of 61 patients with Lyme borreliosis however, the incidence of Lyme carditis was found to be only 1.6%[15]. In the European literature, only retrospective studies on the incidence of Lyme carditis are available and a frequency of 1.6% in 817 German patients[16] and 3.3% in 272 French patients[17] with Lyme borreliosis is reported. Difficulties in confirming Lyme carditis, unfamiliarity with the clinical pattern and accidental cardiac abnormalities in cases with Lyme borreliosis without causal relation to this disease may lead to under- or overestimation of the frequency of cardiac manifestations in Lyme borreliosis[18].

Lyme borreliosis and thus Lyme carditis can be found on every continent and in every country[1,13,19-26], although there are endemic area known.

Lyme borreliosis and -carditis may occur at either sex, although Lyme carditis was found to be more frequent in men (male-female ratio 3:1)[9]. It may occur at any age, even in children[23,24,27].

Clinical features

Lyme carditis may be the only manifestation of Lyme borreliosis. Lyme carditis may present as:

- Conduction disturbances (disturbances of impulse formation and conduction)
- Rhythm disturbances (extrasystoles, brady- and tachyarrhythmias)
- Pericarditis and/or pericardial effusion
- Myocarditis and/or recent onset heart failure
- Dilating cardiomyopathy as a late manifestation

Conduction disturbances. Intra-atrial, AV, and intraventricular conduction disturbances may all occur. A fluctuating degree of AV block is, however, the most frequently observed sign of Lyme carditis. Rapid changes in degree of AV block may occur. In several reports unstable escape foci have been described (Fig 1).

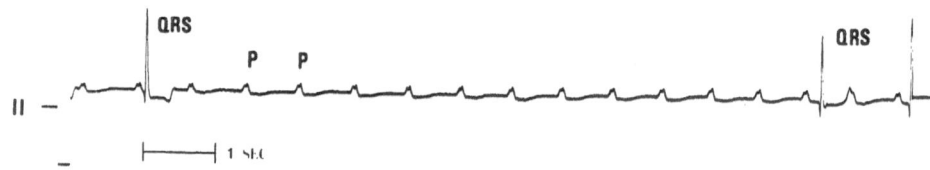

Fig. 1. Rhythm strip (lead II) of a 40-year-old patient with complete AV block due to biopsy proven Lyme carditis. The instability of the escape focus in this patient is demonstrated by the ventricular standstill of approximately 9 seconds (despite infusion of isoproterenol). A temporary pacemaker was inserted.

The block may be restricted to the supra-hisian (i.e., in the AV node) or infra-hisian areas (i.e., in the common His bundle and bundle branches) or the conduction disturbances may be diffuse, in which case the AV node, common His bundle, bundle branches and the Purkinje system may all be affected, with a possible concomitant delay of intra-atrial conduction[7,28].

Rhythm disturbances. Rhythm disturbances in Lyme carditis seem to be less frequent than conduction disturbances, but may include atrial fibrillation or flutter, paroxysmal atrial tachycardia, premature supraventricular and ventricular beats and (sinus)bradycardia[7.9]. In 1991 a 67-year-old white patient was reported with Lyme carditis-related periods of nonsustained and sustained monomorphic ventricular tachycardia with global hypokinesis of a dilated left ventricle and normal coronary arteries. After several anti-arrhythmic drugs, he was treated with ceftriaxone intravenously with good effect on the ventricular tachycardia and a significant improvement in left ventricular ejection fraction after six months[10].

Pericarditis, pericardial effusion, myocarditis and heart failure. Pericarditis and pericardial effusion in Lyme carditis is frequently described. Pericarditis can be found in up to 15% of patients with Lyme carditis and in half of these patients echocardiographically demonstrated pericardial effusion was present. In some cases this was complicated by right-sided heart failure. Up to 8% of the patients with Lyme carditis may experience myopericarditis (pericarditis associated with enzyme changes, but with no evidence of myocardial infarction), in some cases with right sided heart failure[9,28-32]. Overt clinical left heart failure seems to be an infrequent manifestation of Lyme carditis, although it has been described in several reports[9,10,13].

Dilating cardiomyopathy. Until recently, late manifestations of Lyme carditis were unknown. Stanek and coworkers[8] demonstrated that in a group of patients with dilating cardiomyopathy of unknown cause, some patients may have suffered Lyme carditis in earlier years, resulting in this late manifestation. Another case report on this subject was published in the former East Germany[13].

DIAGNOSTIC TOOLS

In short, the diagnostic procedures include a complete history and physical examination with respect to Lyme borreliosis and more specific Lyme carditis. General laboratory tests and specific tests on B. burgdorferi are routine practice. With respect to Lyme carditis, serial surface electrocardiograms (ECG's), chest X-rays, an echocardiogram and Holterscans have to be performed according to the clinical situation. In case of first degree AV block with PR > 300 msec, high degree AV block, rapidly changing AV block, unstable escape foci or hemodynamically significant rhythm disturbances, hospital admission for continuous monitoring is necessary. The same applies for clinical heart failure. Multigated bloodpool scintigraphy can be done on specific indication. In case of diagnostic difficulties, a gallium-67 scintigraphy or an indium-111 scan can be performed to distinguish active myocarditis from primary cardiomyopathies and other non-inflammatory myocardial disorders[21,33-37]. A normal gallium-67 or indium-111 scan however, does not exclude Lyme carditis[7,21,33]. Endomyocardial biopsies are not routinely performed in Lyme carditis, but can differentiate acute endomyocarditis and a more chronic inflammation[33,38]. The presence of spirochetes in endomyocardial tissue may be demonstrated by special staining methods[39,40]. In case of diagnostic difficulties, endomyocardial biopsies may be indicated, possibly guided by a gallium-67 or indium-111 scan. Electrophysiological studies can be useful in Lyme carditis, although they are not fundamental for diagnosis and treatment. Electrophysiological studies in Lyme carditis may be restricted to patients in whom it is already necessary to introduce a temporary pacemaker, in those patients where myocardial biopsies have to be performed and for research purposes in patients who have given special informed consent. In combination with the surface ECG, location and extension of the conduction disturbances (intra-atrial, supra- or infra-his) can be determined[7,41,42].

THERAPY, COURSE AND PROGNOSIS

Therapy. Treatment of Lyme carditis can be divided in two parts. First, the treatment of the manifestations and symptoms of Lyme carditis. In case of asymptomatic and low grade AV block or other mild rhythm or conduction disturbances, an expectative attitude with rhythm monitoring is sufficient. When symptomatic or hemodynamically compromising high degree heart block is present, a temporary pacemaker must be inserted. Retrospective analysis of 105 patients with Lyme carditis showed about 35 % use of temporary pacemakers worldwide[12]. Patience with respect to permanent pacemaker implantation is important. Tachyarrhythmias must be monitored and, if necessary, treated with anti-arrhythmic drugs (caution with negative inotropic drugs). Pericarditis and myocarditis without right or left ventricular heart failure can be treated with a few days of immobilisation. If heart failure is present, additional drugs like diuretics and vasodilators must be applied. Duration and intensity of these therapies depend on the clinical pattern. Secondly, antibiotic therapy should be given as soon as the suspicion of Lyme carditis is likely enough to justify treatment. Usually it is not possible to wait for the definite proof of Lyme carditis. The present-day recommended antibiotic choice is ceftriaxone (1x2 gram intravenously for 10-14 days), cefotaxime (2x2 gram intravenously for 10-14 days) or penicillin G (4x5 million U intravenously for 14 days) in case of clinically significant Lyme carditis. In asymptomatic first degree AV block (PR < 300 msec), tetracycline (3-4 x 500 mg orally for 10-30 days) or doxycycline (2x100 mg orally for 10-30) days, may be sufficient[43]. The application of steroids is controversial. These recommendations are partly based on empirical data and partly on data from the results of treatment of other manifestations of Lyme borreliosis[44,45].

Course and prognosis. In general, Lyme carditis has a favourable course. More than 90% of the documented patients with Lyme carditis had a complete recovery within 2-6 weeks, with or without antibiotic therapy[9,12]. In less than 10% minor conduction disturbances remain and in about 2% a permanent pacemaker had to be implanted[7,21]. Despite these reassuring data, there have been at least 3 patients described with fatal Lyme carditis[20,46,47]. Furthermore, recent reports have demonstrated the possibility of late manifestations of Lyme carditis like dilating cardiomyopathy[8,13].

REFERENCES

1. Steere AC, Batsford WP, Weinberg M, Alexander J, Berger HJ, Wolfson S, Malawista SE. Lyme carditis: cardiac abnormalities of Lyme disease. *Ann Intern Med* 1980;93(1):8-16.
2. Cornuau C, Bernard M, Daumas PL, Oblet B, Poirot G, Valois M. Les manifestations cardiaques de la maladie de Lyme. A propos de deux observations. (Cardiac manifestations of Lyme disease. Apropos of 2 cases). *Ann Cardiol Angeiol (Paris)* 1984;33(6):395-399.
3. Houwerzijl J, Root JJ, Hoogkamp-Korstanje JAA. A case of Lyme disease with cardiac involvement in the Netherlands (letter). *Infection* 1984;12:358.
4. Schott B. Erytheme de Lipschtz avec Atteinte Cardiaque. A Propos d'une Cas: Etiology Méconne de BAV Paroxystique. Synthèse et Rapport avec la Maladie de Lyme aux USA. *Thesis*, Louis Pasteur University, Strasbourg, France, 1985.
5. Bannwarth A. Chronische lymphocytaere Meningitis, entzuentliche Polyneuritis und Rheumatismus. *Arch Psychiatr Nervenkr* 1941;113:284-376.

6. Hopf HC. Acrodermatitis Chronica Atrophicans (Herxheimer) und Nervensystem, *in*: "Monographien aus dem Gesamtgebiete der Neurologie und Psychiatrie", vol 114., Berlin-Heidelberg-New York: Springer, 1966;70-71.

7. Linde van der MR, Crijns HJGM, de Koning J, Hoogkamp-Korstanje JAA, de Graaf JJ, Piers DA, van der Galien A, Lie KI. Range of atrioventricular conduction disturbances in Lyme borreliosis: a report of 4 cases and review of other published reports. *Br Heart J* 1990;63:162-168.

8. Stanek G, Klein J, Bittner R, Glogar D. Isolation of Borrelia burgdorferi from the myocardium of a patient with longstanding cardiomyopathy. *N Engl J Med* 1990;322:249-252.

9. Linde van der MR. Lyme carditis: clinical characteristics of 105 cases. *Scand J Infect Dis (suppl)* 1991;77:81-84.

10. Vlay SC, Dervan JP, Elias J, Kane PP, Dattwyler R. Ventricular tachycardia associated with Lyme carditis. *Am J Med* 1991;121:1558-1560.

11. Jukema JW, Werner HA, Reinders Folmer SCC, Wesdorp JCL. Lyme-borreliose en gedilateerde cardiomyopathie. *Ned Tijdschr Cardiol* 1992;6:16-19.

12. Linde van der MR, Ballmer PE. Lyme carditis. In: Weber K, Burgdorfer W, ed. *Aspects of Lyme Borreliosis*. Berlin-Heidelberg: Springer Verlag, 1993;131-151.

13. Wunderlich E, Graf A, Thess G, Foelske H. Dilatative Herzmukelerkrankung als Folge einer chronischen Lyme-Karditis. *Z Kardiol* 1990;79:599-600.

14. Steere AC, Malawista S, Hardin J, et al. Erythema chronicum migrans and Lyme arthritis: The enlarging clinical spectrum. *Ann Intern Med* 1977;86:685-698.

15. Rubin DA, Sorbera C, Nikitin P, McAllister A, Wormser GP, Nadelman RB. Prospective evaluation of heart block complicating early Lyme disease. *Pace* 1992;15:252-255.

16. Schmidt R, Kabatzki J, Hartung S, Ackermann R. Erythema-migrans-Borreliose in der Bundesrepu blik Deutschland: Epidemiologie und klinisches Bild. *Dtsch Med Wochenschr* 1985;110:1803-1807.

17. Dournon E, Assous M, Fourcade C. Clinical features of Lyme disease in France. In: Abstract Book of the 27th Interscience Conference on Antimicrobial Agents and Chemotherapy (ICAAC), 4-7 Oct 1987, New York, USA, 1987;319:1274.

18. Kaell AT, Volkman DJ, Gorevic PD, Dattwyler RJ. Positive Lyme serology in subacute bacterial endocarditis: a study of four patients. *JAMA* 1990;264:2916-2918.

19. Cox J, Krajden M. Cardiovascular manifestations of Lyme disease. *Am Heart J* 1991;122:1449-1455.

20. Cary NRB, Fox B, Wright DJM, Cutler SJ, Shapiro LM, Grace AA. Fatal Lyme carditis and endodermal heterotopia of the atrioventricular node. *Postgrad Med J* 1990;66:134-136.

21. Artigao R, Torres G, Guerrero A, Jimenez-Mena M, Bayas Paredes M. *Am J Med* 1991;90:531-533.

22. Bianchi G, Rovetta, Monteforte P, Fumarola D, Trevisan G, Crovato F, Cimmino MA. Articular involvement in European patients with Lyme disease. A report of 32 Italian patients. *Br J Rheum* 1990;29:178-180.

23. Slavik Z, Janousek J, Tax P, Chaloupecky V. Cardiac involvement in Lyme disease. Case report. *Cs Pediat* 1990;45:276-278.

24. Patial RK, Kashyap S, Bansal SK, Sood A. Lyme disease in a Shimla boy. *JAPI* 1990;38:503-504.

25. Carlberg H, Naito S. Lyme Borreliosis. A review and present situation in Japan. *J Derm* 1991;18:125-142.

26. Stewart A, Glass J, Patel A, Watt G, Cripps A, Clancy R. Lyme arthritis in the Hunter Valley. *Med J Aust* 1982;1:139.

27. Woolf PK, Lorsung EM, Edwards KS, Li KI, Kanengiser SJ, Ruddy RM, Gewitz MH. Electrocardio graphic findings in children with Lyme disease. *Pediatr Emerg Care* 1990;7:334-336.

28. Kapusta P, Fauchier JP, Cosnau P, Huguet R, Grezard O, Rouesnel P. Troubles conductifs sino-auriculaires et auriculo-ventriculaires de la maladie de Lyme. A propos de deux observations. (Sinoatrial and atrioventricular conduction disorders in Lyme disease. Apropos of 2 case reports). *Arch Mal Coeur* 1986;79(9):1361-1366.

29. Lavaud P, Etienne J, Chamot E et al. Bloc auriculo- ventriculaire aiqu associé une maladie de Lyme. Un cas confirmé par sérologie. (Acute auriculo-ventricular block associated with Lyme disease. A case confirmed by serology (letter)). *Presse Med* 1985;14(39):2020.

30. Allal J, Coisne D, Thomas P et al. Manifestations cardiaques de la maladie de Lyme. (Cardiac manifestations of Lyme disease). *Ann Med Interne* 1986;137(5):372-374.

31. Lorcerie B, Boutron MC, Portier H, Beuriat P, Ravisy J, Martin F. Manifestations pericardiques de la maladie de Lyme (Pericardial manifestations of Lyme disease). *Ann Med Interne (Paris)* 1987;138:601-603.

32. Veyssier P, Davous N, Kalousstian E, Maitre B, Lallement PY, Serret A. Atteintes cardiaques au

cours de la maladie de Lyme. Deux observations. (Cardiac involvement in Lyme disease. 2 cases). *Rev Med Interne* 1987;8:357-360.

33. O'Connell JB, Henkin RE, Robinson JA, Subramanian R, Path MRC, Scanlon PJ, Gunnar RM. Gallium-67 imaging in patients with dilated cardiomyopathy and biopsy-proven myocarditis. *Circulation* 1984;70:58-62.

34. Jacobs JC, Rosen JM, Szer IS. Lyme myocarditis diagnosed by gallium scan. *J Pediatr* 1984;105:950-952.

35. Rienzo RJ, Morel DE, Prager D, Barron L, Post R. Gallium avid Lyme myocarditis. *Clin Nucl Med* 1987;12:475-476.

36. Ponsonnaille J, Citron B, Karsenty B et al. Myocardite aique au cours d'un syndrome de Lyme. Interet de la scintigraphie myocardique au gallium 67. (Acute myocarditis in Lyme's syndrome. Value of myocardial scintigraphy with gallium 67). *Arch Mal Coeur* 1986;79:1946-1950.

37. Linde van der MR, Crijns HJGM, Lie KI. Transient complete atrioventricular block in Lyme disease. *Chest* 1989;96:219-221.

38. Duray PH, Steere AC. The spectrum of organ and systems pathology in human Lyme disease. *Zentralbl Bakteriol Hyg (A)* 1986;263:169-178.

39. Koning de J, Bosma RB, Hoogkamp-Korstanje JAA. Demonstration of spirochaetes in patients with Lyme disease with a modified silver stain. *J Med Microbiol* 1987;23:261-267.

40. Koning de J, Hoogkamp-Korstanje JAA, Linde van der MR et al. Demonstration of spirochetes in cardiac biopsies of patients with Lyme disease. *J Inf Dis* 1989;160:150-153.

41. Dunica S, Piette JC, Nassar N, Beaufils P. Une nouvelle cause de bloc auriculo-ventriculaire aiou transitoire: la maladie de Lyme. (A new cause of acute transitory auriculoventricular block: Lyme disease). *Arch Mal Coeur* 1986;79:1251-1255.

42. Fauchier JP, Cosnay P, Sirinelli A, Moquet B, Rabut H. Myocardite de Lyme sans troubles conductifs auriculo-ventriculaires. *Press Med* 1988;17(38):2036-2037.

43. Mayer-Weber W, van der Linde MR, Hassler D. Therapy of Lyme Carditis. In: Weber K, Burgdorfer W, ed. *Aspects of Lyme borreliosis*. Berlin-Heidelberg: Springer Verlag, 1993;344-349.

44. Dattwyler RJ, Halperin JJ, Pass H, Luft BJ. Ceftriaxone as effective therapy in refractory Lyme disease. *J Infect Dis* 1987;1322-1325.

45. Dattwyler RJ, Halperin JJ, Volkmann DJ, Luft BJ. Treatment of late Lyme borreliosis - randomized comparison of ceftriaxone and penicillin. *Lancet* 1988;ii:1191-1194.

46. Marcus LC, Steere AC, Duray PH, Anderson AE, Mahoney EB. Fatal pancarditis in a patient with coexistent Lyme disease and babesiosis: demonstation of spirochetes in the myocardium. *Ann Intern Med* 1985;103:374-6.

47. Koning de J, Houwerzijl J. Personnal communication.

BORRELIA BURGDORFERI SHOWN BY PCR FROM SKIN BIOPSY SPECIMEN AFTER A FLY BITE

Jarmo Oksi[1,2], Inkeri Helander[3], Heikki Aho[4], Merja Marjamäki[5] and Matti K. Viljanen[5]

Departments of Medical Microbiology[1] and Pathology[4], Turku University; Departments of Medicine[2] and Dermatology[3] Turku University Central Hospital; National Public Health Institute[5], Department in Turku, Finland

CASE REPORT

The patient, 41-year-old physician, had previously been healthy. He recalled no tick bites in his life. After jogging on August 17, 1992, he was bitten by a deer fly. Erythema migrans like lesion developed around the bite area in right lower leg slowly in two weeks. No fever or other symptoms developed. Examination on September 2 disclosed 5 x 12 cm homogenous erythema. (Fig.1.) No itching or other hypersensitivity symptoms were present in addition to the erythema.

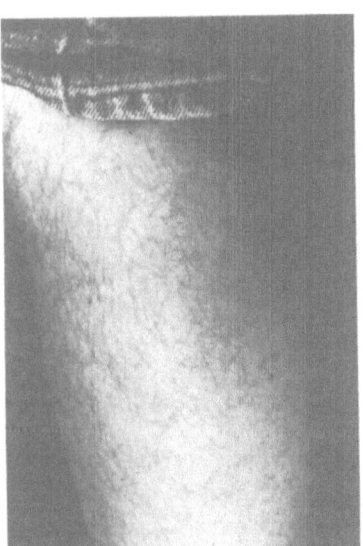

Figure 1. Homogenous erythema migrans in the lower leg two weeks after deer fly bite. (Photo: J. Oksi)

Histological biopsy specimen from the centre of the erythema was cut serially: no pieces from the fly's stinger were found, but there was heavy perivascular lymphoid infiltrate with eosinophils indicating an insect bite reaction.(Fig.2.) Histological sections from the border of the lesion resembled more typical erythema migrans with mainly lymphocytic perivascular infiltrate.

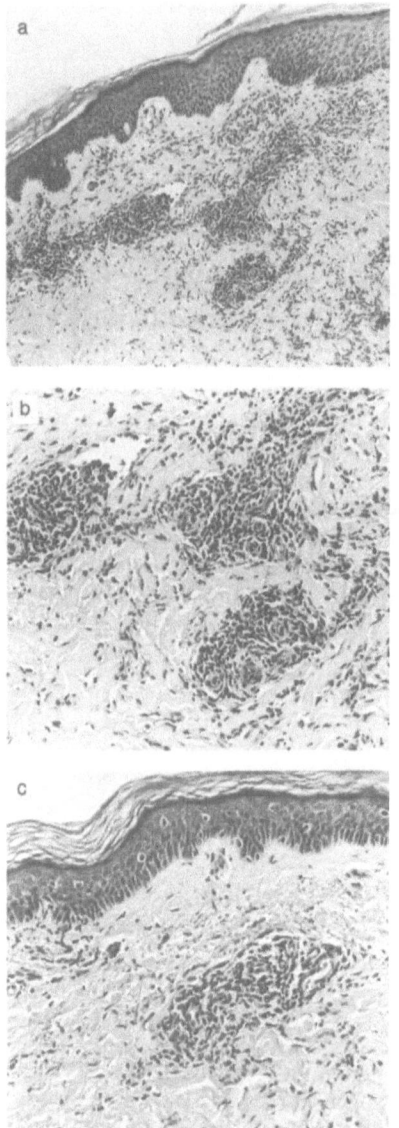

Figure 2. a) Perivascular lymphocytic infiltration around dermal blood vessels. Slight acanthosis and sparse lymphocytic exocytosis is seen in the epidermis. Biopsy from the center of the lesion. Hematoxylin and eosin, x90.

b) The blood vessel endothelium is slightly hypertrophic. Some eosinophils are present among perivascular lymphocytes. The cellular infiltrate extends also between the dermal collagen bundles. Higher magnification of the area seen in a). Hematoxylin and eosin, x190.

c) Some lymphocytes are seen around dermal blood vessels indicating mild reaction of erythema migrans -type. Biopsy from the periphery of the lesion. Hematoxylin and eosin, x105. (Photo: H. Aho)

Cultivation and DNA extraction was performed from other specimen taken from the border of the erythema. The culture remained negative, and so was the dark field microscopying. *Borrelia burgdorferi* DNA could be detected using polymerase chain reaction (PCR).[2-3]

Serum antibodies to *B. burgdorferi* were measured from sera obtained on September 2, September 22, 1992, and January 7, 1993. Seroconversion against native flagellin of *B. burgdorferi* was detected in the second serum, in which the amount of IgM antibodies was on positive level, and again on negative level in the third serum. Antibodies against sonicated whole *B. burgdorferi*[1] were on negative level in all three samples.

Erythema migrans disappeared completely in one week during therapy with amoxicillin 1 g two times a day. The antimicrobial therapy was continued for 25 days. No other symptoms of borreliosis have appeared during follow-up of five months.

METHODS

Cultivation

The skin biopsy specimen was immediately put to a tube containing BSK-II-medium (Kelly's modification) and incubated in 30°C. The tube was macroscopically checked two times a week and new passaging was performed once a week.

Treatment of the skin biopsy specimen

1 ml of physiological NaCl solution containing minced biopsy specimen was centrifuged (13000 rpm, 10 min.), the pellet was washed once with distilled water. DNA was extracted with phenol-chloroform, precipitated with ethanol, and dissolved in water.

Amplification by PCR

PCR was run in two steps, first with external primers, prB31/41-4 and prB31/41-5[2], resulting in a 730 bp PCR product, and then with nested PCR with internal primers, WK1 and WK2 [3], resulting in a 290 bp fragment. The oligonucleotides were synthesized on an automated DNA synthesizer (391 DNA Synthesizer PCR Mate; Applied Biosystems, Inc., Foster City, Calif.). Amplification of *B. burgdorferi* target sequence was carried out in a 50 µl reaction mixture containing 10 mM Tris-HCl (pH 8,8), 1,5 mM $MgCl_2$, 50 mM KCl, 0,1% Triton X-100, 200 µM deoxynucleoside triphosphates, 30 pmol of external or internal primers, 1,0 U of DynaZyme™ DNA polymerase (Finnzymes Oy, Espoo, Finland), and various amounts of DNA extracted from the sera. The reactions were subjected to 35 cycles of amplification using an automated DNA thermal cycler (HB-TR1; Hybaid Ltd., Middlesex, United Kingdom). Each cycle involved heating to 94°C for 1 min (DNA denaturation), cooling to 42°C for 2 min (primer annealing) and again heating to 72°C for 2 min (primer extension).

For nested PCR, 5 µl of the reaction mixture from the first PCR was added to a new PCR mixture containing 30 pmol of the internal primers, and was subjected to an additional 15 cycles. Each cycle in the nested PCR involved heating to 94°C for 1 min, cooling to 50°C for 1 min and again heating to 72°C for 1 min.

Determination of the PCR product was performed by gel electrophoresis on 1,5% agarose gel with ethidium bromide staining.

A positive control (*B. bergdorferi* ATCC 35210 was amplified in every reaction, and the fragment obtained was 290 bp, as expected. As a negative control were all reagents without any added DNA. A negative control was also included in DNA extraction procedure. The negative controls remained negative in each run. Simultaneously were tested also other clinical samples, which remained negative.

Serological studies

One test assessed IgM and IgG antibodies against sonicated *B. burgdorferi*[1]. Interpretation of results was based on the analysis of 110 healthy blood donor sera. The other test was a commercial test kit measuring IgM and IgG antibodies against purified, native 41 kD flagellin of *B. burgdorferi* (Lyme Borreliosis ELISA Kit, 2nd generation. Dako, Copenhagen).

DISCUSSION AND CONCLUSION

This is to our knowledge the first report in which *Borrelia burgdorferi* has been directly demonstrated from erythema migrans skin lesion after a deer fly bite. This patient case further supports the observation reported by Luger[4] that Lyme borreliosis can be transmitted by a biting fly, and that a very short attachment of the fly is enough for the transmission.

The fact, that patients suffering from Lyme borreliosis, do not very frequently recall tick bite in their history, may indicate importance of other vectors[5-6].

ACKNOWLEDGMENTS

This study was supported by the Foundation of Emil Aaltonen, Foundation of Emil and Blida Maunula, Turku University Society and Turku University Foundation.

REFERENCES

1. Viljanen MK, Punnonen J. The effect of storage of antigen-coated polystyrene microwells on the detection of antibodies against *Borrelia burgdorferi* by enzyme immunoassay (EIA). J Immunol Methods 1989;**124**:137-41
2. Wallich R, Moter SE, Simon MM, Ebnet K, Heiberger A, Kramer MD. The *Borrelia burgdorferi* flagellum-associated 41 kilodalton antigen (flagellin) - molecular cloning, expression, and amplification of the gene. Infect Immun 1990;**58**:1711-9
3. Kruger WH, Pulz M. Detection of *Borrelia burgdorferi* in cerebrospinal fluid by the polymerase chain reaction. J Med Microbiol 1991;**35**:98-102
4. Luger SW. Lyme disease transmitted by a biting fly. N Engl J Med 1990;**322**:1752
5. Magnarelli LA, Anderson JF. Ticks and biting insects infected with the etiologic agent of Lyme disease, *Borrelia burgdorferi*. J Clin Microbiol 1988;**26**:1482-6
6. Magnarelli LA, Anderson JF, Barbour AG. The etiologic agent of Lyme disease in deer flies, horse flies, and mosquitoes. J Infect Dis 1986;**154**:355-8

LYME ARTHRITIS: EXPERIENCE FROM SOMERSET, ENGLAND

T G Palferman

Yeovil District Hospital
Higher Kingston
Yeovil
Somerset, BA21 4AT
U.K.

INTRODUCTION

Joint manifestations of Lyme Disease are the rule in the United States of America, but seem to be the exception in the United Kingdom. One report from southern England documented a patient who developed a monoarthritis within 72 hours of a tick bite, a more generalised arthropathy evolving 2 weeks later when erythema chronicum migrans (ECM) was still present[1]. Sixty eight patients were identified as seropositive for Borrelia burgdorferi out of a total of 431 sera tested in a reference laboratory; of these seropositive patients only one reported joint symptoms[2] which took the form of a symmetrical polyarthritis and which resolved without antibacterial therapy.

Whether Lyme arthritis in the UK is rare, misdiagnosed or simply not documented is undetermined. Four cases are now presented followed by a consideration of the distribution of the arthropathy and a discussion about the difficulties of diagnosis and the imprecision of extant laboratory aids to diagnosis.

PATIENTS

Case 1

Male Aged 47 Dairy/Sheep farmer

> Presented October 1986 to allergy clinic.
> History from March/April 1985: evanescent, flitting, mildly pruritic rash described as "heat bumps". Contemporaneous onset of painful, stiff joints affecting small joints of hands and feet plus swelling of left knee. Associated with episodic "flu-like" symptoms and shivering.
> The General Practitioner recorded swelling of first metacarpophalangeal joints and pain and tenderness in heels.

Skin testing for allergens negative. Indirect immunofluorescent antibody test for Borrelia burgdorferi positive 1/128.
December 1986: given two week course of oral penicillin. Referred to Rheumatology clinic.
February 1987 seen by Rheumatologist. Patient significantly improved. Fading, pink, maculopapular rash present over left abdomen. No synovitis discernable.
Rheumatoid factor and antinuclear antibodies negative. Normal acute phase proteins and immunoglobulins. Patient admits to deer on farm land and physical contact with dead fawn, but has no recall of preceding tick bite.
May 1987: No return of rash. One short lived episode of arthralgia in hands. Discharged from clinic.
September 1990: referred back to Rheumatology clinic. Three month history of recurrence of arthralgia affecting small joints of hands and feet; rash – persistent lesion back of right thigh, flu-like symptoms and shivering. On attendance he was into the third week of a month's course of doxycycline.
Small erythematous, macular lesion visible over right iliac crest. No other abnormality.
ELISA – IgG Borrelia burgdorferi – negative.
January 1991: reported short lived episode of arthralgia in fingers of right hand, otherwise asymptomatic.
No subsequent recurrence.

Case 2

Female Aged 47 Farmer's wife

Presented November 1986: emergency admission to hospital. Five day history of polyarthritis involving small joints of hands, left wrist, shoulders, knees and left ankle.
Fourteen months earlier suffered a painful left ankle, there was no preceding trauma and the joint settled spontaneously. Nine months earlier had developed a "frozen" left shoulder.
Examination: temperature 37.5^{0}C. Fine, erythematous, macular, mildly pruritic rash forearms. Synovitis pip, mcp and both wrist joints. Shoulders painful and restricted. Left ankle swollen.
Plasma viscosity 1.84 cp (1.52 – 1.72); wbc 11.2 x 10^{9}/l, CRP 11.7 mg/dl, rheumatoid factor negative, ANA negative, ASOT negative. Full infective screen negative including virology and brucella antibodies.
Admits to deer on farm. No recent recall of preceding tick bite.
Indirect immunofluorescent antibodies to Borrelia burgdorferi positive.
Initially given NSAIDs with minimal improvement. Subsequently prescribed ten days oral penicillin.
Outpatient follow-up for two years. Intermittent arthralgia hands and wrists, persistently painful, restricted left shoulder. Two intra-articular steroid injections were administered four months apart.
November 1988: minimal discomfort left shoulder. Otherwise asymptomatic and no subsequent recurrence.

Case 3

Female Aged 43 Teacher

5th December 1989: presented to General Practitioner with a right Bell's palsy.
27th December 1989: left Bell's palsy – given short course of

Prednisolone 60 mg daily by doctor friend while away from home.
Patient lives in a house in 26 acres of woodland heavily populated
by deer. She admitted to frequent tick bites and suffered an
especially painful one to the right ear-lobe November 1989 with
subsequent facial ECM.
One week later there were 3 episodes of high fever associated with
shivers, profound weakness and impaired memory.
Investigations underway, she was given ten days penicillin
empirically.
ELISA - IgG Borrelia burgdorferi positive 41.5 units. After
treatment 37 units.
Two months from onset of right facial palsy she developed severe
joint pains and stiffness in hands, wrists, elbows, knees, ankles
and feet. Pip joints swollen, unable to her remove rings.
For six months she remained disabled by articular symptoms which
reduced walking and precluded riding. There was pain on weight
bearing and wrists were particularly painful on movement. To a
lesser degree, but still troublesome, these features continued to
October 1992, relieved only by Scotch whisky (but not by any
other alcohol).
Homeopathy started, given "gelsenium". Initially this caused a
mild increase in symptoms and transient return of right Bell's
palsy. Subsequent progress excellent.
May 1993: Occasional ankle discomfort, mild tenderness above right
heel.

Case 4

Male Aged 43 Agricultural engineer

Presented to Rheumatology Clinic April 1992. Four month history of
polyarthralgia in small joints of hands, wrists, shoulders, knees
and feet. Associated episodic "flu-like" symptoms and lethargy.
Examination revealed a restricted right shoulder, effusion in left
knee and tender mtp joints.
Investigations: FBC, PVT, CRP negative. RA latex weakly
positive, ANA weakly positive. X-rays: no erosive changes in
hands or feet.
June 1992: variable joint pains, signs unchanged.
Although he works in forests, there was no recall of recent tick
bite or rash.
ELISA - IgG Borrelia burgdorferi positive 1: 200 - 1: 400.
Western Blot negative.
September 1992: left knee aspirated and injected with
methylprednisolone.
January 1993: knee settled, variable arthralgia hands and wrists
continues, associated with "flu-like" symptoms. After discussion
with Bacteriologist patient given tetracycline 500 mg qid which he
discontinued after two weeks as "flu-like" and joint symptoms
exacerbated.
March 1993: ELISA - IgG Borelia burgdorferi negative. RA latex
borderline, ANA weakly positive. No erosive changes on X-ray.
? Lyme arthritis. ? Early rheumatoid disease. ? Treatment.

DISCUSSION

Different expressions of Lyme disease in the USA and UK are thought to
represent separate strains of Borrelia burgdorferi(3), but host immunity
and possibly the early introduction of antibacterial therapy in Europe
might also be contributory(4,5). There have been increasing reports

throughout Europe of Lyme arthritis where the distribution of joint disease appears to be similar to that seen in patients from the USA(6). Nonetheless case reports from the UK remain exceptional.

Of the four cases presented here, three had rashes, but only one ECM. All had constitutional disturbances, three responded well to antibiotics when other treatments had failed, whereas in the forth patient antibiotics were discontinued after two weeks when symptoms increased. One had neurological disease which preceded the onset of joint symptoms by two months. All four had peripheral polyarthralgia with varying degrees of synovitis, three had persistent single joint involvement, the knee in two patients and a shoulder in a third, all of whom received intra-articular corticosteriods with success. Two of these patients received their intra-articular corticosteriods after antibiotic therapy and this appeared to help in the resolution of the arthritis, unlike the suggestion that intra-articular steriods can prolong joint disease before antibacterial therapy is given(7). The pattern of joint disease conformed to that accepted as occurring in Lyme arthritis, similarly the knee joint was the most commonly affected, in keeping with previous published data(5,6). Figure 1 shows the affected joints in the four subject cases. There was no particular seasonal onset to rheumatic symptoms and the latent period after infection remains undertermined. All four cases presented with the associated "virus-like" syndrome, a feature of early localised Lyme disease (stage 1).

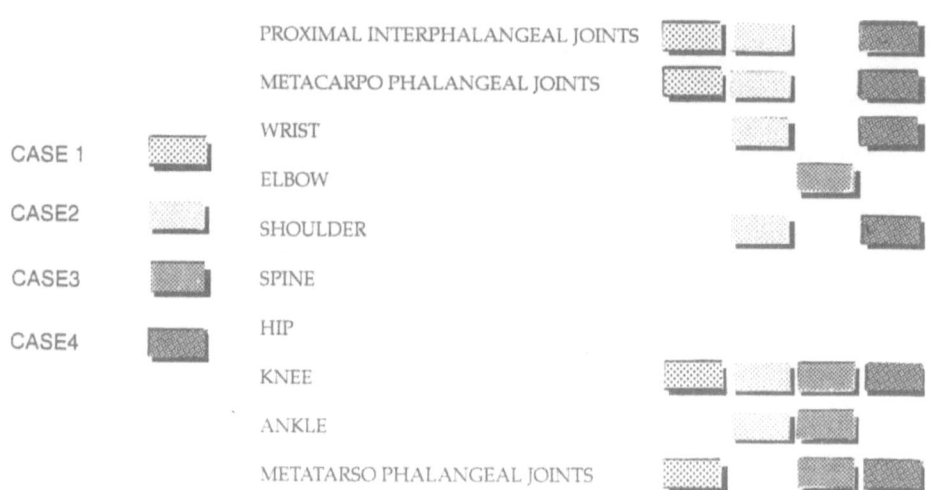

Figure 1. AFFECTED JOINTS: BASED ON SYMPTOMS

The reasons for the difficulties in recognition and diagnosis of the disease are several. Knowledge of infection with Borrelia burgdorferi is poor among Medical Practitioners in the United Kingdom. Even in areas where ticks are endemic awareness of the possibility of infection and the clinical manifestations other than ECM is probably lacking owing to the rarity of multi-system Lyme disease. Of the four patients presented here, Case 1 was referred to a hospital clinic 18 months after the onset of symptoms when the true diagnosis had not been suspected. Case 2 was admitted to hospital as an emergency, again without the diagnosis being suspected. Case 4 was referred to the Rheumatology Outpatients Department having not responded adequately to nonsteriodal anti-inflammatory drugs and when low titres of rheumatoid factor and anti-nuclear antibodies had been detected on routine investigation. The atypical nature of the arthropathy was commented on by the referring family doctor, but Lyme arthritis was not suspected. In contrast Lyme disease was swiftly suspected in Case 3 by an astute General Practitioner in rural practice familiar with the skin manifestations and constitutional upsets characteristic of early localised Lyme disease and cognizant of neuro-Borrellosis.

Attempts to confirm the diagnosis serologically relied on the techniques then available, With cases 1 and 2, investigations in 1986 were of indirect immunofluorescent antibody tests in which the accuracy of the titres with a positive test was uncertain. More recently this has been superseded by an ELISA - IgG technique. The method in use produced either a result expressed as a titre or numerically in units. In selected cases sera are further subjected to Western blotting. The interpretation of the results is hampered further by the uncertain correlation between Lyme arthritis and the degree of seropositivity. Herzer(6) divided patients with a specified minimum degree of seropositivity into those with "definite" or "probable" Lyme arthritis. The former group those with preceding ECM or neurological features while the latter had an undifferentiated arthritis without other features of Lyme Borreliosis. Patients with lower or borderline results and similar joint manifestation are not discussed. Sigal, however, makes the point repeatedly that the diagnosis of Lyme disease is clinical and not serologic(8).

The four patients presented in this paper were all at risk from tick bites, the distribution of the rheumatic symptoms compatible with the series of patients documented from the USA and mainland Europe. Cases 1 - 3 were seronegative for rheumatoid factor and antinuclear antibodies, Case 4 has borderline levels of rheumatoid factor and low levels of antinuclear antibodies present, both these findings previously documented(6,9,10) in Lyme patients. Moreover, this patient more than a year after the onset of his arthropathy has no erosive changes demonstrated radiologically and the joint disease on clinical examination more oligo-articular in type rather than the symmetrical polyarthritis characteristic of rheumatoid disease. Extensive investigations have been performed on all four subjects, reactive seronegative arthropathies in particular have been excluded and in Cases 1 - 3 prompt and significant improvements occurred with antibacterial therapy. In patient 4 the exacerbation of symptoms possibly reflected a modified Jarisch-Herxheimer reaction. Lack of experience in dealing with Lyme arthritis in 1986 meant that opportunities to seek antibodies to Borrelia burgdorferi in synovial fluid were missed. If these cases are representative of Lyme arthritis in the south west of England then it is likely that the disease is overlooked and therefore under reported. A wider appreciation of the possibility of Lyme arthritis is required, particularly by primary care physicians in rural communities when faced with a patient with an atypical seronegative arthropathy. In such circumstances referral to a Rheumatologist with an interest in Lyme disease might be helpful in establishing the diagnosis.

Currently there is too much discrepancy among results from different reference laboratories which leads to mistrust about the accuracy of information provided. There is an urgent need for laboratory investigations to be available which provide results of higher specificity and sensitivity, subjected to quality control with inter-hospital uniformity and reproducibility. These aims are increasingly being met and techniques constantly improving. That of polymerase chain reaction (PCR) holds the promise of a more sensitive diagnostic test, but as yet is still undergoing evaluation and is not widely available.

SUMMARY

Four patients are presented in whome Lyme arthritis has been diagnosed. Their modes of presentation were not uniform, other features of

Lyme Borreliosis not consistently present and the patients presented from 5 days to 18 months after the onset of symptoms. Indirect immunofluorescent antibody and ELISA - IgG tests were employed, all patients were sero-positive, but to a variable degree. Detailed clinical and laboratory investigations failed convincingly to unearth any other form of arthropathy or erosive changes on X-Ray. The outcome in 2 patients has been of complete cure; in 1, mild intermittent arthralgia still infrequently occurs 3 years after the onset; the 4th patient remains under investigation and review, antibiotics having caused an increase in symptoms.

It is clear that Lyme Borreliosis as a cause of synovitis should be more frequently considered in the differential diagnosis of arthropathies. Hitherto joint disease has apparently been exceptionally rare in the UK and more studies are required to determine the frequency of Lyme arthritis. To this end close collaboration between General Practitioners and Rheumatologists is likely to be fruitful.

REFERENCES

1. Macallan DC, Hughes, CA, Bradlow A. Lyme Arthritis in Southern England. Br Med J 1987; **294**; 1062-1063.

2. Muhlemann NF, Wright DGM. Emerging pattern of Lyme disease in the United Kingdom and Irish Republic. Lancet 1987;i: 261-262.

3. Barbour AG, Heiland RA, Howe TR. Heterogeneity of Major Proteins in in Lyme disease borreliae: A molecular analysis of North American and European isolate. J Infect Dis 1985; **152**:478 - 84.

4. Steere AC, Gibofski A, Patarroyo ME, Winchester RJ, Hardin JA, Mlawista SE. Chronic Lyme arthritis: clinical and immunogenetic differentiation from rheumatoid arthritis. Ann Intern Med 1979; **90**:896.

5. Steere AC, Schoen RT, Taylor E. The clinical evolutiion of Lyme Arthritis. Ann Intern Med 1987; **107**:725.

6. Herzer P. Joint Manifestations of Lyme Borreliosis in Europe. Scand J Infect Dis 1991 (suppl); **77**:896.

7. Steere AC, Green J, Schoen RT, Taylor E, Hutchinson GJ, Rahn DW, Malawista SE. Successful parenteral penicillin therapy of established Lyme arthritis. N Engl J Med 1984;**312**:869.

8. Sigal LH. Lyme Disease : Testing and Treatment. In: Sergent JS, Panush RS, Eds. Rheumatic Disease Clinics of North America. Saunders: Philadelphia, 1993; **19:1**; 79 - 93.

9. Levin RE. An Unusual Presentation of Lyme Arthritis. J Rheumatol 1989; **16:11** : 1500 -01.

10. Goebel KM, Kranse A, Neurath F. Acquired transient autoimmune reaction in Lyme arthritis : correlation between rheumatoid factor and disease activity. Scand J Rheum 1988 (Suppl); **75**:314.

NO EVIDENCE TO IMPLICATE *BORRELIA BURGDORFERI* IN THE PATHOGENESIS OF DILATED CARDIOMYOPATHY IN THE UK

David H E Rees,[1] Phillip J Keeling,[2] William J McKenna,[2] and John S Axford[1]

[1]Academic Rheumatology Group
[2]Division of Cardiology
St George's Hospital Medical School
University of London
Cranmer Terrace
London SW17 ORE, UK

INTRODUCTION

Lyme borreliosis (LB) is endemic in several regions of the United Kingdom[1]. The initial manifestations of the disease are the characteristic skin lesion, erythema migrans, and symptoms of general malaise (stage 1). This may be followed weeks to months later by acute articular, cardiac and neurological manifestations (stage II). Months to years later intermittent or continuous attacks of mono/oligoarthritis may occur (stage III). Nearly half of all patients with stage II and III LB however do not report a preceding skin rash or tick bite[2].

Lyme carditis was first described in the USA in 1980[3] and in Europe in 1984[4]. Histological changes have been described[5] and spirochaetes have been isolated from endomyocardial biopsies[6]. Cardiac involvement has mainly been reported as myocarditis, pericarditis and conduction disorders occuring in 1 - 8% of cases of LB[6]. It usually begins three to six weeks after the initial illness, though cardiac manifestations may be the sole presenting manifestation of LB[7], occurring without a prior history of erythema migrans or tick-bite. In a recent review of 105 cases of Lyme Carditis in the USA and Europe[8], 93% had conduction abnormalities (77% had atrioventricular block, 13% bundle branch block or intraventricular block, 3% sinoatrial block), 18% had arrythmias, 16% pericarditis and 13% heart failure. 35% required insertion of a temporary pacing wire and only one patient developed permanent AV block . 94% of all patients made a complete recovery. Only 34% remembered a preceeding tick bite and 67% reported the characteristic rash of erythema migrans, though, of note, this was less common in Europe (58%) than in the USA (82%).

Recently *B.burgdorferi* infection has been implicated in the pathogenesis of some cases of dilated cardiomyopathy (DCM). In 1991 a 54 year old man with established DCM was reported who had raised antibodies to *B.burgdorferi* and typical histological changes of Lyme carditis on endomyocardial biopsy from which Bb was isolated[9]. A subsequent report using Enzyme linked immunosorbent assay (ELISA) to detect antibodies to Bb showed a significantly higher prevalence of raised anti-Bb antibodies in patients with DCM (26.4%) compared to those with coronary artery disease (12.7%) or healthy blood donors (8.2%)[10]. Although, these findings should be interpreted with caution, as raised antibody titres detected by ELISA are not diagnostic of *B.burgdorferi* infection.

As *B.burgdorferi* has been implicated in the pathogenesis of DCM and cardiac manifestations may be the sole presentation of Lyme Borreliosis we felt it important to determine whether there was evidence for infection with *B.burgdorferi* in UK patients with idiopathic DCM.

SUBJECTS

Sera was obtained from 97 consecutive patients with idiopathic DCM (mean age 43 years, range 12-74, 71 male) referred to a St George's Hospital, a tertiary centre in the UK and 77 controls (38 healthy, age and sex matched; 39 household contacts). The diagnosis of DCM was established by WHO criteria, all patients having normal coronary arteriography and no specific heart muscle diseases.

METHODS

The patients notes were reviewed for clinical evidence of Lyme Borreliosis. Antibodies to *B.burgdorferi* were detected by enzyme linked immunoabsorbent assay (ELISA) using sonicated whole cell spirochaetes (strain B31) as antigen and samples with raised antibody levels immunoblotted to further define antibody specificity.

Enzyme Linked Immunosorbent Assay

One half of 96-well Immulon 1 immunoassay plates (Dynatech) were coated with 50 μl of a 5 μg/ml suspension of sonicated spirochaetes in carbonate buffer (pH 9.6) by incubation overnight at 4°C. The plates were subsequently blocked with 3% bovine serum albumin (BSA) in phosphate buffered saline (PBS) for 1 hour at 37°C. Serum samples were diluted 1 in 800 in PBS containing 0.05% Tween-80 (BDH), 3% goat serum and 1% BSA. 100 μl of each sample was added in triplicate to both the antigen coated wells and uncoated control wells and incubated for 1 hour at 37°C. The same sample from a patient with raised antibody levels (positive control) was also added in triplicate to each plate. After washing, the wells were further incubated for 1 hour at 37°C with goat anti-human IgG - alkaline phosphatase conjugate (Sigma Chemicals). Antibody binding was detected using p-nitrophenyl phospate substrate (Sigma Chemicals) and the optical density read at 405nm. Background binding to the uncoated control wells was subtracted from the readings and results expressed as a ratio of the mean optical density of the test sample to the positive control.

Immunoblotting

B.burgdorferi proteins (strain B31) were separated on a 10% polyacrylamide gel as described elsewhere[11] and transferred onto a nitrocellulose membrane by semi-dry blotting. Excess binding sites were blocked by incubation in 5% dried skimmed milk in PBS. Strips cut from the nitrocellulose were incubated for 1 hour at 37°C in serum diluted 1 in 500 in PBS containing 5% dried skimmed milk and 0.2% Tween-80. After washing, the strips were incubated for 1 hour at 37°C with biotinylated goat anti-human IgG (Amersham) and for a further hour in streptavidin horseradish peroxidase (Amersham), both diluted 1 in 1000 with PBS-Tween and antibody binding subsequently detected using the enhanced chemiluminescence system (Amersham). Binding of antibody to any of 14 previously reported *B.burgdorferi* associated protein bands[12] with molecular weights of 17, 22, 29, 31, 34, 39, 41, 46, 55, 60, 66, 75, 83, 94 kilodaltons (kDa) was assessed blindly by an experienced and independant observer.

Detection of Antibodies to Treponema pallidum

Samples with raised anti -*B.burgdorferi* antibody levels by ELISA (n = 9) were screened for binding to *Treponema pallidum* by the *Treponema pallidum* haemagglutination (TPHA) test. One sample was TPHA positive and this was further tested using the venereal disease reference laboratory (VDRL) test.

STATISTICAL ANALYSIS

ELISA data was analysed using a chi-squared test.

RESULTS

Patients

Patients had been symptomatic for a mean duration of 34 ± 48 months, 30 (31%) described an acute viral illness at disease onset and 13 (14%) had either clinical (n=7) or histological (n=6) evidence of myocarditis. None described the characteristic rash of erythema migrans or any other previous clinical illness compatible with Lyme Borreliosis.

Antibody Detection

Analysis of the ELISA data showed the presence of raised antibodies to *B.burgdorferi* in eight of 97 patients with DCM (8.2%) and three of 77 controls (3.9%). Immunoblot analysis however only showed very weak antibody binding to the 60 kilodalton (kd) protein in one DCM patient, 17kd in a further patient and 17 and 20kd in one other. One of the controls had very weak binding to 30 and 41kd and another to the 41kd proteins. These patterns of antibody binding with very weak reactivity are unlikely to result from infection with *B.burgdorferi*.

Antibodies to *Treponema Pallidum*

One sample was TPHA positive, this sample was further analysed and found to be VDRL negative.

DISCUSSION

In this study we have investigated patients with idiopathic DCM for clinical and serological evidence of infection with *B.burgdorferi*. We found no clinical evidence of Lyme borreliosis in this group and although, using ELISA, 7.2% of the patients with DCM tested had raised serum anti-*B.burgdorferi* antibody levels, this was not significantly more than in the control group (p = 0.3) and subsequent immunoblot analysis failed to show antibody binding patterns consistent with *B.burgdorferi* infection in any of the samples. The lack of specificity, using ELISA, is consistent with other data published[13]. Antibody levels to *treponema pallidum*, a potentially cross-reactive infection[14], were only raised in one patient.

We have therefore found no evidence to implicate *B.burgdorferi* in the pathogenesis of idiopathic dilated cardiomyopathy in the UK. In the absence of specific symptoms or likely exposure to *B.burgdorferi* we would not recommend routine serological testing for Lyme borreliosis in this group of patients.

REFERENCES

1. Muhleman MF, Wright DJM. Emerging pattern of Lyme disease in the United Kingdom and Irish Republic. *Lancet* 1987; **1**: 260-262

2. Bianchi G, Rovetta G, Monteforte P, Fumarola D, Trevisan G, Crovato F et al. Articular involvement in European Lyme disease. A report of 32 Italian patients. *Brit J Rheum* 1990; **29**: 178-180

3. Steere AC, Batsford WP, Weinberg M et al. Lyme carditis: cardiac abnormalities of Lyme disease. *Ann Intern Med* 1980; **93**: 8-16

4. Houwerzijl J, Root JJ, Hoogkamp-Korstanje JAA. A case of Lyme disease with cardiac involvement in the Netherlands. *Infection* 1984; **12**: 358

5. Reznick JW, Braunstein DB, Walsh RL et al. Lyme carditis. Electrophysiologic and histopathologic study. *Am J Med* 1986; **81**: 923-927

6. De Koning J, Hoogkamp-Korstanje JAA, Van der Linde MR, Crijns M et al. Demonstration of spirochaetes in cardiac biopsies of patients with Lyme disease. *J Infect Dis* 1989; **160**: 150-153

7. Vlay SC. Complete heart block due to Lyme disease. *N Engl J Med* 1986; **315**: 1418

8. Van der Linde MR. Lyme carditis: Clinical characteristics of 105 cases. *Scand J Infect Dis* 1991; **77** : 81-84

9. Stanek G, Klein J, Bittner R, Glogar D. Isolation of Borrelia burgdorferi from the myocardium of a patient with longstanding cardiomyopathy. *N Engl J Med* 1990; **322**: 249-252

10. Stanek G, Klein J, Bittner R, Glogar D. *Borrelia burgdorferi* as an etiologic agent in chron8ic heart failure. *Scand J Infect Dis* 1991; **77**: 85-87

11. Laemmli UK. Cleavage of structural proteins during the assembly of the head of bacteriophage T4. *Nature* 1974; **227**: 680-685

12. Bingnan MA, Christen B, Leung D, Vigo-Pelfrey C. Serodiagnosis of Lyme Borreliosis by western immunoblot: reactivity of various significant antibodies against *Borrelia burgdorferi*. *J Clin Micro* 1992; **30(2)**: 370-376

13   Berg D, Abson KG, Prose NS. The laboratory diagnosis of Lyme disease. *Arch dermatol* 1991; **127**: 866-870

14.   Magnarelli LA, Miller JN, Anderson JF, Riviere GR. Cross-reactivity of non specific treponemal antibody in serological tests for Lyme disease. *J Clin Microbiol* 1990; **28**: 1276-1279

EVIDENCE FOR LYME DISEASE IN URBAN PARK WORKERS: A POTENTIAL NEW HEALTH HAZARD FOR CITY INHABITANTS

David H E Rees and John S Axford

Academic Rheumatology Group
Division of Immunology
St George's Hospital Medical School
University of London
Cranmer Terrace
London SW17 ORE, UK

INTRODUCTION

Lyme disease is endemic in several rural regions of the UK, including the New Forest (Hampshire), Thetford forest (Norfolk) and parts of Scotland[1]. Recently, Ixodes ticks infected with *B.burgdorferi* have also been found in Richmond and Bushey parks, London[2]. These parks are close to the centre of the city, have a large population of deer and are visited by an estimated 3 million people each year. There are, so far, no reported cases of human Lyme disease resulting from a tick-bite in either of these parks, although there is a confirmed case in a dog, whose exposure to ticks was in Richmond park[2]. To determine whether Lyme disease can be contracted in these parks we have surveyed the park workers clinically and serologically for evidence of infection with *B.burgdorferi* .

SUBJECTS

Park workers (n = 44, 41 male, mean age 44.3 years, range 16-66 years) from Richmond (n = 28) and Bushey (n = 16) parks, who were employed in outside work, were questioned to assess their exposure to tick-bites and possible symptoms of Lyme disease during their employment in the park and asked to provide a blood sample. This was carried out on a voluntary basis and more than 80% of park workers participated in the study.

Comparison was made to healthy individuals (n = 49) who were either employed in a similar occupation or lived in the same geographic locality. These individuals consisted of zoo keepers (n = 27, 19 male, mean age 34.4 years, range 17-58 years) from Whipsnade wildlife park in Bedfordshire, laboratory workers (n = 10, 7 female, mean age 32 years, range 20-48 years) and bone marrow donors (n = 12, 6 male, mean age 30.5 years, range 15-50 years). All healthy controls, apart from the bone marrow donors, were questioned to determine whether tick-bites or symptoms of Lyme disease had ever occured.

To determine whether cross reactive antibodies from patients with other diseases occur, thereby confounding the interpretation of raised ELISA readings, sera was obtained from 21 patients with other infectious and inflammatory diseases: Rheumatoid arthritis (n=5), Osteoarthritis (n=5), Pulmonary tuberculosis (n=5), Reiter's disease (n=3), Sarcoidosis (n=2) and psoriatic arthritis (n=1).

DETECTION OF ANTIBODIES TO *BORRELIA BURGDORFERI*

Antibodies to *B.burgdorferi* were detected by two different, validated ELISAs using sonicated whole cell spirochaetes (strain B31) and purified flagellin, the flagellar protein of *B.burgdorferi* (SIA Lyme disease kit - Sigma diagnostics) as antigens. The specificity of these antibodies was subsequently determined by immunoblotting.

ELISA Using Sonicated *B.burgdorferi* as Antigen

One half of 96-well Immulon 1 immunoassay plates (Dynatech) were coated with 50 µl of a 5 µg/ml suspension of sonicated spirochaetes in carbonate buffer (pH 9.6) by incubation overnight at 4°C. The plates were subsequently blocked with 3% bovine serum albumin (BSA) in phosphate buffered saline (PBS) for 1 hour at 37°C. Serum samples were diluted 1 in 800 in PBS containing 0.05% Tween-80 (BDH), 3% goat serum and 1% BSA. 100 µl of each sample was added in triplicate to both the antigen coated wells and uncoated control wells and incubated for 1 hour at 37°C. A positive control was also added in triplicate to each plate. After washing, the wells were further incubated for 1 hour at 37°C with goat anti-human IgG - alkaline phosphatase conjugate (Sigma Chemicals). Antibody binding was detected using p-nitrophenyl phospate substrate (Sigma Chemicals) and the optical density (OD) read at 405nm after one hour incubation. Background binding to the uncoated control wells was subtracted from the readings and results expressed both as mean OD values and as a ratio of the mean OD of the test sample to the positive control.

ELISA Using Purified Flagellin as Antigen

This ELISA was carried out following the manufacturers instructions (Sigma diagnostics). Briefly, wells precoated with flagellin, were incubated with serum samples diluted 1 in 100 with serum diluent and antibody binding detected using anti-human IgG and IgM alkaline phosphatase conjugate and p-nitrophenyl phosphate substrate and the optical density read at 405nm after 15 minutes incubation. Each sample was analysed in duplicate and antibody binding again expressed both as mean OD values and as a ratio of the mean OD of the test sample to the positive control.

Immunoblotting

Serum antibody binding from park workers and zoo keepers, who, as they worked in a similar outdoor environment, were considered the most comparable control group, was analysed by immunoblotting. *B.burgdorferi* proteins (strain B31) were separated on a 10%

polyacrylamide gel, as described elsewhere[3] and transferred onto a nitrocellulose membrane by semi-dry blotting. Excess binding sites were blocked by incubation in 5% dried skimmed milk in PBS. Strips cut from the nitrocellulose were incubated for 1 hour at 37°C in serum diluted 1 in 200 in PBS containing 5% dried skimmed milk and 0.2% Tween-80. After washing, the strips were incubated for 1 hour at 37°C with biotinylated goat anti-human IgG (Amersham) and for a further hour in streptavidin horseradish peroxidase (Amersham), both diluted 1 in 1000 with PBS-Tween and antibody binding subsequently detected using the enhanced chemiluminescence system (Amersham) with exposure times of 30, 60 and 120 seconds.

Binding of antibody to any of 14 previously reported *B.burgdorferi* associated protein bands[4] with molecular weights of 17, 22, 29, 31, 34, 39, 41, 46, 55, 60, 66, 75, 83, 94 kilodaltons (kDa) was assessed blindly by an experienced, independant observer reviewing all 3 exposures.

DETECTION OF RHEUMATOID FACTOR AND ANTIBODIES TO *TREPONEMA PALLIDUM*

To determine whether raised *B.burgdorferi* antibody levels were due to potentially cross reactive antibodies, serum samples with the highest antibody levels by each ELISA (n = 16) were screened for IgM rheumatoid factor by latex agglutination[5] and for binding to *Treponema pallidum* by the venereal disease reference laboratory (VDRL), *Treponema pallidum* haemagglutination (TPHA) and IgG fluorescent treponemal antibody (FTA) tests.

STATISTICAL ANALYSIS

Differences in mean antibody levels were analysed using the unpaired Student's t-test and antibody levels were considered positive if the OD ratio was >2SDs above the mean of the healthy control group.

RESULTS

Clinical

10 of 44 park workers (23%) reported tick-bites within Richmond and Bushey Parks and none were bitten outside the parks. Three (7%) described an illness compatible with Lyme disease.

The first developed a red, expanding rash on his leg one week after a tick-bite which resolved spontaneously in two weeks, though, as it was clearing he developed headache and vomiting lasting one week. He was treated with antibiotics by his general practitioner. Three weeks later he developed shoulder arthralgia lasting two weeks. The second park worker developed a spontaneous, red, circular rash on his calf, which expanded and cleared from the centre in two weeks. He felt unwell and subsequently developed nausea and headache, which gradually resolved over a month. He was unaware of a preceeding tick-bite. The third park worker had an unexplained monoarthritis of the knee lasting one year and underwent arthroscopy in 1990. His symptoms have since resolved. He has lived in a lodge in Richmond park for 11 years and had noticed several tick-bites prior to his illness. None of the control population questioned reported previous tick-bites or symptoms compatible with Lyme disease.

Rheumatoid factor and antibodies to *Treponema Pallidum*

The 16 park workers with the highest *B.burgdorferi* antibody levels by ELISA were tested for IgM rheumatoid factor activity and antibody reactivity to *Treponema pallidum*. Five were positive by the FTA test (IgG antibodies), however, all these were negative by both VDRL and TPHA. Two had IgM rheumatoid factor detectable at titres of 1/1600 and 1/400 and neither of these described symptoms of Lyme disease. The first had very high antibody levels to *B.burgdorferi* using sonicated B31 as antigen (OD ratio 1.4, >3SDs above control mean) but normal levels using purified flagellin. The other had moderately raised levels only using purified flagellin as antigen (OD ratio 0.086, >2SDs above control mean). Both had one band only on immunoblot. These two park workers were omitted from the subsequent analysis as it was possible that their antibody binding to *B.burgdorferi* on ELISA was due to rheumatoid factor activity. The park workers with positive FTA tests were not omitted as all had negative VDRL and TPHA tests and this pattern of reactivity is consistent with antibodies generated as a consequence of infection with *B.burgdorferi* rather than *T. pallidum* [6].

Enzyme Linked Immunosorbent Assay

Park workers and healthy controls (table 1 and figs 1&2). Significantly raised *B.burgdorferi* antibody levels were found in the park workers when compared with healthy controls, using both the purified flagellin and whole cell sonicate antigen preparations.

7 (17%) and 10 (24%) of the 42 park workers had antibody levels above two SDs of the healthy control mean, using whole cell sonicate and purified flagellin as antigen respectively. Only one of the three park workers who described previous symptoms compatible with Lyme disease, a monoarthritis, had raised antibody levels and these were detected by both ELISAs.

Disease controls. No difference in the mean antibody level was detected between the healthy controls (0.035 ± 0.003, mean OD ± SEM) and disease controls (0.036 ± 0.014) using flagellin as antigen and significantly lower levels were seen in the disease controls (0.022 ± 0.006) compared to the healthy controls (0.045 ± 0.005) using whole cell sonicate as antigen (-50%; $p<0.01$).

Raised anti-*B.burgdorferi* antibody levels were rarely found in the disease controls, only one sample with each ELISA having antibody levels above two SDs of the healthy control mean. One sample was from a patient with Reiter's disease and was detected only using flagellin as antigen and the other from a patient with pulmonary tuberculosis and was detected only using whole cell sonicate as antigen.

Immunoblots (fig 3)

Analysis of the immunoblots from all park workers (n=44) and zoo keepers (n= 27) revealed that, out of a possible 14 *B.burgdorferi* associated protein bands[4], significantly more were detected in the park workers (mean 1.8, range 0-6) than the zoo keepers (mean 0.8, range 0-4; $p<0.001$) and a higher frequency of antibody binding was found to all bands other than the 75 kDa protein in the park workers.

Furthermore, 14 (32%) of the park workers had reactivity with three or more protein bands, whilst only one of the zookeeper group showed this level of antigen binding. Two of the three park workers who described symptoms compatible with Lyme disease had reactivity with three or more bands; one together with a positive ELISA.

Table 1
Antibody levels to *Borrelia burgdorferi* (Mean OD Ratio ± SEM (mean OD range)) were elevated in park workers when compared to other healthy individuals (*whole cell sonicate: p <0.05, **flagellin: p < 0.001)

Subjects	Antibodies against whole cell sonicate	Antibodies against flagellin
Park workers (n = 42)	0.112 ± 0.035 (0.00 - 0.741)	0.054 ± 0.005 (0.01 - 0.110)
Zoo Keepers (n = 27)	0.045 ± 0.005* (0.00 - 0.081)	0.031 ± 0.004** (0.00 - 0.078)
All Healthy Controls (n = 49)	0.045 ± 0.005* (0.00 - 0.081)	0.035 ± 0.003** (0.00 - 0.078)

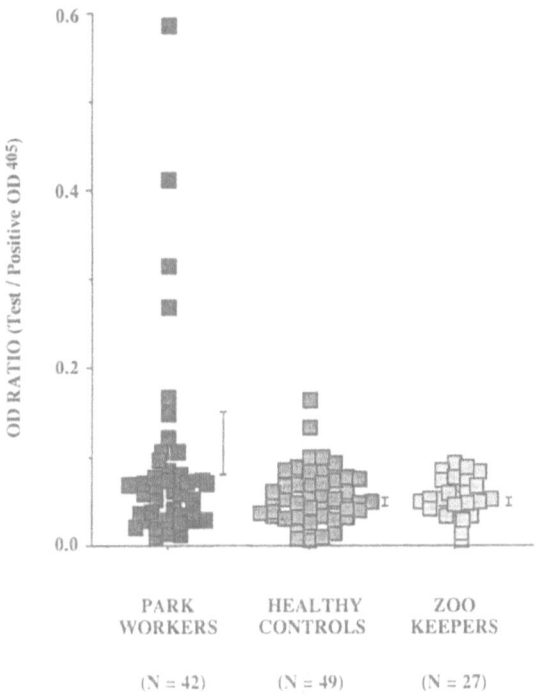

Fig. 1
Serum antibodies to *Borrelia burgdorferi* whole cell sonicate, detected by ELISA, were significantly increased in park workers when compared to zoo keepers (+148%, p<0.05) and all healthy controls (+148%, p<0.05). The squares represent individual OD ratio values and bars represent mean OD ratios ± SEM

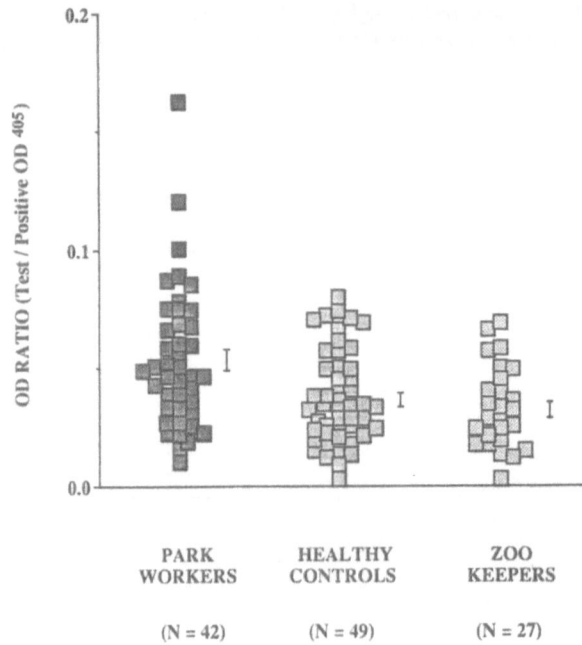

Fig.2
Serum antibodies to *Borrelia burgdorferi* flagellin protein, detected by ELISA, were significantly increased in park workers when compared to zoo keepers (+74%, p<0.001) and all healthy controls (+54%, p<0.001). The squares represent individual OD ratio values and bars mean OD ratios ± SEM

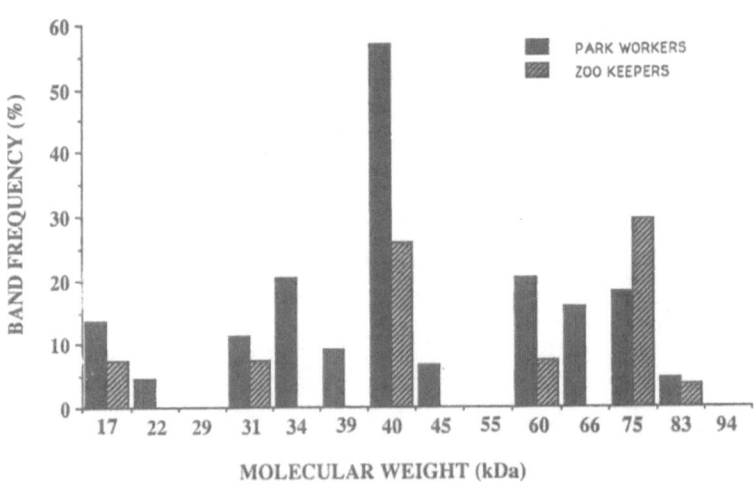

Fig.3
Immunoblot analysis showing the frequency of serum antibody reactivity to *Borrelia burgdorferi* proteins. Significantly more protein bands were detected by serum antibodies from park workers than zoo keepers (p<0.001). Fourteen (32%) of the park workers, including two individuals describing symptoms compatible with Lyme disease, had reactivity with ≥ 3 protein bands, whilst only one of the zoo keeper group showed this level of antibody binding.

DISCUSSION

This study has demonstrated that Richmond and Bushey park workers are exposed to tick-bites, develop clinical symptoms indistinguishable from Stage 1 and 2 Lyme disease and have significantly raised serum *Borrelia burgdorferi* antibody levels when compared to healthy individuals from the same locality and others in a similar occupation. Of the park workers surveyed, 23% reported tick-bites occurring within the park, whilst none had been bitten elsewhere. Two of the individuals, who had noticed tick-bites, developed clinical symptoms compatible with Lyme disease. In one, there was a rash characteristic of erythema migrans, followed by symptoms of meningism and later monoarthralgia and, in the other, an unexplained monoarthritis occurred, which resolved spontaneously. One further park worker, who was unaware of a preceeding tick-bite, described an erythema migrans-like rash, followed by symptoms of meningism. None of the park workers interviewed had current symptoms compatible with Lyme disease.

These clinical findings are unusual, were not found in the randomly selected group of healthy individuals and would strongly suggest previous infection with *Borrelia burgdorferi*. This was supported when serum antibody levels were determined, as they were significantly raised in two ELISAs utilising different *B.burgdorferi* antigen preparations . This was reinforced by the finding of substantially raised antibody levels (above two SDs of the healthy control mean) in 17% and 24% of park workers when tested against the whole cell sonicate and the flagellin antigens respectively. These findings were unlikely to result from antibody cross-reactivity, as less than 5% of samples from patients with a variety of other infectious and inflammatory diseases had raised levels.

An increase in antibody levels to both *B.burgdorferi* preparations was generally not found, as only one park worker had raised antibody levels against both antigen preparations; interestingly this occurred in the park worker with the unexplained monoarthritis. This disparity is not surprising as the immune response in Lyme disease varies with time after infection and is directed against many different protein components of the organism[7]. Importantly, the majority of individuals with raised *B.burgdorferi* antibody levels had no symptoms and this presumably indicates that usually the immune system has little difficulty irradicating this bacterium.

Confirmation of the antibody specificity was obtained by immunoblot analysis, as significantly more park workers (32%) had reactivity with three or more *B.burgdorferi* associated protein bands, when compared to the zoo keepers (4%). A higher frequency of antibody binding was found to all bands, other than the 75 kDa protein, in the park workers and two of the three who described symptoms compatible with Lyme disease had reactivity with three or more bands. That these antibodies were generated in park workers as a result of *B.burgdorferi* infection is likely, as other than geographic location, the characteristics of both groups were similar and such a high frequency of antibody binding is not common in a healthy population, nor in those with other diseases, such as rheumatoid arthritis, in which cross reactive antibodies might be expected[8]. The reason why a higher frequency of reactivity was found against the 75 kDa band in the zoo keepers cannot readily be explained, as, unlike the cross-reactive 41 and 60 kDa proteins, this band is reasonably specific[4]. It may be that these individuals have been infected with an organism which is not, as yet, known to be cross-reactive.

Tests for antibodies to *Treponema pallidum,* a potential cross-reacting organism[9], revealed five park workers had positive FTA tests, but none had antibodies detectable by VDRL or TPHA tests. This pattern of reactivity is consistent with antibodies generated as a consequence of infection with *B.burgdorferi* rather than *T.pallidum* [6], especially as such a high prevalence of syphilis is most unlikely in this population. Furthermore, the raised antibody levels were not due to non-specific binding by rheumatoid factor, as the park workers with both raised *B.burgdorferi* antibody levels and detectable rheumatoid factor were excluded from the analysis. It is, however, possible that the rheumatoid factor detected in these park workers was related to *B.burgdorferi* infection, as there is evidence that Lyme disease may be associated with rheumatoid factor production[10].

These data show, for the first time, that there is evidence for infection with *Borrelia burgdorferi* in workers from these urban parks and suggest that individuals visiting or working in them are at increased risk of contracting Lyme disease. The park workers surveyed, however, are likely to have a higher exposure to tick-bites than those visiting the parks. This may represent a changing pattern of this zoonosis in the UK, but whether the increased risk of infection translates itself into significant clinical disease for the visiting population will only be determined by further studies of patients with symptoms compatible with Lyme disease e.g. oligoarthritis and facial nerve palsies. In the meantime, these findings should alert general practitioners and hospital physicians to the possibility that a person who has visited either of these parks and has an unexplained illness may have Lyme disease. Until the public health aspects these data suggest are further investigated, our advice is to continue visiting these parks, wear clothes that protect against tick-bites when walking through undergrowth, and check for, and remove ticks after such expeditions.

REFERENCES

1. Muhleman MF, Wright DJM. Emerging pattern of Lyme disease in the United Kingdom and Irish Republic. *Lancet* 1987; **1**: 260-262 .

2. Guy EC, Farquhar RG. *Borrelia burgdorferi* in urban parks (letter). *Lancet* 1991; **338**: 253-255.

3. Laemmli UK. Cleavage of structural proteins during the assembly of the head of bacteriophage T4. *Nature* 1974; **227**: 680-685.

4. Bingnan MA, Christen B, Leung D, Vigo-Pelfrey C. Serodiagnosis of Lyme Borreliosis by western immunoblot: reactivity of various significant antibodies against *Borrelia burgdorferi*. *J Clin Micro* 1992; **30(2)**: 370-376.

5. Singer JM, Plotz CM. The latex fixation test: application to the serological diagnosis of rheumatoid arthritis. *Amer J Med* 1956; **21**: 893-898.

6. Magnarelli LA, Anderson JF, Johnson RC. Cross-reactivity in serological tests for Lyme disease and other spirochaetal infections. *J Infect Dis* 1987; **156**: 183-188.

7. Craft JE, Fisher DK, Steere AC. Antigens of *Borrelia burgdorferi* recognised during Lyme disease. *J Clin Invest* 1986; **78**: 934-939.

8. Cooke WD. Prevalence of "Natural antibody" to Borrelia burgdorferi. *Arth Rheum* 1992; **9**: S36.

9. Magnarelli LA, Miller JN, Anderson JF, Riviere GR. Cross-reactivity of non specific treponemal antibody in serological tests for Lyme disease. *J Clin Microbiol* 1990; **28**: 1276-1279.

10. Kajula GA, Steere AC, Davis J. IgM rheumatoid factor in Lyme disease: correlation with disease activity, total serum IgM and IgM antibody to *Borrelia burgdorferi*. *J Rheum 1987*; **14**: 772-776.

MANAGEMENT OF A DEER TICK BITE

Sunil K. Sood, Mark B. Salzman, Lillian Carmody, Lorry G. Rubin, and
Joseph Piesman

Division of Infectious Diseases
Schneider Children's Hospital of Long Island Jewish Medical Center
The Long Island Campus for Albert Einstein College of Medicine
New Hyde Park, NY 11042, USA

and

Lyme Disease Vector Section
CDCP Division of Vector Borne Infectious Diseases
National Center for Infectious Diseases
Ft. Collins, CO 80522, USA

INTRODUCTION

Lyme borreliosis is acquired following the bite of an infected Ixodes tick.
Inoculation of the spirochete into a human may result in a localized rash termed erythema
migrans, dissemination to distant organs through the bloodstream, or both. Because such
organ involvement can be associated with severe and occasionally persistent disease, some
physicians have considered the administration of an antibiotic following an *I. dammini* bite
as a means of preventing the infection. This practice is controversial. The efficacy of such
treatment is uncertain. The incidence of infection following an individual tick bite is low,
even in an endemic area, and needs to be balanced against the incidence of adverse effects
from an antibiotic, which can be high.[1] Shapiro et al conducted a controlled trial of
prophylaxis with amoxicillin versus placebo following a deer tick bite in a highly endemic
area in Connecticut, and found no evidence for its efficacy given the low likelihood of
infection.[2] Conversely, Majid et al. used cost-effectiveness analysis to evaluate the wisdom
of empiric treatment of all persons bitten by a deer tick, and concluded that such a strategy
was indicated in areas where the probability of infection was 3.6% or higher.[3] All
authorities agree that any recommendations to practitioners need to be based on the
consideration of a number of variables that influence the risk of infection.

GEOGRAPHIC LOCATION

The risk of infection varies with geographic location. The prevalence of *I. dammini* infected with *B. burgdorferi* varies significantly even within an endemic area. Surveys of tick infection have shown rates of 47-100% on Shelter Island and Long Island, New York, and 10-35% in Connecticut.[4,5] In contrast, only 1.5% of *I. pacificus* ticks in California were found to be infected. Unfortunately, such information is lacking for most areas.

Information regarding the risk of human infection following an *I. dammini* bite is available from several studies performed in endemic areas.[2,6-8] Despite the high prevalence of infected ticks, this risk is 1.2 to 4.4% (U.S. studies are summarized in the Table).

Table. Incidence of either clinical Lyme borreliosis or seroconversion to *B. burgdorferi* following an *I. dammini* bite.

Location (reference)	Incidence	Erythema migrans with seroconversion	Erythema migrans without seroconversion	"Flu-like" illness with seroconversion	Asymptomatic seroconversion
Westchester, NY (6)	4 of 90 (4.4%)	1		1*	2*
Connecticut (2)	2 of 173 (1.2%)	1	1		
Connecticut (7)	1 of 29 (3.4%)		1		
Long Island, NY (8)	1 of 23 (4.3%)	1			
Long Island, NY (see text)	4 of 98 (4.1%)		2		2
Total	12 of 413 (2.9%)	3	4	1	4

*Small rise in titer by immunofluorescence assay makes it unclear if all had true seroconversion.

In southern Germany, 47 (17%) of 272 *I. ricinus* ticks removed from camp attendees were found to be infected with *B. burgdorferi*, but only 1 of the 47 (2.1%) developed symptomatic Lyme disease.[9] In a rural, forested area of northeastern Italy, 26 of 221 (11.8%) of subjects not reporting a tick bite had serological evidence of antibodies to *B. burgdorferi* by immunofluorescence assay (IFA) compared to 11 of 45 (24.4%) subjects who recalled having a tick bite in the preceding two months.[10] However, the conclusion that the risk of seroconversion to *B. burgdorferi* following a tick bite in the latter endemic area is very high (positive predictive value 24%) is questionable because of the high seroprevalence in controls in this population with a potential for past tick bites, and because confirmation of IFA titers, e.g. by immunoblot, was not performed.

STAGE OF TICK AND TIME OF YEAR

The risk of acquiring *B. burgdorferi* infection also varies according to the developmental stage of *I. dammini* that bites the person. Most cases of infection result from the nymphal stage.[11] The nymph feeds in the spring and early summer when people are outdoors. It is much smaller than an adult tick and often goes unnoticed, resulting in prolonged attachment that may facilitate spirochete transmission.

DURATION OF ATTACHMENT

Another factor which appears to influence the risk of infection is the duration of attachment of the tick. The spirochete resides and replicates in the midgut of the tick, and there is a delay between the onset of feeding and the appearance of spirochetes in the saliva, via which inoculation occurs. In animal studies, the duration of tick attachment correlated closely with the rate of transmission of the spirochete. Laboratory infected *I. dammini* ticks were allowed to feed on the natural host *Peromyscus leucopus* (white-footed mouse) for various lengths of time.[12] With a mean of 15 ticks feeding on each mouse, the transmission rate increased from 7% at 36 hours to 25% at 42 hours and then steeply to 75% at 48 hours. A calculated rate of transmission which assumes that only 1 tick successfully transmitted the infection to each mouse, called the minimal transmission rate (MTR), may be more relevant with regard to tick-to-human transmission. The MTR was 0.5% at 36 hours, 1.5% at 42 hours, and 5.8% at 48 hours. Additionally, ticks fed on infected rodents for various lengths of time were homogenized and injected into mice. Unfed, 24 hour and 48 hour tick homogenates produced infection rates in mice of 0%, 20% and 80% respectively, suggesting that spirochete multiplication occurred during feeding. Thus the occurrence of one or more critical events at about 48 hours of attachment results in a marked increase in the risk of transmission.

Duration of attachment appears to be important in humans as well. Often, a tick that has been attached for several days is visibly engorged. Among 226 persons in Westchester County, New York, all 4 who developed erythema migrans within 6 weeks of the bite had engorged nymphal *I. dammini* removed.[13] In our pilot study at the Lyme Disease Center at Schneider Children's Hospital on Long Island, the only seroconversion among 23 children presenting with an *I. dammini* bite occurred in a child in whom the tick was noted to be engorged.[8]

Prospective study of deer tick bites

We hypothesized that *I. dammini* attachment for 48 hours or more increases the incidence of *B. burgdorferi* infection in humans. A likely temporal threshold for the increased risk is approximately 48 hours based on the animal studies. The present study is being conducted in the New York metropolitan area, which covers several hyperendemic foci of Lyme borreliosis in Westchester County, Long Island, New Jersey and lower Connecticut. Persons reporting a suspected *I. dammini* bite are interviewed. Ticks are inspected to ascertain their likelihood of being *I. dammini*, and other subjects are excluded. Blood is drawn at enrolment and again at 6 weeks, and sera are assayed for antibodies to *B. burgdorferi* by enzyme immunoassay, confirmed by immunoblot. No antibiotics are administered unless the subject develops symptoms.

Because the history of duration of attachment is often unreliable, we use an index of engorgement of the tick known as the scutal index as an objective measure of the duration of attachment.[14] This is calculated by dividing the body length, measured from the basis capitulum to the tip of the opisthosoma, which increases with time of feeding, by the width of the scutum, an inflexible structure. This index has been shown to have a linear correlation with the duration of attachment.[15] The duration of attachment in hours can be calculated from separate regression equations for the adult and nymphal stage, developed by Dr. Richard Falco, Westchester County Health Department.

Ticks are analyzed for carriage of *B. burgdorferi* using the polymerase chain reaction (PCR), in which a flagellin sequence is amplified by a nested primer approach.[16]

Preliminary results

Of 169 enrolled subjects with a possible *I. dammini* tick bite over the first 6 months of the study, 151 had identifiable ticks. Of these, 114 or 75.5% were *I. dammini*. Ninety-eight subjects had complete data. Fifty-two percent were the nymphal stage, 42% were adult female ticks, and 6% were larvae. Thirteen percent were positive for *B. burgdorferi* by PCR.

Four of 98 (4.1%) subjects developed evidence of Lyme borreliosis, 1 of 72 (1.4%) with tick attachment <48 hours, versus 3 of 36 (11.5%) with attachment ≥48 hours, ($p < 0.05$). Two manifested with erythema migrans, and 2 had asymptomatic seroconversion. Tick analysis showed a PCR negative adult female tick in the subject with attachment <48 hours. Two nymphs, both PCR positive, and 1 adult female, PCR negative, were the implicated ticks in the subjects with attachment ≥48 hours.

These limited results suggest that persons with a brief duration of deer tick attachment are at a very low risk of *B. burgdorferi* infection. The risk of infection following 48 hours or longer attachment is eight-fold higher. However, the power of the study is limited, because of the relatively small number of subjects enrolled, and these findings remain tentative.

CONCLUSION

The risk of *B. burgdorferi* infection following a deer tick bite is low, even in an endemic area. A large proportion of tick bites are not deer tick bites, which further lowers the risk. Currently, most experts agree that prophylactic antibiotic treatment is not warranted even in most endemic areas. Identification of persons at much higher risk of acquiring infection may be possible with the help of data on regional prevalence of infected ticks and when further data are available on the relationship of duration of attachment to transmission of infection in humans.

ACKNOWLEDGEMENTS

We thank Doctors Vincent Bonagura, Eileen Hilton, Kevin Feig and Martin Mayer for their assistance with the study.

REFERENCES

1. R.B. Nadelman, S.W. Luger, E. Frank, et al., Comparison of cefuroxime axetil and doxycycline in the treatment of early Lyme disease, *Ann Intern Med.* 117:273-80 (1992).

2. E.D. Shapiro, M.A. Gerber, L.P.N. Holabird, et al., A controlled trial of antimicrobial prophylaxis for Lyme disease after deer-tick bites, *N Engl J Med.* 327:1769-73 (1992).

3. D. Magid, B. Schwartz, J. Craft, and J.S. Schwartz, Prevention of Lyme disease after tick bites -- a cost-effectiveness analysis, *N Engl J Med.* 327:534-41 (1992).

4. J.F. Anderson, Mammalian and avian reservoirs for *Borrelia burgdorferi. Ann NY Acad Sci.* 539:180-91 (1988).

5. W. Burgdorfer, Vector/host relationships of the Lyme disease spirochete, *Borrelia burgdorferi, Rheum Dis Clin North Am.* 15:775-87 (1989).

6. F. Agre and R.M. Schwartz, The value of early treatment for the prevention of Lyme disease [abstract], *Am J Dis Child.* 145:391 (1991).

7. C.M. Costello, A.C. Steere, R.E. Pinkerton, and H.M. Feder, A prospective study of tick bites in an endemic area for Lyme disease, *J Infect Dis.* 159:136-9 (1989).

8. M.B. Salzman, L.G. Rubin, and S.K. Sood, Incidence of seroconversion to *Borrelia burgdorferi* by immunoblot after a tick bite in an endemic area for Lyme disease [abstract]. *Am J Dis Child.* 146:515 (1992).

9. H. Paul, H.J. Gerth, and R. Ackermann, Infectiousness for humans of *Ixodes ricinus* containing *Borrelia burgdorferi, Zentralbl Bakteriol Mikrobiol Hyg. [A]* 263:473-6 (1986).

10. R. Milanese, M. Marin, A. Antonini-Canterin, et al., Borreliosis risk after tick bite in north-eastern Italy, *Microbiologica* 14:357-9 (1991).

11. J. Piesman, T.N. Mather, G.J. Dammin, S.R. Telford, C.C. Lastavica, and A. Spielman. Seasonal variation of transmission risk of Lyme disease and human babesiosis, *Am J Epidemiol.* 126:1187-9 (1987).

12. J. Piesman, Dynamics of *Borrelia burgdorferi* transmission by nymphal *Ixodes dammini* ticks, *J Infect Dis* 167:1082-5 (1993).

13. R.B. Nadelman, G. Forseter, H. Horowitz, et al., Natural history of patients bitten by *I. dammini* in Westchester County, New York, USA, *in*: "Program and Abstracts of the Fifth International Conference on Lyme Borreliosis, 1992," Federation of American Societies for Experimental Biology, Bethesda, MD, A80, abstract (1992).

14. J. Piesman and A. Spielman. Human babesiosis on Nantucket Island: Prevalence of *Babesia microti* in ticks, *Am J Trop Med Hyg.* 29:742-746 (1980).

15. R.C. Falco, D. Fish, and J. Piesman. Duration of *Ixodes dammini* attachment to humans, *in*: "Program and Abstracts of the Fifth International Conference on Lyme Borreliosis, 1992," Federation of American Societies for Experimental Biology, Bethesda, MD, A46, abstract (1992).

16. B.J.B. Johnson, C.M. Happ, L.W. Mayer, and J. Piesman, Detection of *B. burgdorferi* in ticks by species-specific amplification of the flagellin gene, *Am J Trop Med Hyg.* 47:730-741 (1992).

LYME BORRELIOSIS IN AUSTRALIA

Richard D. Barry,[1] Bernard J. Hudson,[2] Darren R. Shafren[1]
and Michelle C. Wills[1]

[1]Faculty of Medicine, The University of Newcastle, NSW, Australia
[2]Royal North Shore Hospital, Sydney, NSW, Australia

SUMMARY

Since the first clinical description of the Lyme syndrome in Australia in 1982, there have been continuing efforts, so far with little success, to demonstrate that a genuinely indigenous Lyme borreliosis exists in the country. *Ixodes sp.* ticks are abundant throughout coastal regions of the eastern seaboard and while they have been found to contain fastidious, slowly growing borreliae, none have proved so far to be *B. burgdorferi*. Neither has *B. burgdorferi* been isolated nor detected in humans or other vertebrates.

On the assumption that Lyme borreliosis in Australia will be similar to its northern hemisphere equivalent we are assembling a register of likely chronic, late stage LB patients. Representative clinical and serological data for five candidate patients are provided in this presentation. As well as excluding other likely causes of disease, we have tested these patients for evidence of antibody specific to the *B. burgdorferi* antigens 41 kDa (flagellin) 34 kDa (OspB) and 31 kDa (OspA) of each of the three strains B31, ACA-1 and NBS-16, using an immunoblotting technique. Two patients were antibody positive for all flagellin antigens, and for the OspA of ACA. We conclude that the illness experienced by these patients is highly likely to have been caused by exposure to a strain of *B. burgdorferi* that has antigenic epitopes in common with European rather than US strains. The antibody response of the remaining patients is varied, but all had antibody to the OspA antigen of at least one test strain; no reaction to OspB was detected.

These results persuade us that at least one, and perhaps several, types of *B. burgdorferi* are circulating in Australia. The exact nature of these agents awaits their successful isolation, in the meantime it may be possible by further immunoblotting, using a variety of other *B. burgdorferi* isolates, to develop an identikit image of the antigenic makeup of the cause of Australian LB.

INTRODUCTION

Despite much conjecture and abundant anecdotal reports of tick bite associated "Lyme"-like illness in eastern coastal Australia, and an ever-increasing number of requests to diagnostic laboratories for Lyme serology, it is not known currently whether Lyme borreliosis (LB) exists at all, or to any significant extent on the Australian continent. There are no published reports or reviews of the likely symptomatology, severity, incidence or prevalence of suspected LB, nor are there considerations of its differentiation from other endemic vector borne illnesses. Only three sporadic case reports, published in the early to mid 1980s (Stewart *et al.*, 1982; Lawrence, Bradbury & Cullen 1986; McCrossin 1986) hint at the possible existence of LB but no confirmatory laboratory data were obtained.

Nevertheless, there appear to be many hundreds of patients, particularly in tick infested areas of coastal New South Wales, who are chronically ill with what could be late stage LB. Their illnesses are usually of at least several years duration and associated with antecedent tick bite. Many recall or report EM (often recurrent) and at present have chronic arthritis or fibromyalgia, cranial nerve palsies and/or a variety of other central and peripheral nervous system problems; on subjective assessment, neurological rather than rheumatological problems seem to predominate. There is need for the establishment of criteria appropriate for Australia on which a presumptive clinical diagnosis of LB can be based.

One of the major reasons for lingering uncertainty as to the reality of LB in Australia is that *Borrelia burgdorferi* or related Borreliae have yet to be isolated (or detected by PCR technology).

Another difficulty is that the arthropod vector, presumably an *Ixodes* species tick, has not been identified with certainty. There are reasons for believing that both the natural history of indigenous LB in Australia will prove to be quite different from that occurring elsewhere and that there will be major biological and structural differences in the causal Borreliae. An explanation such as this may account for the apparently disappointing outcome of serological testing. Using *B. burgdorferi* B31 as antigen in either IFA or ELISA tests, only 38 (3.1%) positive patients were detected out of 1296 likely LB sera tested (Russell *et al.*, 1991).

Establishing a definitive diagnosis of LB in Australia is complicated not only by the absence of a suitable serodiagnostic confirmatory test, but by the co-existence in likely endemic areas of other vector borne illnesses. Most common is epidemic polyarthritis (EP) or Ross River disease, a mosquito borne arbovirus with which physicians are quite familiar and whose chronic manifestations include a severe monoarticular or oligoarticular arthritis that may persist for years and could be mistaken for Lyme arthritis. The disease manifestations of rickettsial spotted fever (*Rickettsia australis*), Q Fever (*Coxiella burneti*) have also to be considered and the sero cross-reactivity of other spirochaetal (leptospirosis and syphilis) and viral (EBV) illnesses have also to be taken into account.

To determine whether LB exists to any significant extent in Australia we are working towards two objectives, namely, a case definition that might be helpful in the differentiation of LB from other chronic diseases, and the development of a suitable serodiagnostic test that will confirm the diagnosis of LB.

We have set about finding likely LB patients. Not only is a detailed clinical history obtained, but subjects undergo thorough clinical assessment and serology for exclusion of EP and other infectious and non-infectious chronic illness. Because of the well-documented shortcomings of IFA and ELISA tests for LB diagnosis, we have investigated the usefulness of immunoblotting as a confirmatory test, concentrating particularly on the detection of antibody to the 41kDa flagellin polypeptide and to the

34kDa and 31kDa outer surface proteins, the OspA and OspB polypeptides; three strains of *B. burgdorferi* were used as antigen - B31 (US origin), ACA-1, NBS-16 (European origin).

THE SEARCH FOR BORRELIAE

Since 1990 several Australian laboratories have been collecting and testing likely tick vectors for the presence of *B. burgdorferi* or related spirochaetes by culture in BSK medium (Wills & Barry, 1991; Russell *et al.*, 1991); findings so far are virtually identical. In our hands we find that characteristically a positive spirochaete culture requires up to 8 weeks for the optical detection of the microbe, by which time the culture is grossly contaminated and the number of spirochaetes/culture remains low until becoming senescent. Filtration and/or selective antibiotic inhibition has a very low success rate for eliminating contaminants and does not enhance the growth of spirochaetes. Of 167 processed ticks, we obtained 70 positive cultures (42%), but only 4 have been concentrated and purified to the point of further characterization. Acrylamide gel electrophoresis indicates that these isolates have a polypeptide profile typical of *Borrelia burgdorferi*; they react weakly with MAB H5332 which recognises an OspA epitope. In our experiments, the presence of flagellin has been detected in several isolates by PCR, and a morphological similarity to *B. burgdorferi* is detected when stained for immunofluorescence using a flagellin specific MAB (Fig. 1). Russell *et al* (1991) report PCR amplification products of the size expected for the OspA gene in some of their isolates. Thus *B. burgdorferi*-like spirochaetes are relatively common in *Ixodes* and several other tick species collected from regions of Australia where LB is suspected to occur. It is highly likely that the cause of LB in Australia will ultimately be detected from among these microbes.

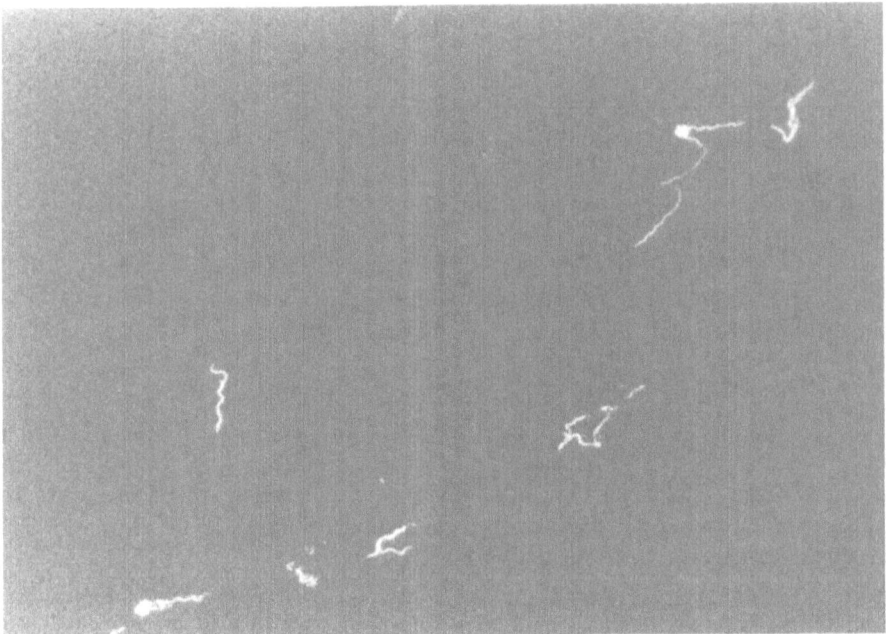

Figure 1. *Borrelia* isolate H-51 obtained on BSK culture of *Ixodes holocyclus* adult female tick. Immunofluorescence x 40 following reaction with MAB H5332.

Findings are presented for five patients who collectively exemplify the spectrum of illness and *B. burgdorferi* serology of an evergrowing number of subjects whom we have examined and tested. Each patient lives in heavily tick infested regions of coastal New South Wales, either to the north or south of Sydney, and with one exception, has never travelled outside Australia. Each patient is seronegative for EP, Q-fever, leptospirosis, *Rickettsia australis*, syphilis and rheumatic fever. Details for one case (No. 3) are provided below, while the clinical findings and exposure histories for the other four patients are summarised in Table 1.

Table 1. Pattern of illness of LB patients

Case	Age/Sex	Illness	Duration of illness	Where infected	Overseas travel
1	36/M	Rash, Fibromyalgia CFS[1]	5 years	Wingham[3]	N
2	47/M	Rash, Fibromyalgia, CFS[1]	12 years	Grafton[3]	N
3	9/M	Monoarthritis Fever	4 months	Avalon[4]	N
4	60/M	Rash, arthritis Fibromyalgia, facial nerve palsy	3 years	Tea Gardens[3]	Y
5	52/F	EM[2] Fibromyalgia	2.5 years	Nowra[3]	N

[1]Chronic fatigue syndrome
[2]*Erythema migrans*
[3]Coastal town of New South Wales, Australia
[4]Beach suburb of Sydney

Case Description (No. 3)

Presentation: A 9 year old male from Avalon, a suburb in the Northern Beaches Shire of Sydney, presented in May 1992 for evaluation of an illness of 4 months duration. The illness was characterised by recurrent fevers up to 39°C, left-sided headaches, with photophobia in the left eye, persistent lethargy, dyspnoea on exertion, swelling of left knee joint, with development of a popliteal fossa cyst. His performance at school had deteriorated over the previous 18 months.

History: There was no history of overseas travel. He had sustained many tick bites around his home in Avalon where he had lived for 9 years. He had many episodes of rash related to these which were labelled as hypersensitivity rections to tick bites. A history of typical EM rash was not elicited.

Management: A 14 day course of oral erythromycin (400mg daily) was prescribed resulting in an improvement in most symptoms, including resolution of fever, but these recurred within 4-6 weeks. He was re-treated with the same course of erythromycin again but also relapsed with further fevers and recurrence of knee joint swelling. In September 1992 he was treated with ceftriaxone, 1g intravenously, for 28 days. During the second week of this treatment he became most unwell with high fevers up to 39°C, increased swelling and pain in his left knee joint (still with popliteal fossa cyst) and photophobia. Between days 21-28 of treatment his symptoms resolved, including the knee joint swelling. His general condition improved dramatically following this treatment and he has remained well (as at February 1993). His teacher and parents have witnessed a dramatic improvement in his scholastic performance since October 1992. He has returned to his full sporting activity level.

SEROLOGICAL TESTING FOR LYME BORRELIOSIS

Immunoblotting in diagnosis

The prominent immunodominant antigens of *B. burgdorferi* are the 94kDa, 60kDa, 41kDa (flagellin) 34kDa (OspB), 31kDa (OspA), 30kDa, 21kDa (OspC) and 17/18kDa proteins (Wilske *et al.*, 1990). In the WB technique it has been found the 60kDa, 41kDa and 34kDa polypeptides reveal marked cross reactivity both with other spirochaetes and a variety of unrelated bacteria (Bruckbauer *et al.*, 1992); the early immune response in LB is triggered mainly by the 41kDa flagellin (Craft *et al.*, 1986). In late stage disease, antibodies to various polypeptides are present, particularly those in the range of 21-39kDa; among the latter, OspA seems to be one of the specific, if not unique, structural antigens (Bruckbauer *et al.*, 1992). OspA is immunodominant for the late immune response in American patients (Barbour, 1988) but antibodies to this protein are rarely detected in European patients (Wilske *et al.*, 1990).

The sera of the five patients presented in Table 1 (all of whom we consider to be likely late stage LB patients) were tested by immunoblotting for the presence of antibody both to the group specific 41kDa flagellin antigen, and to the species-specific 31kDa OspA antigen. Three strains of *B. burgdorferi* were used as antigen - the US isolate B31 and the European isolates ACA-1 and NBS16. The results obtained with ACA are shown in Figure 2. Patient 4 (lane 3) and Patient 3 (lane 7) are both strongly positive for both flagellin and OspA antigens, while patient 1 is positive for OspA only. Antibody responses of all five patients to each of the three test strains is shown in Table 2. If OspA reactivity is used as the criterion for likely LB positivity, all five patients are positive to at least one strain. By this standard, however, B31 is the least sensitive detector of reactivity.

Immunoblotting for screening

The results presented above suggest that European strains of *B. burgdorferi* might be useful for the detection of likely LB patients. To check how well the detection of antibody to both flagellin and OspA correlates with a clinical diagnosis of LB, the serological response of 23 likely LB patients to these antigens was determined and compared to the response of two other target groups - apparently healthy individuals and those suffering from other chronic illnesses. The healthy volunteers were young adults, mostly city dwelling students, who were volunteers in an influenza vaccine trial. The 56 subjects comprising the chronic illness group consisted predominantly of patients with rheumatic arthritis, SLE and EBV mononucleosis. The results are presented in Table 3.

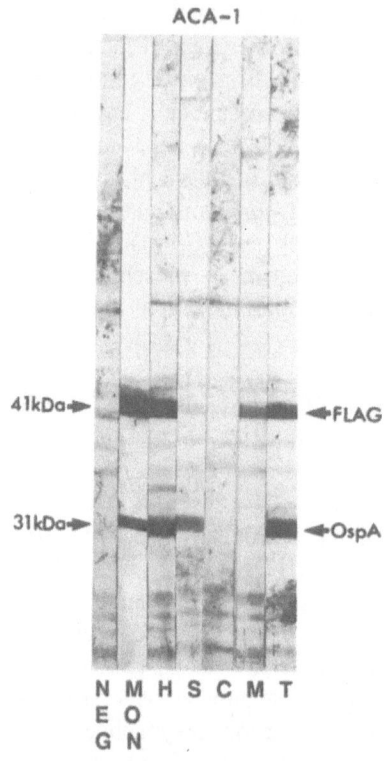

Figure 2. Western blot analysis of sera from 5 patients suspected to have LB (Table 1). Lane 1 - negative control serum; Lane 2 - MABs H9724 (41kDa) and H5332 (31kDa); Lane 3 - patient 4; Lane 4 - Patient 1; Lane 5 - Patient 2; Lane 6 - Patient 5; Lane 7 - Patient 3.

Table 2. Antibody response of candidate LB patients to the flagellin and OspA antigens of 3 strains of *B.burgdorferi*.

Antigen	Case No.				
	1	2	3	4	5
B31					
OspA	−	+	−	-	−
Flag.	−	−	+	+	−
ACA					
OspA	+	−	+	+	−
Flag.	−	−	+	+	−
NBS-16					
OspA	+	+	−	−	+
Flag.	−	−	+	+	−

Table 3. Reactivity of sera from candidate LB patients, other patients and healthy volunteers when immunoblotted against Swedish isolates of *B. burgdorferi*

	Number tested	Number positive	Percentage positive
A Healthy Volunteers	92	2	2.2
B Chronic illness other than Lyme disease	56	2	3.3
C Candidate Lyme disease patients	23	13	55.0

A,B not significant
A,C $P < 0.01$
C,B $P < 0.05$
AB,C $P < 0.01$

DISCUSSION

The diagnosis of LB is based on clinical presentation confirmed usually by positive serology. In the present uncertain state of knowledge about the nature and causation of LB in Australia, we decided that the relatively unreliable IFA and ELISA procedures are inappropriate for individual patient diagnosis and began evaluating the usefulness of immunoblotting. On the assumption that Australian borrelia will be similar, yet distinct, from B31 and possibly more closely related to European or Asian types of *Borrelia burgdorferi*, we have begun our comparison by examining the serological status of our five candidate LB patients (Table 1) when tested against B31, ACA-1 and NBS-16.

All patients had illnesses consistent with the clinical spectrum of LB. The likelihood that illness was due to causes other than *B. burgdorferi* is greatly reduced by the serological exclusion of epidemic polyarthritis, and other spirochaetal and rickettsial illnesses; other Borrelial illnesses, such as relapsing fever, do not occur in Australia. Thus, a cross-reactive response to the 41kDa flagellin, commonly associated with *B. hermsii* infection, is less likely in Australian than US patients.

Of the five patients in Table 1, two (Nos. 3 and 4) were antibody positive both to the flagellin (41kDa) and OspA (31kDa) of the Swedish human *B. burgdorferi* isolate ACA-1. It is interesting to note that while both patients also react with the 41kDa antigen of B31 and NBS-16, neither react with the OspA of those strains. The other three patients (Nos. 1,2,5) do not have flagellin reactions, but have OspA reactions to only one or two of the test strains (e.g. No. 1 - ACA-1 and NBS-16). Since OspA is one of the species specific (and possibly a unique) antigen (Bruckbauer *et al.*, 1992), we consider all patients as likely positives, with Nos. 3 and 4 being highly likely. It is paradoxical that Australian chronic LB patients have strong anti-Osp reaction while European patients do not. It is noteworthy that only 1/5 patients had antibody to B31 OspA, while 3/5 had antibody to the OspA of either NBS-16 or ACA-1. In our hands B31 rarely detects likely LB patients.

Preliminary data (Table 3) suggest that the "background" prevalence of LB in the community at large may be about 2-3%, although some of this may be attributable to cross-reactivity. Patients who, on clinical grounds, are likely to have LB have a $> 50\%$

chance of being sero-positive to certain European *B. burgdorferi*. Other Eurasian strains of *B. burgdorferi* are currently being evaluated.

Antimicrobial treatments of patients 3 and 4, including one treated with intravenous ceftriaxone, resulted in complete recovery with resolution of symptoms, a finding that is also consistent with a diagnosis of LB. Patient 3 was also strongly positive for LB when tested against commercially available IFA and ELISA reagents (data not presented). This patient has never travelled outside Australia, which suggests that his Borrelia-like infection has been acquired locally. This leads to the speculation that an indigenous cycle of LB exists in Australia, in which any or all of its main components - strain of *Borrelia burgdorferi*, arthropod vector, host reservoir - may differ significantly from their counterparts in the Northern hemisphere.

ACKNOWLEDGEMENTS

We are most grateful to Professor A.G. Barbour, Dr. V. Bundoc and Dr. D. Thomas of University of Texas Health Sciences Center at San Antonio for kindly providing strains of *B. burgdorferi*, monoclonal antibodies H5332, H6831 and H9724, and for the immunofluorescence result shown in Figure 1.

Generous financial assistance has been provided by Dr. S. Buckingham and Mr. A. Seeney, Roche Products Australia, and by the Arthritis Foundation of Australia.

REFERENCES

Barbour, A.G., 1988, Laboratory aspects of Lyme borreliosis, *Clin. Microbiol. Rev.* 1:1711-1719.

Brückbauer, H.R., Preac-Mursic,V., Fuchs, R. and Wilske, B. 1992, Cross-reactive proteins of *Borrelia burgdorferi*, *Eur. J. Clin. Microbiol. Infect. Dis*, 11:224-232.

Craft, J.E., Fisher, D.K., Shimamoto, G.T. and Steere, A.C., 1986, Antigens of *Borrelia burgdorferi* recognised during Lyme disease, *J. clin. Invest.* 78:934-939.

Lawrence, R.H., Bradbury, R. and Cullin, J.S., 1986, Lyme disease on the central coast, *Med. J. Aust.*, 145:364.

McCrossin, I., 1986, Lyme disease in the NSW south coast, *Med. J. Aust.*, 11:724-725.

Russell, R.C., Doggett, S., Dickeson, D., Hunt, C., Munro, R., 1991, Lyme disease in Australia: the current situation, *Proc. 3rd Ann. Aust. Trop. Hlth. & Nutr. Conf.* 265-279.

Stewart, A., Glass, J., Patel, A., Wyatt, G., Cripps, A. and Clancy, R. 1982, Lyme arthritis in the Hunter Valley, *Med. J. Aust.* 1:139.

Wills, M.C., and Barry, R.D., 1991, Detecting the cause of Lyme disease in Australia, *Med. J. Aust.*, 155:275.

Wilske, B., Preac-Mursic, V., Fuchs, R., Bruckbauer, H., Hofmann, A., Zumstein, G., Jauris, S., Sontschek, E., Motz, M., 1990, Immunodominant proteins of *Borrelia burgdorferi*: implications for improving serodiagnosis of Lyme borreliosis *in* "New antibacterial strategies", H.C. Neu, ed.,Churchill Livingstone, Edinburgh, 47-63.

GEOGRAPHIC DIVERSITY OF LYME BORRELIOSIS

Gerolamo Bianchi, Guido Rovetta

DIMI-Division of Rheumatology, Institute E. Bruzzone
Rheumatological Centre, University of Genova and U.S.L. XVI
Genova, Italy

INTRODUCTION

Lyme Borreliosis (LB) is a multi-system disease which, like syphilis, is characterized by early and late phases and by protean clinical manifestations that may involve different organ systems.

Symptoms of this infection, primarily affecting the skin and nervous system, were first described in Europe at the beginning of the century (1,2). However, the full clinical picture and worldwide distribution have been recognized only recently after the American reports of arthritis and carditis in patients with erythema migrans (EM), the dermatological hallmark of LB (3).

Shortly after, attention was focused on different prevalence figures of clinical manifestations of LB, especially for articular involvement, which seemed to be more frequent in American than in European patients (4). However, the issue of geographic diversity of LB is still a matter of debate and it is not yet clear whether these reported differences are real (5).

The following paper tries to review this topic with focus on Italian data.

GEOGRAPHIC DIVERSITY OF LB: FACT OR ARTIFACT?

Firstly, some differences between USA and Europe seem to exist as regards the infection and manifestation rate of LB (table 1).

Infection rate of vector ticks with *Borrelia burgdorferi* seems to be a little lower in Europe than in the USA, even if a great variability has been reported in the different studies. The variability in the infection rates in each area must be evaluated with regard to the existence of hyper-endemic areas that have been found both in the USA and Europe.

Manifestations rate of LB in individuals bitten by infected ticks is reported to be 8.5 % in the USA (6), but only 0.5 % in a study from Germany. In Germany the seroconversion rate in people bitten by an infected tick was as high as 29 %, thus leading the authors to

Lyme Borreliosis, Edited by J.S. Axford and
D.H.E. Rees, Plenum Press, New York, 1994

suggest that in Europe a great majority of Bb infection in humans might follow an asymptomatic course (7).

Geographic diversity of LB was suggested for the prevalence of the clinical manifestations of LB in different countries.

We have analyzed data from serological- and clinico-epidemiological studies presented at the V International Conference on Lyme Borreliosis (8), and we have the impression that prevalence data from North American studies are substantially homogeneous, whereas European data are more differentiated.

Table 1. Infection and manifestation rate of Lyme borreliosis

	USA	EUROPE
Infection rate	17-50	3-35
Manifestation rate	8.5	0.5

In Europe there is a great variation for the reported prevalence of skin involvement, with a higher prevalence in North European countries. There is no substantial difference between the North and South of Europe for the prevalence of nervous system involvement. and there is a higher prevalence in the Southern European countries for the prevalence of articular involvement. As far as the cardiac manifestations, these are less frequent with minor differences between European countries.

As regards Italy, where LB is endemic in two areas (Liguria and Friuli-Venezia-Giulia), epidemiological data were collected by the Italian Group for the Study of Lyme Borreliosis (IGSLB) (9). The cumulative Italian prevalence rates of clinical manifestations of LB in Italy are shown in table 2, together with the estimated prevalence rates of clinical manifestations in the USA, Northern and Southern Europe.

Table 2. Prevalence rates of clinical manifestations of LB in different areas

	Skin	Nervous system	Joints	Heart
United States	76	11	46	4
North Europe	60	33	7	1
South Europe	44	29	21	6
Italy	54	21	30	2

So, Italian data suggests a prevalence of arthritis lower than in the USA but higher than in other European countries. A prevalence of nervous system involvement higher than the USA but lower than in the rest of Europe. A prevalence of skin involvement the same as in Europe and a bit lower than that reported in American patients.

These data also seems to confirm, as previously reported, that cutaneous manifestations and arthritis are less common in European than in American patients and that, on the other hand, neurological manifestations are more frequent in Europe than in the USA.

Geographic diversity of LB not only is concerned with the absolute prevalence of involvement of any different organ system, but also with the pattern of clinical manifestations.

A milder, prolonged course of EM is more common, and multiple skin lesions are less frequent in European patients, while EM is thought to precede more often nervous system involvement in North America (10,11). Acrodermatitis chronica atrophicans (ACA) and lymphocytoma cutis appear to be almost uniquely reported in Europe, though ACA is almost exclusively present in North and Central Europe (4,12). We point out that ACA is extremely rare in Italy.

Other sclerotic and atrophic lesions, such as morphea, lichen sclerosus et atrophicans, and atrophoderma of Pasini-Pierini have been seen late in the course of European disease (14). A dermatomyositis-like syndrome has recently been described in Europe as caused by Borrelia burgdorferi infection (15).

As regards joint manifestations, the pattern of articular involvement seems to be similar in Europe and in North America, even if European patients seem to develop arthritis earlier in the course of the disease (16,17). However, in Europe diffuse hand and finger swelling has been noted as a striking feature of early Lyme arthritis (18). In North America chronic Lyme arthritis appears to have an immunogenetic basis, linked to HLA-DR4 or -DR2 specificities (19), but this has not been confirmed in European patients.

Neurologic involvement is more frequent in Europe, where authors have emphasized clinical phenomena, such as painful radiculitis (Garin-Bujadoux-Bannwarth syndrome) and progressive borrelia encephalomyelitis, and it has been pointed out that intra-techal antibodies production is a feature of European patients (20,21,22). Meningitis is more frequent in the USA, and North American patients seem to develop milder forms of nervous system involvement. Lyme encephalopathy, primarily manifested in memory, mood, and sleep disturbances, is a common late neurologic manifestation in these patients (23). The neuropathy associated with ACA, seen in European patients, seems to be quite similar, to the distal axonopathy described in American patients.

Transient atrio-ventricular block is the most frequent manifestation of Lyme carditis, with a similar prevalence of complete heart block, and myocarditis has less frequency in both areas (24). However, in Europe it has been further suggested an association or even an etiological role for Bb in dilated cardiomyopathy (25).

Moreover, in European patients there is more frequently only one clinical manifestation of Bb infection, and there is a tendency to the occurrence of fewer clinical manifestations of the disease (26), though further clinico-epidemiological studies are certainly needed for confirmation.

GEOGRAPHIC DIVERSITY OF LB: POSSIBLE EXPLANATIONS

Various explanations for the geographic diversity of LB may be suggested. However, one must consider that ascertainement bias might play a role.

Bias may be due to selection of patients by different specialists (dermatologist, neurologist, rheumatologist) with possible different levels of awareness of each clinical manifestation. To the frequent presence, in Europe, of only one clinical manifestation, and to the occurrence of mild forms of the disease, such as first degree atrio-ventricular block or rapidly disappearing erythema migrans, which patients and physicians fail to observe.

Besides, currently used diagnostic criteria are not completely satisfactory in terms of specificity and sensitivity. For these reasons, many cases of LB may not be identified and the casistics may not correctly represent the true spectrum of the disease.

Furthermore laboratory confirmation of LB is mainly based on serologic techniques and these are at present unsatisfactory. Problems in Lyme serology arise from different methods used (and their sensitivity and specificity), intra- and inter-laboratory variation, and the difficulty of defining cut-off values, as in endemic areas seropositivity in normal subjects ranges between ten and thirty percent. To assess whether a positive serology is due to asymptomatic seroconversion, active infection, or to latent infection is very difficult (27).

Besides, in some individuals, most likely immuno-suppressed, no antibodies may be found. Finally, the ability of strains to undergo antigenic variation may further confuse the serology, even in individual cases (28).

For all these reasons LB can be both overdiagnosed, mainly due to background seropositivity present in endemic areas, and, on the other hand, it can be underestimated because of symptoms which may not correctly be related to LB by general practitioners and specialists (29,30).

The first explanation (table 3) for geographic diversity of LB may be that different strains of Bb could have a variable pathogenetic capacity. The recent delineation of three genospecies of Bb, among isolates from Europe, North America and Japan (Bb sensu stricto, B garinii, and group VS461), which tended to correlate with different clinical manifestations, seems to confirm this possibility (31).

Recent findings from Italy, indicate also that strains isolated from a very restricted area can be genetically very heterogeneous (32).

Table 3. Possible explanations for geographic diversity of LB.

- strain-related clinical polymorphism
- variations in HLA-related host response
- influence of antibiotic treatment

Secondly, the difference may be due, at least for some of the late manifestations, to variations in HLA-related host response: for instance in North America chronic arthritis appears to have an immunogenetic basis, being associated with HLA-DR4 or -DR2 specificities (19), but this is not the case in Europe, where increased frequencies of HLA-A2 and Cw3 were reported in patients with Bb infection (33).

Finally, European physicians have been treating patients affected with EM or Garin-Bujadoux-Bannwarth syndrome with antibiotic therapy, since previous reports suggested the efficacy of antibiotic therapy (34,35), and this may well have changed the incidence and/or the pattern of subsequent clinical manifestations of LB. Besides, in endemic areas, the use of antibiotics, effective against Bb, in the therapy of other microbial diseases, could also play a role in modifying the incidence of clinical patterns of Bb infection. For example, an enthusiastic use of penicillin therapy in children supposed to have rheumatic fever because of arthralgias or arthritis. Another example is what happen in our region, where Rickettsia conorii is endemic, and general practitioners routinely treat subjects reporting a tick bite with tetracycline.

CONCLUSIONS

In conclusion, present data do not allow us to state definitely whether the geographic diversity of LB is indeed real, even if we feel that such a diversity seems likely to exist.

For sure, some issues will have to be addressed to answer the question whether geographic diversity of LB is a fact or an artefact.

Epidemiological studies with a larger number of people and a more accurate selection of patients are needed. Improvement in diagnostic tests will have to be made, and more data on the follow up of patients are necessary.

Finally, clinical scientists are challenged to identify other clinical manifestations that could perhaps be part of the spectrum of Borrelia burdorferi infection.

Probably more researchers are needed to clarify these points and we should also try to increase the attention paid to LB by many other medical specialties.

REFERENCES

1. Afzelius A. Verhandlungen der dermatologischer Gesell-schaft zu Stockholm. Arch Dermatol Syph 1910; 101:404.
2. Garin CH, Bujadoux M. Paralysie par les tiques. J Med Lyon 1922; 71:765.
3. Steere AC, Malawista SE, Snydman DR, Shope RE, Andiman WA, Ross MR, Steele RM. Lyme Arthritis:an epidemic of oligoarticular arthritis in children and adults in three Connecticut communities. Arthritis Rheum 1977;20:7-13.
4. Stanek G, Wewalka G, Groh V, Neumann R, Kristoferitsch W. Differences between Lyme disease and European arthropod-borne Borrelia infections. Lancet 1985; 1:401.
5. Dattwyler RJ, Volkman DJ, Luft BJ, Halperin JJ. Lyme disease in Europe and North America Lancet 1987; 1:681.
6. Shapiro ED, Gerber MA, Persing D, Berg AT,Feder H. Prevention of Lyme Disease: a randomized clinical trial of antimicrobial prophylaxis for people bitten by a deer tick.Proceedings and abstracts of the Fifth International Conference on Lyme Borreliosis, Washington,D.C.,May 30-June 2,1992. Bethesda,Md: Fed. Am. Soc.Exp. Biol. 1992; A273.
7. Paul H, Ackermann R, Gerth HJ. Infection and manifestation rate of European Lyme Borreliosis in humans. Zbl Bakt 1989; suppl 18:44-49.
8. Proceedings and abstracts of the Fifth International Conference on Lyme Borreliosis, Epidemiology and control II, Washington, D.C., May 30-June 2, 1992. Bethesda,Md: Fed. Am. Soc. Exp. Biol. 1992; A53-A62.
9. Rovetta G, Trevisan G, Cinco M, Crovato F, Parodi A, Mancardi GL, Fumarola D, Bianchi G. Italian Group for the Study of Lyme Borreliosis. Epidemiology of Lyme borreliosis . Proceedings and abstracts of the Fifth International Conference on Lyme Borreliosis, Washington,D.C.,May 30-June 2,1992. Bethesda,Md: Fed. Am. Soc.Exp. Biol. 1992; A340.
10. Asbrink E, Howmark A. Early and late cutaneous manifestations in Ixodes-borne borreliosis (erythema migrans borreliosis,Lyme borreliosis). Ann NY Acad Sci 1988;539:4-15.
11. Weber K, Neubert U. Clinical features of early erythema migrans disease and related disorders. Zbl Bakt Hyg 1986; A 263:377-388.
12. Howmark A, Asbrink E, Weber K, Kaudewitz P. Borrelial lymphocytoma. In Weber K, Burgdorfer W eds. Aspects of Lyme Borreliosis. Berlin: Springer-Verlag 1993: 122-130.
13. Danda J. Die Weltfrequenz der Akrodermatitis chronica atrophicans. Hautarzt 1963;14:337-40.
14. Aberer E, Stanek G. Histological evidence for spirochetal origin of morphea and lichen sclerosus et atrophicans. Am J Dermatopathol 1987; 9:374-379.
15. Detmar U, Maciejewski W. Borrelial dermatomyositis-like syndrome In Weber K, Burgdorfer W eds. Aspects of Lyme Borreliosis. Berlin: Springer-Verlag 1993: 259-265.
16. Steere AC, Malawista SE, Hardin JA, Ruddy S, Askenase PW, Andiman WA. Erythema chronicum migrans and Lyme arthritis: the enlarging clinical spectrum. Ann Intern Med 1977;86:685-698.
17. Bianchi G, Rovetta G, Monteforte P, Fumarola D, Trevisan G, Crovato F, Cimmino MA. Articular involvement in european patients with Lyme disease. A report of 32 italian patients. Br J Rheum 1990; 29:178-180.
18. Herzer P. Joint manifestations of Lyme borreliosis in Europe. Scand J Infect Dis suppl 1991;77:55-63.
19. Steere AC, Dwyer E, Winchester R. Association of chronic Lyme arthritis with HLA-DR4 and HLA-DR2 alleles. N Eng J Med 1990; 323: 219-23.
20. Ackermann R., Gollmen E., Relse-Kupper B. Progressive Borrelien - Encephalomyelitis. Dtsch Med. Wochenschr 1985; 110:1039 - 1042.
21. Kristoferitsch W., Spiel G., Wessely P. Zur Meningopolyneuritis (Garin-Bujadoux, Bannwarth). Klinik und Laborbefunde. Nervenarzt 1983; 54: 640-646.
22. Hansen K., Rechinitzer C., Pedersen NS, Arpi M, Jessen O. Borrelia meningitis in Denmark. Zbl. Bakt Hyg 1986; A263: 348-350.
23. Halperin JJ. North American Lyme neuroborreliosis. Scand J Infect Dis suppl 1991;77:74-80.
24. Van der Linde MR. Lyme carditis: clinical characteristics in 105 cases. Scand J Infect Dis suppl 1991;77:81-4
25. Stanek G, Klein J, Bittner R et al. Borrelia burgdorferi as an etiologic agent in chronic heart failure? Scand J Infect Dis suppl 1991;77:85-7.

26. Weber K. Clinical differences between European and North-American Lyme borreliosis. A review. Zbl Bakt 1989; suppl 18:146:155.
27. Wilske B, Preac-MursicV. Microbiological diagnosis of Lyme borreliosis. In Weber K, Burgdorfer W eds. Aspects of Lyme Borreliosis. Berlin: Springer-Verlag 1993: 267-299.
28. Craft JE, Fischer DH, Shumamoto GT, Steere AC. Antigens of Borrelia burgdorferi recognized during Lyme disease. Appearance of a new immunoglobulin M response and expansion of the immunoglobulin G response late in the illness. J Clin Invest 1986;78:934-939.
29. Matteson EL, Beckett VL, O'Falllon M, Melton LJ, Duffy J. Epidemiology of Lyme disease in Olmsted County, MN, 1975-1990. J Rheumatol 1992; 19:1743-5.
30. Blauuw AAM. Lyme arthritis in the Netherlands. A clinical and epidemiological study (Thesis). Maastricht: Universitaire Pers Maastricht 1993.
31. Baranton G, Postic D, Saint Girons I, Boerlin I, Piffaretti JC, Assous M, Grimont PAD. Delineation of Borrelia bugdorferi sensu stricto, Borrelia garinii sp. nov. and group VS461 associated with Lyme borreliosis. Int J Syst Bacteriol 1992; 42:378-383.
32. Cinco M, De Giovannini R, Fattorini P, Florian F, Graziosi G. Classification of Italian isolates of Borrelia burgdorferi in three genomic groups. 1993 Submitted.
33. Wokke JHJ, van Doorn PA, Brand A, Schreuder GMT, Vermeulen M. Association of HLA-DR antigen with serum IgG antibodies against Borrelia burgdorferi in Bannwarth's syndrome. 1988 J Neurol 235; 415-417.
34. Hollstrom E. Successful treatment of erythema migrans Afzelius. Acta Dermatolo Venereol (Stockh) 1951; 31:235-243.
35. Weber K. Eryhtema-chronicum-migrans-Meningitis eine bakterielle Infektionskrank-heit? Munch Med Wochenschr 1974; 116:1993-1998.

PARRALLELISM BETWEEN LYME AND TICK-BORNE ENCEPHALITIS (TBE)

SERO-EPIDEMIOLOGY FOLLOWING OCCUPATIONAL EXPOSURE

IN S. GERMANY

Jan Clement[1], Herwig Leirs[2], Paula McKenna[1],
Violet Armour[3], David Ward[3], Jan Groen[4],
Ab Osterhaus[4], and Christian Kunz[5]

[1]Belgian Zoonosis Workgroup, Queen Astrid
 Military Hospital, Brussels, Belgium
[2]State University of Antwerp (RUCA)
 Antwerp, Belgium
[3]10th US Medical Laboratory, Landstuhl, Germany
[4]Institute of Public Health and Environmental
 Protection, Bilthoven, Holland
[5]Institute of Virology, University of Vienna,
 Austria

INTRODUCTION

In Europe, the TBE-flavivirus (particularly the Western subtype CEE or Central European encephalitis-virus) and the Borrelia burgdorferi (Bb)-spirochete are maintained in the same wild rodent reservoirs (Clethrionomys glareolus [red bank vole], Apodemus sylvaticus [woodmouse] and Apodemus flavicollis [yellow necked field mouse] (de Boer et al, 1993),(Rodhain et al, 1987). These viral or bacterial entities are also transmitted to man by the same vector, namely a tick (Ixodes ricinus). Persons involved in open field activities, particularly in forested regions are at potential risk of infection with both of these illnesses since contact with ticks is more probable.

In a previous study by McNeil et al, (1985), conducted upon US troops with intensive open field activities in South-Bavaria, TBE IgG screening after the summer training period yielded a seroprevalence of only 1.5% (3/194) with a solid - phase CEE (Hypr strain) ELISA. None of the seropositives reported typical TBE symptoms or a previous tick bite. To date no further data has been available concerning the Lyme epidemiology in US troops in Germany.

Hence a prospective 2-sample (Spring & Autumn) serosurvey was performed to follow US military with frequent outdoor activities in Southern Germany, to detect Lyme and/or TBE seroprevalence as well as potential seroconversions.

Lyme Borreliosis, Edited by J.S. Axford and
D.H.E. Rees, Plenum Press, New York, 1994

MATERIALS AND METHODS

A total of 1,125 serum samples from 614 US soldiers, selected for training in outdoor environments, were collected in 1989. The majority of these sera were paired, taken from the same 511 individuals with an interval of about six months.

All of the subjects participating in the study were male, except one. These individuals were active duty US soldiers with a mean age of 25.9 years (range 18-52 years), stationed between 1 to 3 years in Southern Germany. The locations of the units under study (Baumholder, Bamberg, Bad Tölz) are situated in a region endemic for Lyme disease (ie. the whole of Southern Germany), but outside of the foci known so far for TBE.

In order to try to correlate Lyme or TBE seropositivity with typical symptoms, a history of tick bite or different outdoor activities, a detailed computer questionnaire was completed by the study groups in parallel with each blood sampling.

Both IgG and IgM antibodies to Borrelia burgdorferi were screened for using a commercially available Human Lyme ELISA kit (Cambridge Bioscience, Worcester, MA, USA), employing a whole cell ultrasonicate of American B31 borrelia strain as antigen.

For the detection of TBE specific IgG antibodies, successive screening was performed with a commercially available Immunozym FSME IgG ELISA assay with CEE Neudorfl strain antigen, a home made ELISA kit incorporating CEE Hypr strain as antigen (Bilthoven) and finally a reference ELISA and haemagglutination inhibition (HI) test, using highly purified antigen of a Neudorfl strain (Vienna). Only the confirmed results of the Vienna reference assay were used in the current study, with OD. readings between 63 and 126 Vienna units (VIEU/ml) considered as borderline, and OD.>126 VIEU/ml considered as positive results.

RESULTS

From the 511 individuals participating in the 2-sample study, 40 (7.8%) were found to possess antibodies to Bb and a seroconversion rate of 3.7% was also recorded in this group (Table 1).

An overall clearcut TBE seropositivity (OD.>126VIEU/ml) of 5.7% (35/614) was recorded in a total of 614 study subjects. If borderline seropositives were included, this TBE-seroprevalence augmented to even 7.17% (44/614). Data of seropositivity and -conversion in the 511 subjects with paired serum samples is described in Table 1.
Three subjects had antibodies against both Bb and TBE.

The supplementary information provided by the computer questionnaires allowed us to make the observation that in Bb and/or TBE subjects (even in sero-converters), no typical or general clinical symptoms (fever, fatigue, muscle pain etc.) were documented and no hospitalizations occurred during the 6 to 9 months study period.

In addition, no significant correlation could be made between Bb or TBE-seropositivity and a positive history of previous tick bite(s) or different forms of outdoor activities and/or exposure such as sleeping on ground, digging, clearing brush, wading etc.

DISCUSSION

The overall Bb antibody prevalence of 7.8% and the asymptomatic Lyme seroconversion rate of 3.7% recorded in US troops under study is much higher than the 2.9% (33/1,126) prevalence, respectively 1.49% (3/200) conversion rate found in a similar study in young (mean age 20.9 years +/- 4.2, range 18-28 y) Belgian paratroopers with frequent open air activities (Lefevre et al, 1990). The prevalence is lower however than the 14 to 25% rates found in other professions with a high risk of tick-bites (ie. forestry workers, farmers and track-finders) in Switzerland (Fahrer et al, 1991), UK (Morgan et al, 1989), and Bavaria (Münchhoff et al, 1986). Neither in the current study, nor in the former Belgian serosurvey (Lefevre et al, 1990) intense open field military activities, at least during the limited study period of 6 to 9 months, appeared to be linked to higher Bb antibody prevalences or -seroconversions.

A similar study, ie. a prospective 2 year serosurvey determining the prevalence rates of both Lyme disease and TBE was performed in Lisö, south of Stockholm, Sweden, with a known endemicity for both diseases (Gustafson et al, 1992).

As in our study, the reported seroprevalence rate of TBE appeared to be lower than that for Bb and asymptomatic seroconversions for TBE were also noted. The incidence of both infections was however much higher than the values we recorded, with a seroprevalence of 25.7% for Bb and 11.6% for TBE. In addition, clinical symptoms were also associated with some of the seroconvertors in the Swedish investigation.

Further parallels can be drawn between the prevalence of Lyme specific IgG antibodies and the incidence of asymptomatic Lyme borreliosis seroconversions which we noted in our population study and that recorded by Fahrer et al (1991), in a group of Swiss orienteers (26.1% [248/950] Lyme IgG seroprevalence and 8.1% [45/558] seroconversion rate) within a 6 month period.

In the present study, troops arrived mainly in the previous 1 to 3 years from N.America, where TBE is not prevalent. In this group the rate of TBE prevalence is surprisingly high (5.7%, or even 7.17% when borderlines are considered), especially upon comparison with the low prevalence (1.5%) in the same region in comparable US troops in 1984 (Mc Neil et al, 1985). In view of the fact that 17/37 (46%) of the TBE seropositive subjects had visited Flavivirus endemic regions prior to the onset of the current study, and since possible cross reactions exist between these viruses, the authenticity of the TBE seroprevalence rates remains questionable. Only with the aid of labour intensive neutralization tests can cross reactions with Yellow Fever

Table 1. Comparison of Lyme and TBE serological data in 511 U.S. soldiers participating in a 2-sample study.

	LYME				TBE			
	seropositivity		seroconversion		seropositivity		seroconversion	
Localisation	N	%	N	%	N	%	N	%
Baumholder	29/352	8.2	15/339	4.4	28/352	7.95	13/337	3.8
Bamberg	7/89	7.9	1/82	1.2	4/89	4.5	0/85	0
Bad Tölz	3/65	4.6	2/64	3.1	4/65	6.2	1/62	1.6
Other	1/4	(25)	0/3	0	0/4	0	0/4	0
TOTAL	40/511	7.8	18/488	3.7	36/510	7.0	14/488	2.9

N = Number of samples

(YF) or YF vaccines, dengue or dengue virus vaccine, Japanese B encephalitis virus vaccine, St. Louis or Powassan encephalitis be ruled out.

In conclusion, a striking parallelism would appear to exist in this US population between Bb and TBE seroprevalence and -conversion rate, albeit without any clinical implications to date. A further point which should be considered is that similaritities in the clinical presentation of both these illnesses occur in the early stages of disease. Hence, TBE should be warranted in the differential diagnosis of Lyme disease in Southern Germany especially since the methods of treatment are very different.

ACKNOWLEDGEMENTS

This study was made possible with financial support from the Belgian ministry of defence JSM R&D: Study G51, and also a grant from USAMRID, USA.

REFERENCES

de Boer R., Hovius KE., Nohlmans MKE., Gray JS. 1993, The woodmouse (Apodemus sylvaticus) as a reservoir of tick-transmitted spirochetes (Borrelia burgdorferi) in the Netherlands. Zbl Bakteriol (in press)

Fahrer H., Sjef M., van der Linden SM., Sauvain MJ., Gern L., Zhioua E., Aeschlimann A. 1991, The prevalence and incidence of clinical and asymptomatic Lyme Borreliosis in a population at risk. J Infect Dis 163:305-310.

Lefevre A., Ramon A., Clement J., Lievens M., Vranckx W., Maes L., Kreuger GRF., Fain A., Ackermann R. 1990, De seroprevalentie van Lyme Borreliose in Belgische militairen. Ann Med Mil Belg 4:171-177.

Mc Neil JG., Lednar WM., Stansfield SK., Prier RE., Miller RN. 1985, Central European tick-borne encephalitis : assessment of risk for persons in the armed services and vacationers. J Inf Dis 152:650-651.

Morgan-Capner P., Cutler SJ., Wright DJM. 1989, Borrelia burgdorferi infection in UK workers at risk of tick bites. Lancet 1:789-790.

Münchhoff P., Wilske B., Preac-Mursic V., Scierz G. 1986, Antibodies against Borrelia burgdorferi in Bavarian forest workers. Zentribl Bakteriol Mikrobiol Hyg (A) 263:412-419.

Rodhain F., Woehl-Kremer B., Perret C., Wiederkehr JL., Perez C., Hannoun C. 1987, Nouvelle observation d'encéphalite à tiques en Alsace. Médecine et Maladies Infectieuses 35-37.

INVESTIGATION OF GENETIC CHANGES ASSOCIATED WITH ATTENUATION OF *BORRELIA BURGDORFERI* BY *IN VITRO* CULTIVATION

A. Romina Emilianus, Kathryn J. Cann, David J. M. Wright,
Leonard C. Archard

Departments of Biochemistry and Microbiology
Charing Cross and Westminster Medical School
St. Dunstan's Road
London, W6 8RF

INTRODUCTION

Repeated *in vitro* cultivation of wild type strains of *Borrelia burgdorferi* sensu lato results in attenuation. Changes in (i) the presence, apparent size and antibody rectivity of surface proteins (1, 2, 3, 4), (ii) plasmid profiles (1, 5, 6, 7) and (iii) infectivity in animal models (1, 5, 6) have been observed during laboratory passage. Preferentially expressed proteins of infective strains and the genes which encode them represent potential virulence factors. Study of these factors will lead to a more clear understanding of the pathogenicity of *B. burgdorferi*.

Two biologically and geographically distinct strains, NCH-1, a North American human skin biopsy isolate and German tick, an *Ixodes ricinus* tick isolate were initially chosen for passage experiments. Infective, low passage populations of each strain have been subcultured in liquid medium. At each fifth passage the infectivity of each strain has been assessed in a mouse model using two methods; (i) culture and (ii) polymerase chain reaction (PCR) of mouse tissue. In parallel with these assays, characterizations of total protein and genomic profiles have been undertaken.

After twenty serial passages no protein, genomic or restriction fragment length polymorphisms (RFLPs) were observed for NCH-1 or German tick strains. NCH-1 was passaged a further twenty times. Detailed analysis of surface proteins and their immunogenicity has revealed a low passage-associated immunogen of NCH-1 P5 which is apparently reduced or absent from P40. However, no differences in linear and supercoiled plasmids of these passages have been observed. Infectivity of NCH-1 P40 is currently being assessed.

METHODS

Strains and *In Vitro* Cultivation

(i) NCH-1 strain was originally isolated from a skin biopsy of a patient with erythema migrans (EM) in Wisconsin (8). The strain, kindly provided by C. A. N. Hughes, was obtained in its third *in vitro* passage (P3) after its infectivity had been proven by reisolation from hamster spleen culture. (ii) German tick strain, originally isolated from *Ixodes ricinus* ticks was kindly provided by S. W. Barthold. An isolate from BALB/c urinary bladder was used to produce a pure culture in its second *in vitro* passage (P2). Each strain was cultivated at 34°C in a particular modification of BSKII liquid medium (9).

Lyme Borreliosis, Edited by J.S. Axford and
D.H.E. Rees, Plenum Press, New York, 1994

Mouse Model

Ten inbred BALB/c mice (15-25g, 4-6 weeks old) were inoculated intraperitoneally with 10^9 phosphate-buffered saline (PBS)-washed cells of NCH-1 P5, P10, P15, P20 and German tick P3, P5, P10, P15 and P20 in 5g/l glucose/PBS suspension. Four uninoculated mice served as controls. At seven weeks postinfection the animals were sacrificed. Tissues were removed aseptically for analysis by culture and PCR.

Culture. Half urinary bladder tissues were freshly homogenized in sterile saline, inoculated into 6ml BSKII medium with and without antibiotics (50µg/ml colistin sulphate, 50µg/ml rifampicin, 100µg/ml 5-fluorouracil) and incubated at 34°C. Aliquots were examined for the presence of spirochaetes by darkfield microscopy every three to four days for four weeks, and weekly thereafter. Presence of spirochaetes in a culture would indicate establishment of persistent infection in the mouse. Hence, that tested passage population could be denoted infective or noninfective. Strain NCH-1 P5 was also used to infect ten adult male CBA mice in the manner described above.

PCR. Half urinary bladders and hearts stored at -70°C were minced prior to DNA extraction and quantification (10). 2µg of DNA per sample was included in 50µl PCR reactions. The primer pair 5' GCG ATG GAT CTG GAA AAG CT and 5' TTG AGT CGT ATT GTT GTA C was designed against the *ospA* region of *B. burgdorferi* sensu stricto type strain B31 by K. J. Cann. Each primer, 0.5µmol, was combined with deoxyribonucleotides, 2mmol and *Taq* polymerase, 2.5U, to a final magnesium concentration of 1.5mM. Purified sample DNA was added finally before layering mineral oil above the reaction. Reactions were performed on an automated DNA thermal cycler (Hybaid) with denaturing at 94°C for 45s, annealing at 50°C for 90s and extension at 72°C for 3min. After 30 cycles, 10µl volumes of each reaction in loading buffer were run on 2% agarose gels at 100V. Each PCR run included a negative control for each sample and positive control NCH-1 or German tick DNA (10fg).

Protein Analysis

NCH-1 and German tick passage populations were used to produce whole cell lysates (11). The surface proteins of NCH-1 were also analysed to identify any change in their presence and immunogenicity during passage to P40. NCH-1 P5 and P40 were enriched for membrane proteins (12) and were treated with proteinase K (0.5 mg/ml) (1, 3, 13) in order to identify surface-located proteins.

Samples were resuspended in sample loading buffer (75mM tris-HCl, 100mM dithiothreitol, 2% wt/vol SDS, 10% vol/vol glycerol, 0.01% wt/vol bromophenol blue) to a final protein concentration of 1.3mg/ml (14). After boiling for five minutes samples (10µg per lane) were loaded onto 12.5% polyacrylamide gels in a vertical electrophoresis system (Biorad) for separation by sodium dodecyl sulphate-polyacrylamide gel electrophoresis (SDS-PAGE) (11). Rainbow protein molecular weight markers (Amersham) were used for calibration. Separated proteins were (i) stained with 0.0125% (wt/vol) Coomassie brilliant blue R-250 (Biorad) and destained, or (ii) electroblotted to nitrocellulose membranes for Western analysis. Membranes were incubated with dilutions of LD patient sera (1:200). Bound antibodies were detected using goat anti-human IgG labelled with peroxidase (15).

Nucleic Acid Analysis

Total DNA. Pulsed-field gel electrophoresis (PFGE) using a contour-clamped homogeneous electric field (CHEF) (16) was performed in a Pulsaphor apparatus (Pharmacia LKB) with a hexagonal electrode. Molecular weight markers consisted of lambda DNA concatemers and lambda *Hind*III digests (Promega). Genomic DNA in low melting temperature agarose inserts was prepared as described by Smith *et al.* (17).

Two-dimensional gel electrophoresis (2DGE) allows separation of linear plasmids from supercoiled circular DNA molecules, the latter lagging the linear molecules in the second dimension. In genomic insert DNA shearing is minimal and therefore more likely to represent genomic DNA in its native form in cells. Separation uses PFGE in the first dimension, then a high voltage constant field (6V/cm) in a perpendicular direction.

Restriction enzyme digests. As the genome of *B. burgdorferi* is AT-rich (68-73mol%) (18), restriction enzymes recognizing GC-rich sequences would be expected to produce a limited number of fragments. *Not*I, *Nae*I and *Eco*RI recognise the sequences GC GGCCGC, GCC GGC and G AATTC, respectively. Insert DNA was digested with 10U of the restriction endonucleases *Nae*I, *Not*I, *Eco*RI (NBL) for 16 to 24hrs as recommended by the manufacturer. Restriction fragments were resolved by PFGE.

The largest linear plasmid of NCH-1 was isolated in low melting temperature agarose after its resolution from the rest of the NCH-1 genome by PFGE. *In situ* digestion used *Bgl*II, *Dra*I, *Eco*RI and *Eco*RV, all having A/T-containing 6bp long recognition sites.

RESULTS

In Vitro Cultivation

In their first few subcultures both NCH-1 and German tick strains grew slowly and exhibited a tendancy to form large clumps at the bottom of culture tubes. Darkfield microscopy revealed these not to be simple sediments but aggregates of spirochaetes joined at one or both ends to form balls of cells. This aggregation phenomenon was seen throughout the passaging of the German tick strain to P20, but had much diminished after P18 of NCH-1 through to P40.

Assays of Infectivity

Culture. Mouse urinary bladder culture of each of ten BALB/c mice per tested passage was unsuccessful. However, NCH-1 P5 inoculated into CBA mice was positive in nine out of ten mice. Urinary bladders from control uninfected mice were negative for culture.

PCR. Urinary bladder and heart tissue samples from each of passages NCH-1 and German tick up to P20 were positive for *B. burgdorferi* DNA. Figure 1 shows the results of PCR of each of nine urinary bladder tissues from mice infected with NCH-1 P20. Urinary bladder and heart tissues from uninfected control mice were PCR negative for *B. burgdorferi* DNA.

Protein Analysis

SDS- PAGE protein profiles of whole cell lysates of German tick and NCH-1 strains did not reveal any changes up to passage P20 and, revealed no apparent change in whole cell lysate (WC) profiles of NCH-1 P5 and P40 (Figure 2). Membrane protein extracts (MP) of NCH-1 P5 and P40 revealed enrichment of 41, 38.2 and 32.7kDa proteins, supporting their abundance in the NCH-1 membrane. However 93.1, 71.3 and 28.9kDa proteins in NCH-1 P5 MP were not present in P40. Proteinase K-treated (PK) profiles revealed degradation of most of the major proteins, in particular 38.2 and 32kDa proteins, present in WC profiles. However, 41 and 27.6kDa proteins were still apparent in PK profiles, indicating some intracellular location of these proteins.

In Figure 3, Western blot analysis using serum from a patient presenting with Bell's palsy and confirmed LD (by enzyme-linked immunosorbent assays and immunoblots carried out in the LD Reference Laboratory, Charing Cross Hospital, London) revealed the immunogenicity of the NCH-1 proteins shown in Figure 2. Major proteins of NCH-1 P5 and P40, of apparent sizes 41, 38.2, 32, 28.9 and 22.1kDa reacted with the sera. Interestingly, of the three proteins observed only in the NCH-1 P5 MP profile in Figure 2, 93.1 and 71.3kDa proteins were not significantly immunoreactive in the Western blot, whereas the 28.9kDa protein does exhibit immunogenicity.

Genomic Profiles

Total DNA. The plasmid profiles of NCH-1 and German tick P5 to P20 (shown in Figure 4) were unchanged during twenty *in vitro* passages. 2DGE revealed both strains to contain linear and supercoiled circular plasmids. After twenty passages the linear and supercoiled circular plasmid profiles of NCH-1 and German tick were unchanged. Indeed,

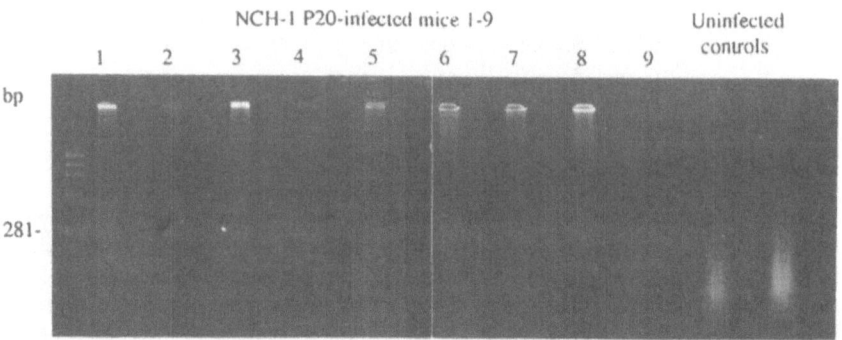

FIGURE 1. PCR products amplified from BALB/c mouse urinary bladder tissue infected with NCH-1 P20. The products were visualized after 2% agarose electrophoresis and ethidium bromide staining. Molecular size standards are shown on the left.

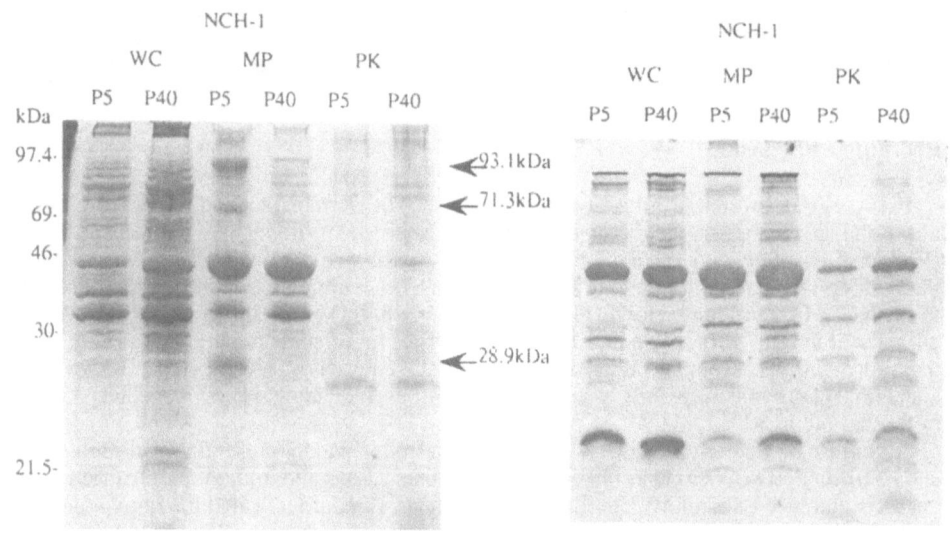

(i) FIGURE 2.
stained with Coomassie brilliant blue.

(ii) FIGURE 3.
Western blotted using a human LD serum.

Whole cell lysates (WC), enriched membrane proteins (MP) and proteinase K-treated cells (PK) of NCH-1 P5 and P40 were separated by SDS-PAGE and

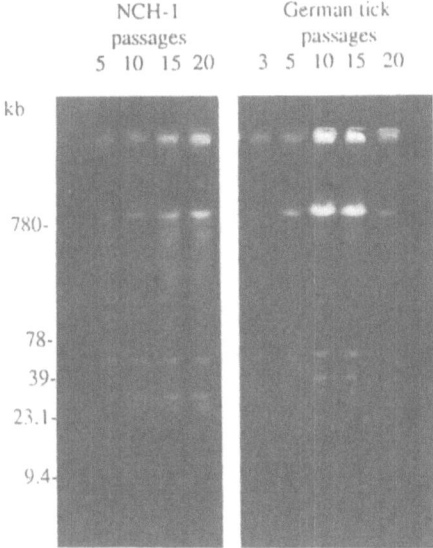

FIGURE 4. *In situ* preparations of genomic DNA from NCH-1 and German tick, separated by PFGE using stepped 1s, 2s, 5s, 10s, 25s and 50s pulses for 3hrs each. Molecular size standards are shown on the left.

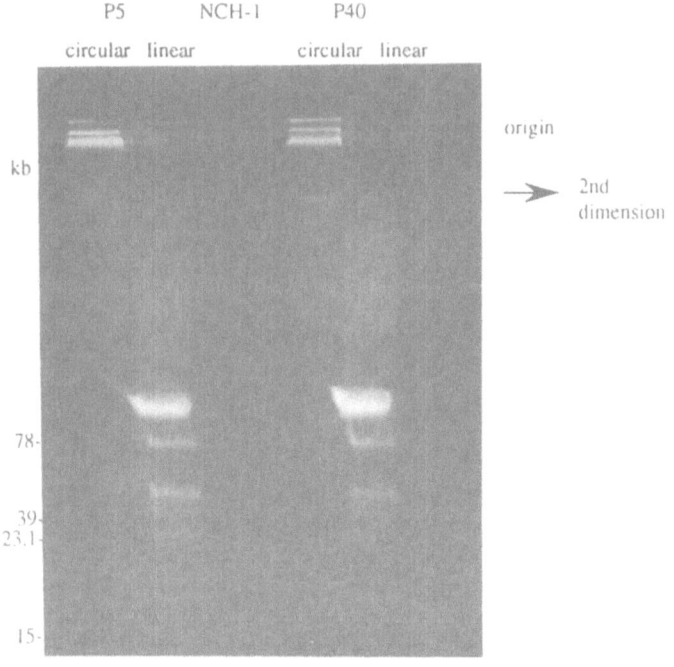

FIGURE 5. In situ preparations of genomic DNA from NCH-1 P5 and P40 were separated by 2DGE: PFGE used 1s pulses for 16hrs, then constant field electrophoresis at 90V for 2hrs at a right angle. Molecular size standards are shown on the left.

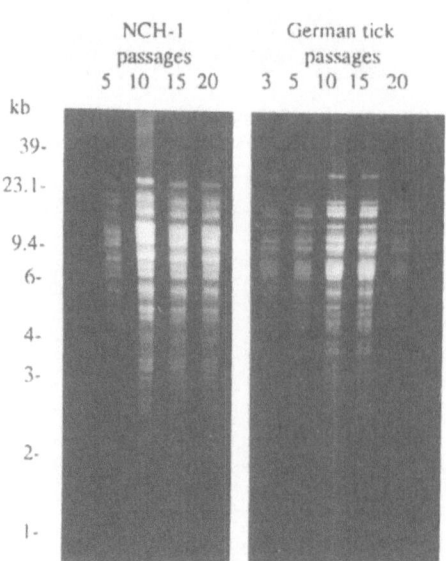

FIGURE 6. *In situ* preparations of genomic DNA from NCH-1 and German tick, digested with *Eco*RI and separated by PFGE using stepped 1s, 2s, 5s and 10s pulses for 3hrs each. Molecular size standards are shown on the left.

Table 1. NCH-1 plasmids.

Type of plasmid		Apparent size of plasmid (kb)					
Linear		76.5		47.5	30	22	17.4
Circular	>700		70				

2DGE revealed that after forty passages of NCH-1 the profile was similar to that of NCH-1 P5 (Figure 5). The full complement of NCH-1 plasmids is given in Table 1. The values given were derived from their apparent mobilities under the conditions used in Figure 5.

Restriction analysis. *Not*I does not appear to cut NCH-1 or German tick DNA. *Nae*I produced at least eighteen discreet fragments for the passages of NCH-1 and also eighteen fragments for the German tick strain passages (data not shown). The profiles of each of the strains NCH-1 and German tick were distinct from each other, however no differences were seen between isogenic passages. The results of *Eco*RI digestion (Figure 6) show this to be true, with the production of a large number of fragments.

Physical mapping. NCH-1 P5 was cultured from CBA mouse urinary bladder and represents an infective, low passage isolate. The largest linear plasmid of NCH-1, apparent size 76.5kb, hybridizes specifically to *ospA* and linear plasmid telomeric sequences from strain B31 sensu stricto (data not shown). These probes serve as markers for restriction mapping of this plasmid. *In situ* digestion used *Bgl*II, *Dra*I, *Eco*RI and *Eco*RV. The resulting fragments were separated by PFGE and their apparent sizes are given in Table 2.

Table 2. Restriction fragments of the largest linear plasmid of NCH-1.

Restriction endonuclease	Apparent sizes of fragments (kb)								Total (kb)
*Bgl*I		15	12.7	9.3	9	8	7.5	5.8 5.3	70.5
*Dra*I	muliple fragments								------
*Eco*RI	16.7	15	12.5			8.7	8.1		72.6
*Eco*RV		15.7	13.5 10.5	9.5	9.1			6.5 5.7	70.7

Mean = 71.3

DISCUSSION

Two infective, low passage isolates of *B. burgdorferi* have been passaged in liquid BSKII. Both strains exhibited a tendency to aggregate in culture, but for NCH-1 this feature was less apparent after P18 through to P40.

Isolation of spirochaetes by culture is a low yield, time-consuming procedure, but is the most definitive method for confirming infection in an animal model. The lack of positive urinary bladder culture in this study may reflect technical failure rather than failure of the BALB/c model itself. BALB/c mice have used successfully for culture of *B. burgdorferi* (4), for *in vivo* passage of *B. hermsii* (19) and for demonstrating a *B. burgdorferi*-associated cystitis (20). Urinary bladder tissue from rodents has been effective for the reisolation of *B. burgdorferi* (20, 21, 22). Indeed, *B. burgdorferi* has been observed within the bladder wall (24) exhibiting persistence (21) and possible tropism for this tissue (21, 24).

That *in vitro* passage of NCH-1 and German tick strains was initiated from infective populations is supported by the successful detection of *B. burgdorferi* DNA in infected mouse tissues by PCR (Figure 1). This indicates persistence of *B. burgdorferi* DNA in mouse tissue up to P20 and suggests maintenance of NCH-1 and German tick strain infectivity after twenty serial passages. PCR assays have reportedly comparable or slightly greater sensitivities than culture for low passage isolates (22, 23). Here, PCR has proven far more sensitive than culture of BALB/c tissue. However, it is not known whether PCR is much more sensitive than culture with higher passage isolates.

Whole cell lysates (WC) of NCH-1 and German tick strains have clearly distinct protein profiles (Figure 2). They do not, however, reveal any differences between the passages. By preparing membrane extracts (MP) surface proteins are enriched (12) and can be identified. Supporting evidence is gained by proteinase K-treatment of intact cells (PK), since surface proteolysis would cause degradation of surface-exposed proteins (1, 3, 13). MP and PK profiles of NCH-1 P5 and P40 revealed 93.1, 71.3 and 28.9kDa proteins to be preferentially present in NCH-1 P5 (Figure 2). This was not evident in WC profiles, possibly because the surface protein bands are obscured by similar-sized proteins from other cellular compartments. Western analysis of these samples (Figure 3) has revealed significant immunoreactivity of the 28.9kDa protein with serum from a human LD patient. It is therefore possible that this protein represents a low passage-associated, surface-exposed antigen. Interestingly, a 28kDa protein, OspD, which was low passage-associated and surface-exposed in strain B31 (1) was not essential for infectivity in mammals, though its involvement in pathogenesis was not ruled out.

Genomic analysis by PFGE and 2DGE has not revealed differences in plasmid profiles among the passages of NCH-1 and German tick strains to P20 (Figure 4). 2DGE analysis of NCH-1 P40, for which there was evidence of changes in surface proteins, has not revealed differences in the linear and supercoiled plasmid profiles with respect to NCH-1 P5 (Figure 5). Restriction analysis also reveals no change among isogenic populations of NCH-1 and German tick strains to P20 (Figure 6). More detailed studies to investigate possible RFLPs among plasmids of NCH-1 are in progress.

Future Studies

The problem of negative culture in the BALB/c model will be addressed. Alternative mouse strains, inoculum sizes and routes of inoculation may be investigated. Assessments of the infectivity of NCH-1 P40 can then follow.

Once attenuation has been achieved and identified, differences between isogenic populations of infective and attenuated *B. burgdorferi* can be determined. Fine mapping of the largest linear plasmid of NCH-1 P5 and P40 will be completed using *ospA* and telomere-specific probes. Mapping of the other plasmids can similarly be performed. Thus, rearrangement within the genome of NCH-1 during passage from P5 to P40 may be revealed. Studies to identify genes associated with infective passages of *B. burgdorferi* will continue.

REFERENCES

1. Norris S.J., Carter C.J., Howell J.K., Barbour A.G. Low-passage-associated proteins of *Borrelia burgdorferi* B31: characterization and molecular cloning of OspD, a surface-exposed, plasmid-encoded lipoprotein. *Infect Immun* 1992; **60**: 4662-4672.
2. Schwan T.G., Burgdorfer W. Antigenic changes in *Borrelia burgdorferi* as a result of *in vitro* cultivation. *J Infect Dis* 1987; **156**: 852-853.
3. Bundoc V.G., Barbour A.G. Clonal polymorphisms of outer membrane protein OspB of *Borrelia burgdorferi*. *Infect Immun* 1989; **57**: 2733-2741.
4. Schwan T.G., Karstens R.H., Schrumpf M.E., Simpson W.J. Changes in antigenic reactivity of *Borrelia burgdorferi*, the Lyme disease spirochete, during persistent infection in mice. *Can J Microbiol* 1991; **37**: 450-454.
5. Schwan T.G., Burgdorfer W., Garon C.F. Changes in infectivity and plasmid profile of the Lyme disease spirochete, *Borrelia burgdorferi*, as a result of *in vitro* cultivation. *Infect Immun* 1988; **56**: 1831-1836.
6. Simpson W.J., Garon C.F., Schwan T.G. Analysis of supercoiled circular plasmids in infectious and non-infectious *Borrelia burgdorferi*. *Microbial Pathogenesis* 1990; **8**: 109-118.
7. Stalhammar-Carlemalm M., Jenny E., Aeschlimann A. Plasmid analysis and restriction fragment length polymorphisms of chromosomal DNA allow a distinction between *Borrelia burgdorferi* strains. *Genetic diversity of Borrelia burgdorferi (Verlag)* 1990: 28-40.
8. Hughes C.A.N., Kodner C.B., Johnson R.C. DNA analysis of NCH-1, the first Northcentral U.S. Human Lyme Disease isolate. *J Clin Micro* 1992; **30**: 698-703.
9. Barbour, A.G. Isolation and cultivation of Lyme disease spirochetes. *Yale J Biol Med* 1984; **57**: 71-75.
10. Maniatis T., Fritch E.F., Sambrook J. Molecular cloning. A laboratory manual. Cold Spring Harbor Laboratory, New York 1982.

11. Laemmli U.K. Cleavage of structural proteins during the assembly of the head of bacteriophage T4. *Nature* 1970; **227**: 680-685.
12. Brandt M.E., Riley B.S., Radolf J.D., Norgard M.V. Immunogenic integral membrane proteins of *Borrelia burgdorferi* are lipoproteins. *Infect Immun* 1990; **58**: 983-991.
13. Finlay B.B., Falkow S. Common themes in microbial pathogenicity. *Microbiol Rev* 1989; **53**: 210-230.
14. Bradford M.M. A rapid and sensitive method for the quantitation of microgram quantities of protein utilizing the principle of protein-dye binding. *Anal Biochem* 1976; **72**: 248-254.
15. Barbour A.G., Tessier S.L., Todd W.J. Lyme disease spirochetes and ixodid tick spirochetes share a common surface antigenic determinant defined by a monoclonal antibody. *Infect Immun* 1983; **41**: 795-804.
16. Chu G., Vollrath D., Davis R.W. Separation of large DNA molecules by contour-clamped homogeneous electric fields. *Science* 1986; **234**: 1582-1585.
17. Smith C.L., Warburton P.E., Gaal A., Cantor C.R. Analysis of genomic organisation and rearrangements by pulsed-field gel electrophoresis. *Genetic Engineering* 1986; **8**. Plenum Press, New York.
18. Hyde F.W., Johnson R.C. Genetic relationship of Lyme disease spirochetes to *Borrelia, Treponema* and *Leptospira* spp. *J. Clin Micro* 1984; **20**: 151-154.
19. Stoenner H.G., Dodd T., Larsen C.. Antigenic variation of *Borrelia hermsii*. *J Exp Med* 1982; **156**: 1297-1311.
20. Czub S., Duray P.H., Thomas R.E., Schwan T.G. Cystitis induced by infection with the Lyme disease spirochaete, *Borrelia burgdorferi*, in mice. *Am J Path* 1992; **141**: 1173-1179.
21. Goodman, J.L., Jurkovich, P., Kodner, C. Persistent cardiac and urinary tract infections with *Borrelia burgdorferi* in experimentally infected Syrian hamsters. *J Clin Micro* 1991; **29**: 894-896.
22. Lebech A.M., Hindersson P., Vuust J. Comparison of *in vitro* culture and polymerase chain reaction for detection of *Borrelia burgdorferi* in tissue from experimentally infected animals. *J Clin Micro* 1991; **29**: 731-737.
23. Pachner A.R., Ricalton N., Delaney E. Comparison of polymerase chain reaction with culture and serology for diagnosis of murine experimental Lyme borreliosis. *J Clin Micro* 1992; **31**: 208-214.
24. Schwan T.G., Burgdorfer W., Schrump M.E., Karstens R.H. The urinary bladder, a consistent source of *Borrelia burgdorferi* in experimentally infected white-footed mice (*Peromyscus Leucopus*). *J Clin Micro* 1988; **26**: 893-895.

PHYTOECOLOGICAL MAPPING OF <u>IXODES</u> <u>RICINUS</u> AS AN APPROACH

TO THE DISTRIBUTION OF LYME BORRELIOSIS IN FRANCE

B. Gilot[1], C. Guiguen[2], B. Degeilh[2], B. Doche[3],
J. Pichot[4], and J.C. Beaucournu[2]

[1]INSERM, Laboratoire de Parasitologie
   Faculté de Médecine - Aix-Marseille II
[2]Laboratoire de Parasitologie et Zoologie appliquée
   Faculté de Médecine - Rennes I
[3]Laboratoire d'Ecologie végétale - Grenoble I
[4]Laboratoire de Parasitologie
   Faculté de Pharmacie, Université Claude Bernard - Lyon

INTRODUCTION

<u>Ixodes ricinus</u> is considered as the main vector of Lyme Borreliosis in western Europe. As the involved bacterium, <u>Borrelia burgdorferi</u>, seems to be present in that tick species wherever it is found, the distribution of tick populations is of considerable interest from an epidemiological point of view. A common idea is that the tick is a common, widely distributed species, but, with the exception of some small countries (like Switzerland) or countries with relatively homogeneous conditions, the distributional records appear to be poor and too localized with regard to the ecological diversity of the countries concerned.

In France, the small scale distribution of the tick which was published in 1972 by Rageau was drawn up from some collections made by different authors and was not suitable for epidemiological purposes. A more detailed study, but restricted to the French Alps and their periphery was published by Gilot in 1985. The aim of the present study was to ascertain the presence of this tick species in the different ecological parts of France and to draw up a small scale map in order to have an overview of the epidemiological risk linked to the species in relation to the environmental conditions.

MATERIAL AND METHODS

We used a "stratified random sampling method" (Daget and Godron, 1982). The sampling procedure used in this study was described by Doche et al., 1993.

1) Phytoecological zoning

A zoning was first carried out in order to define phytoecological and geographical units. Thus fifty four geographically separate and ecologically distinct units were defined.

For the purpose of the study, we used a lot of vegetation maps (scale 1/200.000) which cover practically the whole French territory (66 sheets have been published) and a synthetic work published by Dupias and Rey in 1985.

2) Collecting methods

Tick surveys were conducted in 1992 by three teams of coworkers. Each team studied more especially one quadrant of France (i.e. south-western, north-western, eastern). South-eastern France had already been studied by Gilot (1985). Regions more than 1,000 metres high and the Mediterranean area were neglected, as previous work had evidenced that the Ixodes species is absent or very scarce in such territories. In France, as the tick lives more particularly in wooded areas, only the different types of forest were investigated ; the sites to be prospected were pointed out by an "ecology-specialist" without any a priori idea with regard to their ability to harbour the tick species. The number of forests prospected in each unit depended on its area and its ecological diversity.

In a range of 5 to 15 forest experimental sites, the flagging method was used for one hour's time to take samples of the tick during its period of activity (from late April to the end of October). A total of 309 sites was investigated (figure 1).

Figure 1. Repartition of the sampling sites

3) Tick identifications

We used classic works, such as Pomerantsev's (1959), Arthur's (1963), Morel and Vassiliades's (1962). Because Ixodes ricinus is not the only Ixodes species to be detected by the flagging method (a few adult and preimaginal specimens of Ixodes frontalis and Ixodes ventalloi were detected by this technique ; Gilot, 1984) we very carefully identified the ticks belonging to this genus.

RESULTS

1) 5,348 specimens and eight tick species were collected during the present survey (larvae were not counted). Ixodes ricinus represented 95 per cent of all ticks recovered (larvae were not counted). The results for each quadrant are presented in table 1.

Table 1. Ixodes ricinus collected in each quadrant : total numbers and different stages (n : nymph ; M : male ; F : female).

	n	M	F	
North-western quadrant	1966	437	537	3040
Eastern quadrant	942	210	180	1332
South-western quadrant	500	99	205	804
	3308	746	922	5176

Table 2. Percentage of zones and sites with Ixodes ricinus in the three quadrants investigated.

	ZONES			SITES		
	n°samples	n°positive	%	n°samples	n°positive	%
North-western quadrant	16	16	100%	80	62	77,50%
Eastern quadrant	20	19	95%	100	56	56%
South-western quadrant	17	15	88%	100	42	42%

2) Ixodes ricinus was found in all ecological units except three (95 %). The percentage of negative units varies from one quadrant to another (table 2). The area of negative units is very limited never more than 1/20 of the territory studied.

3) Most of the plant groupings ("dynamics series") which were sampled during the survey harboured Ixodes ricinus : the three exceptions were the followings : alder series (Alnus glutinosa) in flood plains, oak-woods with Quercus pubescens (supra-meditarranean level), Pine-woods with Pinus sylvestris (montane level in the Massif Central).

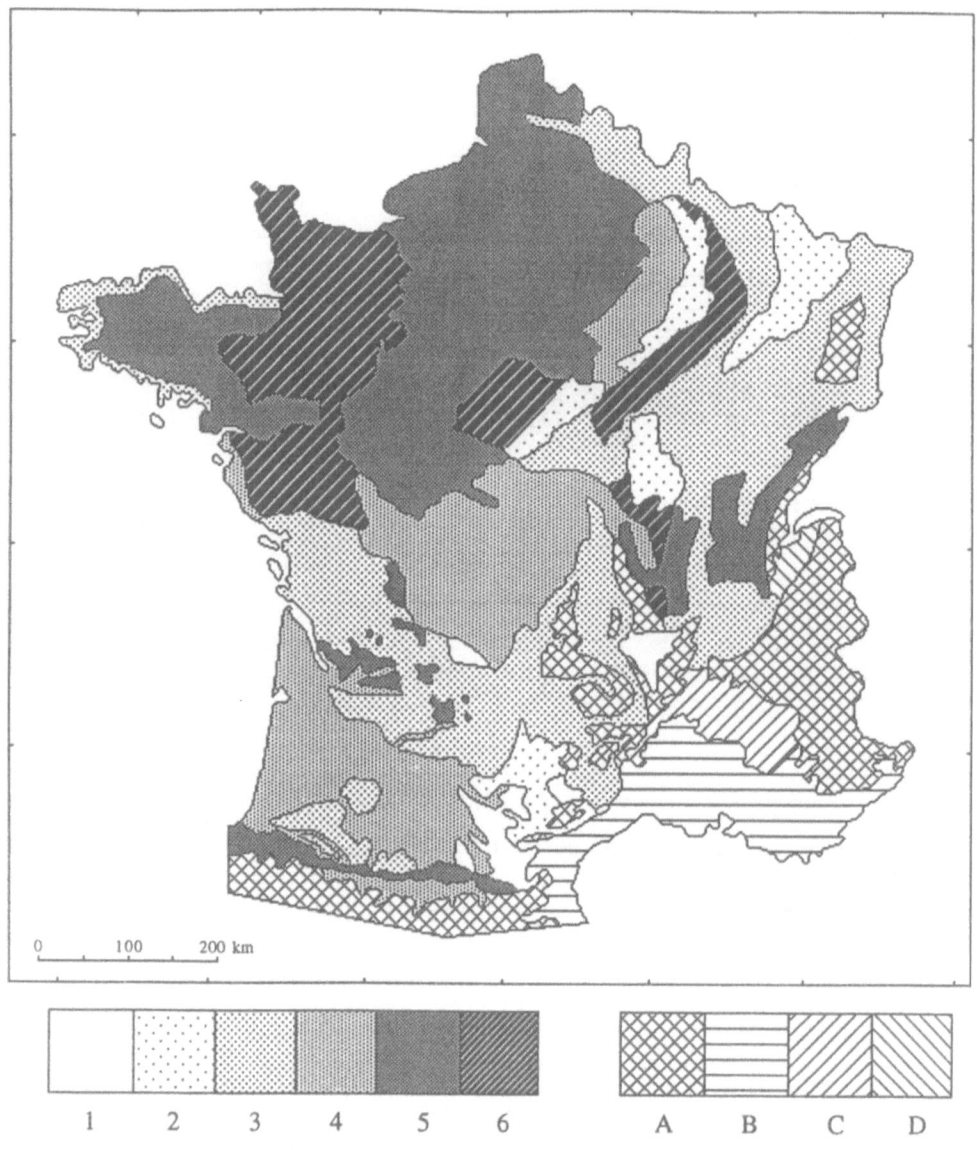

Figure 2. Map visualizing units of "similar potentiality with regard to the <u>Ixodes ricinus</u> frequency

Tick frequencies : 1 : 0 % ; 2 : 1-19 % ; 3 : 20-49 % - 4 : 50-79 % ; 5 : 80-99 % ; 6 : 100%.
Areas not sampled in the present study : A : regions more than 1,000 metres high ; B : mediterranean area ;
C : supramediterranean area ; D : collinean level.

4) Based on these findings, a small scale map can be drawn in order to give an immediate picture of the distribution of <u>Ixodes ricinus</u> in France and to show units of similar potentiality with regard to the tick vector (figure 2).

5) The other tick species which were collected in the biotopes prospected were found either alone (1), or together with <u>Ixodes ricinus</u> (2) (table 3).

Table 3. Other species present in the biotopes. (1) alone, (2) with <u>Ixodes ricinus</u>.

	D.r.	D.m.	H.p.	H.c.	H.i.	I.f.	R.t.
North-western quadrant	1,2					1,2	1
Eastern quadrant	1,2						
South-western quadrant	1,2	1,2	1,2	2	2	1,2	1

D.r. : Dermacentor reticulatus ; D.m. : D. marginatus ;
H.p. : Haemaphysalis punctata ; H.c. : H. concinna ;
H.i. : H. inermis ; I.f. : Ixodes frontalis ;
R.t. : Rhipicephalus turanicus.

DISCUSSION

The present study showed that <u>Ixodes ricinus</u> is a widespread tick-species in France, an assumption put forward by many authors, but, which have never been proved before. In our opinion, this result is enhanced by the sampling procedure. Until now, the spatial distribution in France was based on a few collections made, at random, especially on wild and domestic hosts, in different areas of the French territory (cf. Morel, 1965). Another data source was provided by the scattered collections made for virological and bacteriological purposes, but has the aim of such studies is to evidence pathogenic agents in significant batches of ticks, these ticks are usually collected only in biotopes where their activity was previously established or which are characterized by some plant communities known to be favourable for the species. These two approaches did not take enough into account the geographical and ecological diversity of the country.

From our results we can assume the very large repartition in France of <u>Ixodes ricinus</u> in so far as we can use the different plant groupings as ecological indicators. The use of such indicators is stressed by Ozenda (1986) : whenever a correlation has been established in a given phytoecological territory, between an animal population and a given vegetation series, there is a high likelihood of finding the same population in the whole area covered by the series within the territory. The entire range of a series has thus a single "biological potential" in regard to the tick populations. Such a concept has been taken as a basis for the setting up of medium scale maps (1/50.000 ; 1/200.000 : Gilot <u>et al.</u>, 1979, 1981) the aim of which was to obtain a more precise distribution of various species of ticks in territories showing great disparities from an ecological point of view.

The finding of the present study permites us to forecast to what extent a phytoecological zone could be colonized by the tick vector is new possible. For instance, the likelihood of encountering the tick vector is very high in a phytoecological unit of south-western France, called "Les Landes" : in this area, forests cover over 50 % of the land, and the pine forests of Pinus maritimus which is the prevailing series is according to the present study very favourable for the tick (tick frequency : 70 %). Therefore, the results enable us to assume a large distribution of Lyme borreliosis in this unit in so far as the bacterium is wide spread in the Ixodes population. Another unit, part of the Bassin Aquitain, subjected to a mediterranean influence, is characterized by another type of vegetation (oak-woods of supramediterranean level with Quercus pubescens) which covers only a small area(5 %) and is quite unfavourable for the tick (tick-frequency : 0 %). Consequently, such a unit is likely to be unfavourable for the zoonosis (Gilot et al., submitted paper). The fact that the tick exhibits, in the present study, a very large pattern of distribution gives us an acarological basis to argue a very large distribution of Lyme Borreliosis in France if one can prove a similar extension of the bacterium involved. Until now the evidence of Borrelia in Ixodes ricinus populations has been demonstrated in rather limited areas, but we must stress the fact that the presence of the bacterium has been ascertained wherever it was looked for, and actually in each quadrant investigated here for the tick distribution in France (North-Western : Doby et al., 1989 a et b ; South-Western : Macaigne, 1991 ; North-Eastern : Kremer et al., 1985).

Although the tick is widespread, and present almost in all the sampled vegetation series, it is absent from some plant communities, which are either too hygrophilic or too xerophilic. Alder woods, which are present in most river valleys (Alnetum glutinosae) belong to the former unfavourable type. The pubescent oak formations which belong to a supramediterranean level are linked with ecological conditions which do not agree with the tick requirements (too much drought during the summer months). Similarly, some Pine forests (Pinus sylvestris located on the Eastern border of the Massif Central, which are established on soils exhibiting an insufficient retention power for raining waters, are unable to create the hygrophilic conditions required by the tick.

Although Ixodes ricinus distribution in France is very large, due to the diversity of the climatic conditions, it is far from homogeneous. The reason is that the climate of the southern part of the country is influenced by the mediterranean climate which is not propitious for this Ixodes. It is worth noting that a northern country such as Great Britain, with relatively homogeneous climatic conditions exhibits a homogeneous pattern of distribution of the tick (Mac Leod, 1962 ; Barnett, 1974). Varma stressed the fact in 1965 : "The actual species composits of the vegetation cover seems to matter little. In fact, almost all the rough grazings and moorlands are suitable habitats for the ticks in that they provide the necessary microenvironment "On the contrary, on the southern boundary of the tick distribution, in southern countries, like Morocco or Algeria, Ixodes ricinus is known to have a patchy distribution (Morel, 1965, 1969, Bailly-Choumara et al., 1980) and its presence is linked with a few plant groupings. According to the scarce works published, although the tick is found in larger areas in Spain than in Morocco its distribution is patchy (cf. Gil Collado, 1938, 1948 ; Gilot et al., 1976 ; Estrada-Pena et al ., 1992). Correlatively the epidemiological work which has been done in that country show a quite unhomogeneous distribution of the zoonosis (Oteo et al., 1991, 1992 ; Gomes et al., 1990 ; Anda et al., 1992). The present study shows for France an intermediate statute. In the Northern areas (for example in Brittany), the climatic conditions are homogeneous ; in that case, Ixodes ricinus populations are distributed rather regularly and all the vegetation series are favourable for the tick species (Degeilh et al, in press) ; in southern France, Ixodes ricinus is scarce or absent (Gilot, 1985). The present study shows that the tick populations are less and less frequent as one goes from the areas situated in the Mediterranean Basin and subjected to a Mediterranean climate to areas characterized by a true Atlantic climate.

ACKNOWLEDGEMENTS

This work was supported by a grant from "Institut national de la Santé et de la Recherche Médicale" (contrat n° 900801).

REFERENCES

Anda, P., Rodriguez, I., Asia de la Loma, Fernandez, M.V., Lozano, A., 1993, A serological Survey and Review of Clinical Lyme Borreliosis in Spain, *Clin. Infect. Dis.* 16 : 310-319.

Arthur, D.R., 1963, British Ticks, Butterworth, London, 213 pp.

Bailly-Choumara, H., Morel, P.C., Perez, C., 1980, Deuxième contribution à l'étude des tiques du Maroc (Acari, Ixodoidea), *Bull. Inst. Scient. Rabat.* n° 4, 1-12.

Barnett, S.F., 1974, Economical aspects of Tick-born diseases control in Britain, *Bull. Off. int. Epizoot.* 81, 1-2: 167-182.

Degeilh, B., Guiguen, C., Gilot, B, Doche, B., Pichot, J., Beaucournu, J.C., 1994, Répartition d'Ixodes ricinus (Linné, 1758) (Acarina : Ixodoidae) dans les groupements forestiers du Massif Armoricain, *Acarologia*. 35, 4 (in press).

Daget, Ph., Godron, M., 1982, Analyse de l'écologie des espèces dans les communautés, ed. Masson, Paris.

Doby, J.M., Imbert-Hameurt, C., Jeanne, B., Chevrier, S., 1989 a, Infection de Ixodes ricinus (Acarina, Ixodidae) par Borrelia burgdorferi, agent des spirochétoses à tiques (Maladie de Lyme et autres formes cliniques) dans l'Ouest de la France. I - Résultats globaux de l'examen de 2320 tiques, *Bull. Soc. fr. Parasitol.* 7 : 111-125.

Doby, J.M., Chevrier, S., Couatarmanac'h, A., Imbert-Hameurt, C., 1989 b, Infection de Ixodes ricinus (Acarina, Ixodidae) par Borrelia burgdorferi, agent des spirochétoses à tiques (Maladie de Lyme et autres formes cliniques) dans l'Ouest de la France. II - Résultats détaillés et commentaires, *Bull. Soc. fr. Parasitol.* 7 : 277-289.

Doche, B., Gilot, B., Degeilh, B., Pichot, J., Guiguen, C., 1993, Utilisation de l'indicateur végétal pour la cartographie d'une tique exophile à l'échelle d'un pays : l'exemple d'Ixodes ricinus en France, *Ann. Parasit. hum. comp.* (in press).

Dupias, G., Rey, P., 1985, Document pour un zonage des régions phytoécologiques. Centre d'Ecologie des Ressources renouvelables, C.N.R.S., Toulouse, 40 pp.

Estrada-Pena, A., Osacar, J.J., Gortazar, C., Calvete, C., Lucientes, J., 1992, An account of the ticks of the northeastern of Spain (Acarina ; Ixodidae) des Monts Cantabriques (Espagne), *Ann. Parasit. hum. comp.* 51: 241-254.

Gil Collado, J., 1938, Los Acaros (Ixodoideos) de Espana. *Datos actuales a su distribution, Broteria (Cienc. Nat.).* 7 : 107-118.

Gil Collado, J., 1948, Acaros Ixodoideos de Espana, *Rev. San Mig. Publ. Madrid.* 22 : 389-440.

Gilot, B., 1984, Premier inventaire des Ixodidae (Acarina, Ixodoidea) parasites d'oiseaux dans les Alpes françaises et leur avant-pays. Données écologiques sur quelques espèces peu connues, *Riv. di parassitol.* 1 : 303-318.

Gilot, B., 1985, Bases biologiques, écologiques et cartographiques pour l'étude des maladies transmises par les tiques (Ixodidae et Argasidae) dans les Alpes françaises et leur avant-pays, Thèse Doctorat ès-Sciences, Grenoble, 535 pp.

Gilot, B., Pautou, G., Gosalbez J., Moncada, E., 1976, Contribution à l'étude des Ixodidae (Acarina, Ixodoidea) des Monts Cantabriques (Espagne), *Ann. Parasit. hum. comp.* 51 : 241-254.

Gilot, B., Pautou, G., Lachet, B., 1981, La cartographie des populations de tiques exophiles à visée épidémiologique. Application à la fièvre boutonneuse méditerranéenne. Essai à 1/200.000 dans la basse vallée du Rhône, *Doct. Cart. Ecol.* (Grenoble). 24 : 103-111.

Gilot, B., Pautou, G., Moncada E., Lachet, B., Christin, J.G., 1979, La cartographie des populations de tiques exophiles par le biais de la végétation : bases écologiques, intérêt épidémiologique, *Doct. Cart. Ecol.* (Grenoble). 22 : 65-80.

Gilot, B., Doche, B., Degeilh, B., Guiguen, C., Pichot, J., Bases acarologiques pour l'étude épidémiologique de la borréliose de Lyme : les populations d'Ixodes ricinus (Linné, 1758) du Sud-Ouest de la France, *Acarologia* (submitted for publication).

Gomes, J.M., Sanchez, A., Lozano, F., Lopes, L., 1990, Enfermedad de Lyme en la provincia de Sevilla, *Med. Clin.* 6 : 236-237.

Kremer, M., Rocheteau, A., Bonnat, A., 1985, Note préliminaire sur la mise en évidence en Alsace chez Ixodes ricinus de Borrelia immunologiquement affines de B. burgdorferi, agent de la "Lyme disease", *J. Méd. Strasbourg.* 16 : 590.

Macaigne, F., 1991, Données épidémiologiques de la maladie de Lyme dans le département de la Gironde (France), D.E.A. Université de Bordeaux II, 43 pp.

Mac Leod, J., 1962, Ticks and disease in domestic stock in Great Britain, *Symposia Zool. Soc. London.* 6 : 29-50.

Morel, P.C., 1965, Les tiques d'Afrique et du Bassin méditerranéen, Maison Alfort, (I.E.M.T.Y.). 695 p. (doc. ronéo).

Morel, P.C., Vassiliades, G., 1962, Les Rhipicéphales du groupe sanguineus : espèces africaines (Acariens : Ixodidea), *Rev. Elev. Méd. Vét. Pays Trop.* 15 : 343-386.

Oteo, J.A., Estrada-Pena, A., 1991, Ixodes ricinus, vector comprobado de Borrelia burgdorferi en Espana, *Med. Clin.* 15 : 599.

Oteo, J.A., Martines de Artola, V., Casas, J., Lozano, A., Fernandez Calvo, J.L., Grandival, R., 1992, Epidemiology and prevalence of seropositivity against Borrelia burgdorferi antigen in La Rioja, Spain. *Rev. Epid. et Santé Publ.* 40 : 85-92.

Ozenda, P., 1986, La cartographie écologique et ses applications, Masson éd., Paris, 140 pp.

Pomerantzev, E.I., 1950, Fauna of U.S.S.R. Arachnida. Ixodid Ticks (Ixodidae), Academy of Sciences U.S.S.R. Moscou-Leningrad (vol. IV n° 2), 199 pp. (Translated into english by Alena Elbl Ed. American Institute of Biological Sciences, Washington, 1959).

Rageau, J., 1972, Répartition géographique et rôle pathogène des tiques (Acariens: Argasidae et Ixodidae) en France, *Wiad. Parazytol.* 18 : 707-719.

Varma, M.C.R., 1965, The distribution of Ixodes ricinus in Britain in relation to climate and vegetation, *Proc. Symp. Theor. Nat. Foci Dis.* 301-313.

THE ECOLOGY OF LYME BORRELIOSIS IN SWEDEN

T G T Jaenson[+], S Bergström[*], H A Mejlon[+], L Noppa[*], B Olsén[*]§,
L Tälleklint[+]

[+]The Department of Zoology, University of Uppsala, Villavägen 9, S-
752 36 Uppsala, Sweden
[*]The Department of Microbiology, and § The Department of Infectious
diseases, University of Umeå, S-910 87 Umeå, Sweden

The geographical distribution of Lyme borreliosis (Lb) in the North European
countries appears to coincide with the geographical distribution of the principal
vector, the common tick *Ixodes ricinus*. We have found that in Sweden this tick species
occurs in the southern and south-central parts of the country and along the coast of
northern Sweden. This area corresponds with the distributional area of Lyme
borreliosis[1 2]. *I. ricinus*, and thus also *Borrelia burgdorferi s.l.*, are in general not
present in the interior of North Sweden, presumably because the climate is too harsh
for the vector.

Gustafson and co-workers[3] detected antibody to *B. burgdorferi s.l.* in 26% of
humans living on the island Lisö, which is within an area endemic for Lb, located near
Stockholm. In contrast, only 2% in a group of healthy blood donors from a non-
endemic area were seropositive. In the former group 12% gave a history of previous
Lb, mainly ECM, while less than 1% had had arthritis. This suggests that subclinical
infections are common[3].

B. burgdorferi s.l. has in Sweden only been detected in *I. ricinus*[4 5] and *I. uriae*[6].
Other potential vectors of Lb which we have recorded from Sweden include *I.
hexagonus, I. persulcatus, I. trianguliceps* and *Haemaphysalis punctata*.

Spirochaetal infection in tick larvae in Sweden has not been found suggesting
that transovarial transmission of *B. burgdorferi s.l.* is rare. Infection prevalence in
nymphs, the principal vector stage, may differ between years and sampling sites[7]. In
our studies the prevalence was greater in adults (14-29%) than in nymphs (7-13%) but
similar in male and female ticks[7]. The seasonal host-seeking activity of both larval
and nymphal *I. ricinus* in southern Sweden is generally bimodal with peaks in May-
June and August-September, the greatest density being in mixed forest vegetation[7].

However, the main period of transmission of Lb is during the latter part of the summer and early part of the autumn because at this time both tick activity and human activity - the latter in the form of berry- and mushroom-picking and hunting - are high in those areas where ticks normally occur.

In order for Lb to be maintained for a prolonged period of time in an ecosystem two types of hosts should be present; "reproduction hosts" which provide blood to the female ticks for their reproduction and "reservoir hosts" which are capable to infect feeding ticks with *B. burgdorferi s.l.* Some mammals, such as hares, function as both reproduction and reservoir hosts. Maintenance of Lb is complex in most mainland ecosystems but appears relatively simple in island ecosystems where few host species occur.

Our data suggest that roe deer (*Capreolus capreolus*) and moose (*Alces alces*) are very important as reproduction hosts for *I. ricinus* but these mammals appear incapable to infect this tick with *B. burgdorferi s.l*[8].

In 1987 we isolated *B. burgdorferi s.l.* from the yellow-necked field mouse (*Apodemus flavicollis*) and the bank vole (*Clethrionomys glareolus*)[9]. More recently we have demonstrated that the following mammal species are competent Lb reservoirs: shrews (*Sorex araneus, S. minutus*), small rodents (*A. flavicollis, A. sylvaticus, C. glareolus*) and hares (*Lepus europaeus, L. timidus*)[10]. Small mammals do not usually serve as blood hosts for *I. ricinus* females and are therefore not reproduction hosts but only reservoir hosts. In contrast, the hares are usually infested with all stages of *I. ricinus*. On isolated islands in the Baltic Sea, e.g. Gotska Sandön and Stora Karlsö, the varying hare (*L. timidus*) is the only terrestrial mammal present. Our data suggest that dense populations of *Borrelia*-infected *I. ricinus* are maintained on these islands mainly because of the presence of the hare populations.

A different situation for the maintenance of Lb has been revealed on the isolated island, Bonden. Large seabird colonies are breeding on this island located in the northern Baltic Sea. In July 1991 we collected seventeen *I. uriae* ticks from two juvenile razorbills (*Alca torda*). By phase-contrast microscopy spirochaetes were detected in three of the ticks. After using immunofluorescence techniques and PCR DNA-amplification followed by sequencing parts of the flagellin and the ospA genes we concluded that the spirochaetes belong to *B. burgdorferi s.l.* Since there are no mammals on the island nor any *I. ricinus* population present it is likely that the spirochaetes at Bonden are cycling between seabirds, principally auks, and *I. uriae*. This tick species is associated with colony nesting seabirds in both the northern and southern hemisphere. It infests many species of seabird including far-ranging species such as the wandering albatross, fulmars and gannets. This suggests that seabirds may transmit Lb between the northern and southern hemisphere or vice versa and that hitherto unknown foci of Lb may exist.

ACKNOWLEDGEMENTS

Thomas Jaenson is funded by Magnus Bergvall's Foundation, The Swedish Agency for Research Cooperation with Developing Countries (SAREC) and The Swedish Natural Science Research Council (NFR). Sven Bergström is supported by grants from the Swedish Medical Research Council, the Swedish Council for Forestry and Agricultural Research, the Swedish Research Council for Engineering Sciences, the Medical Faculty of the University of Umeå, and the J. C. Kempes and Hierta Retzius Foundations.

REFERENCES

1 Hederstedt B, Hovmark A, Stiernstedt G, Åsbrink E. Borrelia-diagnos aktuell året om visar serologisk undersökning av 1985 års fall. *Läkartidningen* 1986; **83**: 3987-9.

2 Jaenson T G T, Bergström S, Burman N, Jonsson M, Chirico J. Spirochaete-infected ticks (*Ixodes ricinus*) - risk of infection even in Northern Sweden. *Läkartidningen* 1989; **86**: 2584 [In Swedish].

3 Gustafson R, Svenungsson B, Gardulf A, Stiernstedt G, Forsgren M. Prevalence of tick-borne encephalitis and Lyme borreliosis in a defined Swedish population. *Scand J Infect Dis* 1990; **22**: 297-306.

4 Åsbrink E. Erythema chronicum migrans Afzelius and acrodermatitis chronica atrophicans. Early and late manifestations of *Ixodes ricinus*-borne *Borrelia* spirochaetes. *Acta Dermatol Venereol Suppl* 1985; **118**: 1-63.

5 Stiernstedt G. Tick-borne *Borrelia* infection in Sweden. *Scand J Infect Dis Suppl* 1985; **45**: 1-69.

6 Olsén B, Jaenson T G T, Noppa L, Bunikis, J, Bergström S. A Lyme borreliosis cycle in seabirds and *Ixodes uriae* ticks. Nature 1993; **362**: 340-342.

7 Mejlon H A, Jaenson T G T. Seasonal prevalence of *Borrelia burgdorferi* in *Ixodes ricinus* in different vegetation types in Sweden. *Scand J Infect Dis* 1993; in press.

8 Jaenson T G T, Tälleklint L. Incompetence of roe deer (*Capreolus capreolus*) as reservoirs of the Lyme disease spirochete. *J Med Entomol* 1992; **29**: 813-7.

9 Hovmark A, Jaenson T G T, Åsbrink E, Forsman A, Jansson E. First isolations of *Borrelia burgdorferi* from rodents collected in Northern Europe. *Acta Pathol Microbiol Immunol Scand* 1988; **96**: 917-20.

10 Tälleklint L, Jaenson T G T. Maintenance by hares of European Lyme disease in ecosystems without rodents. *J Med Entomol* 1993; **30**: 273-276.

LYME DISEASE IN SCOTLAND

RESULTS OF A SEROLOGICAL STUDY IN SHEEP

G.B.B. Mitchell and I.W. Smith

Scottish Agricultural College
Veterinary Services
Auchincruive
Ayr
Scotland
KA6 5AE

Lyme Disease is a recently described zoonosis (Burgdorfer et al, 1982) caused by the spirochaete _Borrelia burgdorferi_ and transmitted in Europe by the sheep tick _Ixodes ricinus_, which was added to the list of notifiable diseases in Scotland on 1st January 1990.

The condition, first recorded in Britain in the late 1970s and since increasingly observed in Scotland, England and Northern Ireland, is characterised in humans by a specific skin lesion, erythema chronicum migrans (E.C.M.), followed by an influenza-like systemic illness, arthritis and neurological signs, in some cases clinically and pathologically resembling multiple sclerosis (Gay and Dick, 1986).

In Europe, infection with _Borrelia burgdorferi_ has been reported in farmed cattle, sheep and deer, as well as horses, cats, dogs and free living mammals (Hovmark et al, 1986; Muhlemann and Wright, 1986; Parker and White, 1992). Reports of clinical disease however, in domestic species in Europe due to _Borrelia burgdorferi_ are uncommon. Hovmark et al, 1986, described arthritis in sheep due to _B. burgdorferi_ but, Ellis, 1991 although able to demonstrate the organism in ovine and bovine foeti and adult bovine blood was not able to produce disease experimentally.

The purpose of the present study was to examine sera from different age groups of sheep naturally infested with _Ixodes ricinus_ for evidence of _Borrelia burgdorferi_ infection. Nine commercial hill/upland farms were used for the study in tick infested areas throughout Scotland. Lambs were normally put on to tick infested hill pastures at 2-4 weeks of age and remained there throughout their productive lives. Serological and haematological evidence of tick-borne fever (T.B.F.) was present on all trial farms. Blood samples from tick infested lambs, hoggs (one year old sheep) and ewes, collected during spring and summer were therefore screened using two Elisa techniques incorporating commercially available flagellar ('F') and whole cell ('W.C.') antigens. 'F' (Dako Limited) and 'W.C.' (Biogenesis Limited) antigens were used with rabbit antisheep IgG conjugated to alkaline phosphatase, using 'P' nitrophenyl phosphate (P.N.P.) substrate. In addition engorged and semi-engorged _Ixodes ricinus_ ticks were removed and screened by a polymerase chain reaction (P.C.R.) technique for evidence of _Borrelia burgdorferi_ infection within the tick (Carey, D., pers. comm.).

Lyme Borreliosis, Edited by J.S. Axford and
D.H.E. Rees, Plenum Press, New York, 1994

Table 1
Lyme Disease Elisa O.D. Ratios

Age	No.	"F" Antigen Mean S.D.		"WC" Antigen Mean S.D.		No. Positive Samples		% Positive Samples	
						"F"	"WC"	"F"	"WC"
4-8 Weeks	37	0.469	0.026	0.448	0.032	1	1	2.7	2.7
						"F"	"WC"	"F"	"WC"
1-1.5 Years	45	0.494	0.070	0.470	0.028	18	11	40	24.4
						"F"	"WC"	"F"	"WC"
2-5 Years	88	0.426	0.023	0.454	0.023	5	0	5.7	0

KEY

"F" = Flagella Antigen

"WC" = Whole Cell Antigen

SD   = Standard Deviation

The Elisa results are tabulated in Tables 1 and 2 and illustrated in Figs. 1 -3. Results are expressed as O.D. ratios with reference to a known positive hyperimmune serum and are compared with sera obtained from housed, tick free sheep from a lowground farm. "Cut off" values have been calculated at 2 standard deviations (S.D.) above the mean of the negative control samples to assess positivity. The percentage of positive samples varied from 2.7 (lambs), 40 and 24.4 (hoggs) to 5.7 and 0 (ewes) using 'F' and 'W.C.' antigens respectively. Elisa values of ewe and lamb "tick" samples did not differ significantly from negative control sera, while values of hogg samples were significantly higher ($p < 0.001$) than negative controls, using analysis of variance.

Table 2
Lyme Disease Elisa O.D. Ratios

			0.40	0.42	0.44	0.46	0.48	0.50	0.52	0.54	0.56	0.58	0.60	0.62	0.64
No. of Lambs	"F"	"Tick"	0	2	11	9	9	3	3	0	0*	0	0	0	-
		"Non-Tick"	5	6	11	6	4	5	0	1	3	0	0	1	-a
	"WC"	"Tick"	0	10	18	4	2	1*	0	0	2	-	-	-	-
		"Non-Tick"	0	16	22	4	1	0	0	0	0	-	-	-	-a
No. of Hoggs	"F"	"Tick"	8	0	8	0	6	0*	12	0	4	0	5	0	2
		"Non-Tick"	8	5	13	10	3	3	1	0	0	0	0	0	0b
	"WC"	"Tick"	0	3	12	13	9	4*	2	1	-	-	-	-	-
		"Non-Tick"	0	11	17	10	5	1	0	0	-	-	-	-	-b
No. of Ewes	"F"	"Tick"	0	14	41	28	4	0*	0	0	0	1	-	-	-
		"Non-Tick"	0	5	32	37	9	1	3	0	0	1	-	-	-a
	"WC"	"Tick"	26	32	22	6	2	0*	0	0	0	0	-	-	-
		"Non-Tick"	30	15	20	18	3	0	0	1	0	1	-	-	-c

KEY

"WC" = Whole Cell Antigen
"F" = Flagellar Antigen
* = "Cut-off" Point
a = NS
b = p < 0.001
c = p = 0.01

119

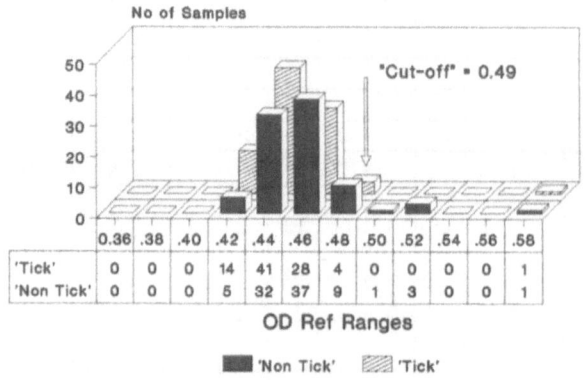

Lyme Disease Elisa
'F' Antigen : EWES

No of Samples

"Cut-off" = 0.49

	0.36	.38	.40	.42	.44	.46	.48	.50	.52	.54	.56	.58
'Tick'	0	0	0	14	41	28	4	0	0	0	0	1
'Non Tick'	0	0	0	5	32	37	9	1	3	0	0	1

OD Ref Ranges

■ 'Non Tick' ▨ 'Tick'

'p'= NS

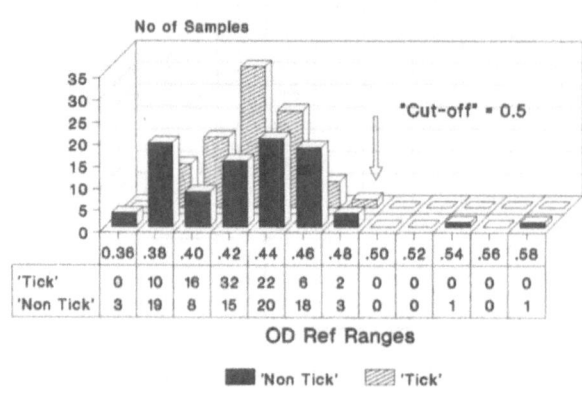

Lyme Disease Elisa
'WC' Antigen : EWES

No of Samples

"Cut-off" = 0.5

	0.36	.38	.40	.42	.44	.46	.48	.50	.52	.54	.56	.58
'Tick'	0	10	16	32	22	6	2	0	0	0	0	0
'Non Tick'	3	19	8	15	20	18	3	0	0	1	0	1

OD Ref Ranges

■ 'Non Tick' ▨ 'Tick'

'p'=0.01

Fig. 1 : Elisa Titres (Lambs).

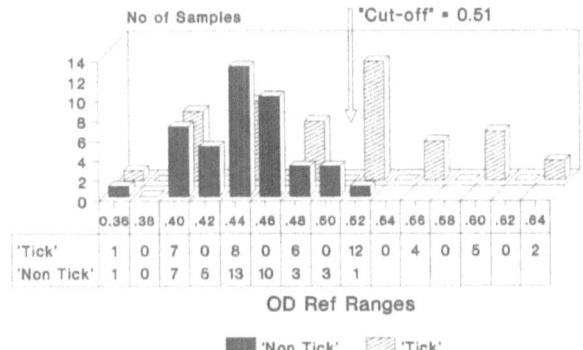

Lyme Disease Elisa
'F' Antigen : Hoggs

No of Samples   "Cut-off" = 0.51

	0.36	.38	.40	.42	.44	.46	.48	.50	.52	.54	.56	.58	.60	.62	.64
'Tick'	1	0	7	0	8	0	6	0	12	0	4	0	5	0	2
'Non Tick'	1	0	7	5	13	10	3	3	1						

OD Ref Ranges

■ 'Non Tick'   ▨ 'Tick'

'p'<0.001

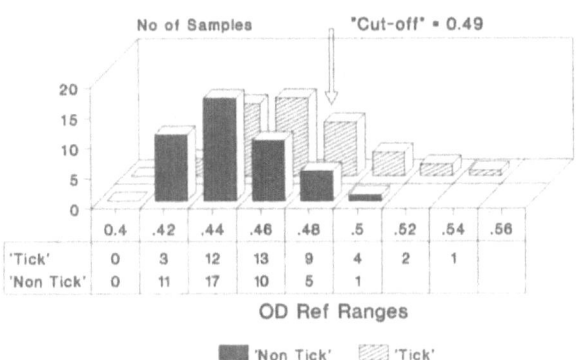

Lyme Disease Elisa
'WC' Antigen : Hoggs

No of Samples   "Cut-off" = 0.49

	0.4	.42	.44	.46	.48	.5	.52	.54	.56
'Tick'	0	3	12	13	9	4	2	1	
'Non Tick'	0	11	17	10	5	1			

OD Ref Ranges

■ 'Non Tick'   ▨ 'Tick'

'p'<0.001

Fig. 2 : Elisa Titres (Hoggs).

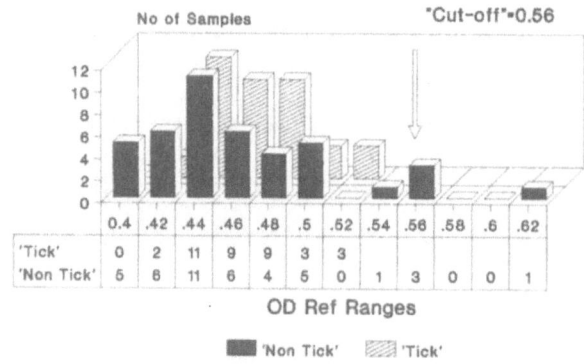

Lyme Disease Elisa
'F' Antigen : Lambs

No of Samples

"Cut-off"=0.56

	0.4	.42	.44	.46	.48	.5	.52	.54	.56	.58	.6	.62
'Tick'	0	2	11	9	9	3	3					
'Non Tick'	5	8	11	6	4	5	0	1	3	0	0	1

OD Ref Ranges

■ 'Non Tick'   ▨ 'Tick'

'p'=NS

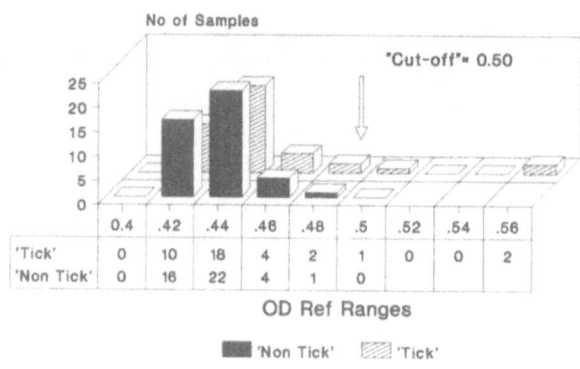

Lyme Disease Elisa
'WC' Antigen : Lambs

No of Samples

"Cut-off"= 0.50

	0.4	.42	.44	.46	.48	.5	.52	.54	.56
'Tick'	0	10	18	4	2	1	0	0	2
'Non Tick'	0	16	22	4	1	0			

OD Ref Ranges

■ 'Non Tick'   ▨ 'Tick'

'p'=NS

Fig. 3 : Elisa Titres (Ewes).

Preliminary results from P.C.R. testing revealed evidence of *Borrelia burgdorferi* in 4.2% of ticks examined, indicative of infection on 42.9% of participating farms.

The study therefore produced compelling evidence of exposure to *Borrelia burgdorferi* infection in hoggs infested with *Ixodes ricinus*, with very limited evidence of individual ewes and lambs being infected. A recent serosurvey of sheep in Norway (Fridriksdottir et al, 1992) gave similar results with the highest prevalence (10%) in the 1.5 year age group and no infection recorded in lambs less than 6 months old. This latter finding, common to both studies, may well reflect lack of immunocompetence of the 4-8 week old lambs sampled in the present study.

Although 75-100% of sheep sampled in the present study harboured significant numbers of *Ixodes ricinus* at any one time neither tick infestation per se nor tick infestation and seropositivity to *Borrelia burgdorferi* were correlated with clinical or sub-clinical disease. Tick numbers were routinely controlled on trial farms using "pour-on" insecticide preparations.

On the basis of the above results, it would be useful to extend the study to tick populations throughout Scotland and to assess the possible role of sheep as a maintenance host of this human pathogen.

ACKNOWLEDGMENTS

The authors are grateful to Mrs. L. Sommerville for skilled technical assistance, and colleagues at St. Boswells, Thurso and Perth Veterinary Centres for assistance with field work.

S.A.C. receives financial assistance from the Scottish Office Agriculture and Fisheries Department.

REFERENCES

Burgdorfer, W., Barbour, A.G., Hayes, S.F., Benach, J.L., Grunwaldt, E. and David, T.P. (1982) Science, 216, 1317.

Ellis, W.A. (1991), Proc. Lyme Disease, Workshop, 19th April 1991, Aberdeen.

Fridriksdottir, V., Nesse, L.L., Gudding, R. (1992), J. Clin. Microbiology, 30, 5, p 1271.

Gay, D. and Dick, G. (1986) The Lancet, 20th September 1986, p 685.

Hovmark, A.E., Asbrink, E., Schwan, O., Hederstedt, B. and Christensson, D. (1986), Acta Vet. Scand., 27, 479-485.

Muhlemann, M.F., Wright, D.J.M. and Alack, C. (1986), The Lancet, 8th March, p 553.

Parker, J.L. and White, K.K. (1992), Cornell Vet., 82, 253.

THE ECOLOGY OF LYME BORRELIOSIS IN THE UK

Patricia Nuttall[1], Sarah Randolph,[2] Dorothy Carey[1], Noel Craine[1,2], Anne Livesley[1], and Lise Gern[3]

[1]NERC Institute of Virology & Environmental Microbiology, Mansfield Road, Oxford, OX1 3SR; [2]Department of Zoology, University of Oxford, South Parks Road, Oxford; [3]Institute of Zoology, University of Neuchâtel, Switzerland

INTRODUCTION

Comparatively little is known of the ecology of *Borrelia burgdorferi* (the aetiological agent of Lyme borreliosis) in the UK. Although ecological similarities with continental Europe are apparent, the physical isolation of the UK (as an island), and the different host preferences of *Ixodes ricinus* (the principal European vector of *B. burgdorferi*) suggest likely differences. Indeed, *I. ricinus* is known as the sheep tick in the UK whereas on the Continent it is called the wood tick.

UK *BORRELIA BURGDORFERI*

An essential step in understanding the ecology of Lyme borreliosis is defining the properties of the aetiological agent. This is important for selecting suitable diagnostic reagents and in understanding the infection characteristics that influence transmission dynamics. Isolation of UK *B. burgdorferi* has proved difficult and to date only one isolate obtained by Dr S. Cutler (Charing Cross Hospital, London) is freely available for study.

We succeeded in obtaining only one isolate from 85 tick pools (representing 504 questing *I. ricinus* nymphs and adults) collected in the UK. In contrast, using identical conditions, *B. burgdorferi* was isolated from one of 7 tick pools (87 ticks) from Switzerland, and a single pool of 10 ticks from Slovakia.[1] Examination of 100 questing *I. ricinus* nymphs from a UK Lyme disease focus, using specific immunofluorescence, revealed 8 ticks positive for *B. burgdorferi* spirochaetes. Of these, 6 nymphs contained 1 to 10 spirochaetes and only one nymph had more than 100 spirochaetes. The low numbers of spirochaetes compared with those recorded on the Continent[2] partly explain the poor success rate in isolating UK *B. burgdorferi*. However, even if the one successful isolation was due to the presence of a highly infected tick, we would expect to have obtained a greater number of UK *B. burgdorferi* isolates.

Lyme Borreliosis, Edited by J.S. Axford and
D.H.E. Rees, Plenum Press, New York, 1994

To examine the problem further, the isolation procedures were monitored by the polymerase chain reaction (PCR using a nested set of primers specific for the OspA gene of *B. burgdorferi*[3]). Eleven of 12 tick samples were PCR positive after 2 weeks in culture but only one sample was positive after 4 weeks and motile spirochaetes were not detected by dark field microscopy.[1] The results indicate that UK *B. burgdorferi* did not adapt to the culture conditions. This observation, together with the comparative ease with which Swiss and Slovakian *B. burgdorferi* were isolated using identical conditions, suggest that the growth requirements of UK *B. burgdorferi* differ significantly from those of other *B. burgdorferi* strains. One such requirement could be related to the diversity in fatty acid profiles of *Borrelia*.[4,5]

GEOGRAPHICAL DISTRIBUTION OF INFECTED TICKS

To assess the geographical distribution of *B. burgdorferi* in the UK, a network of tick collectors has been established with the help of Dr B. Staines (ITE, Banchory) and Mr R. Youngson (Red Deer Commission). Infected ticks have been identified by PCR.[3] The results of this study have been circulated in two Lyme Disease Newsletters. Although the distribution map (Fig. 1) is influenced by the activities of the tick collectors, infected ticks are obviously common and widespread. The distribution of PCR positive ticks broadly corresponds with the recorded distribution of *I. ricinus*.[6]

Figure 1. Distribution of *Ixodes ricinus* ticks collected during 1990 to 1992 and identified as PCR positive for *Borrelia burgdorferi*. Collection sites of positive ticks are indicated by a star.

TICK VECTOR SPECIES

Although *I. ricinus* was by far the most prevalent species collected and diagnosed PCR positive (Fig. 1), other ixodid species found in the UK were identified as positive (Table 1). The full range of potential vector species must be taken into account in studies of the ecology of Lyme borreliosis. For example, *I. hexagonus* has been shown experimentally to be a competent vector of *B. burgdorferi*[7] and in the UK, PCR positive *I. hexagonus* have been found in areas (e.g. Oxfordshire) where *I. ricinus* is not recorded. Thus *I. hexagonus* may extend the range of *B. burgdorferi* beyond that defined by the geographical distribution of *I. ricinus*.

Table 1. Tick species collected in the UK and diagnosed as PCR positive for *Borrelia burgdorferi*.

Tick species	Location	Host associations
Ixodes ricinus	most of UK	numerous vertebrate spp.
Ixodes hexagonus	England	hedgehogs, dogs, cats
Ixodes uriae	Scotland	seabirds
Haemaphysalis punctata	coastal Wales	sheep

Analysis of the PCR results for *I. ricinus* collected in the UK revealed a higher percentage of positives among unfed ticks compared with engorged ticks of the same stage (Fig. 2). This is consistent with reports that constituents of the bloodmeal inhibit the PCR reaction.[8] There was little difference in the proportions of positive unfed adults and nymphs suggesting that most of the infections in the tick population were acquired by feeding larvae and maintained trans-stadially. The detection of two PCR positive pools of questing larvae indicate that transovarial transmission of *B. burgdorferi* occurs from the infected female adult to the succeeding tick generation, as described by other workers.[9] Immunofluorescence assay of unfed *I. ricinus* larvae derived in the laboratory from engorged females collected in a UK focus of Lyme disease revealed that 1.5% larvae were infected.

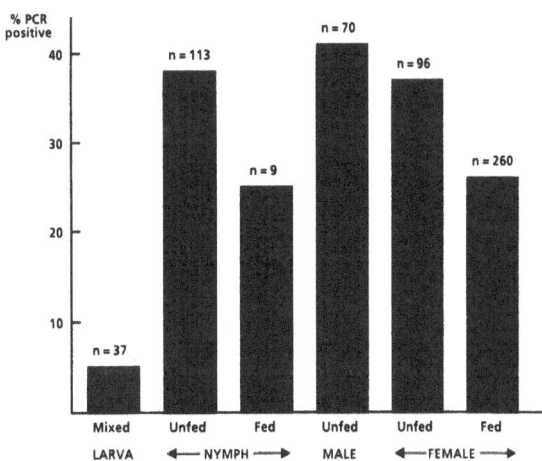

Figure 2. *Ixodes ricinus* ticks collected in Lyme disease foci and identified as PCR positive for *Borrelia burgdorferi*. n=total number examined; mixed=fed and unfed larvae.

VERTEBRATE HOSTS

Most of the feeding *I. ricinus* ticks collected in the UK (Fig. 1) have been from red deer (*Cervus elaphus*) and roe deer (*Capreolus capreolus*) which we assume reflects a collection bias rather than a specific host preference. In addition, PCR positive ticks have been collected from several other wild and domestic vertebrates, *viz.* sheep, cattle, horses, dogs, cats, pheasants (*Phasianus colchicus*), squirrels (*Sciurus carolinensis*), fieldmice (*Apodemus sylvaticus*), and bank voles (*Clethrionomys glareolus*). The susceptibility of these species to infection with UK *B. burgdorferi* is as yet largely undetermined, but even species resistant to infection by *B. burgdorferi* may have profound effects on the prevalence and transmission dynamics of *B. burgdorferi* through their influence on the tick vector population dynamics.

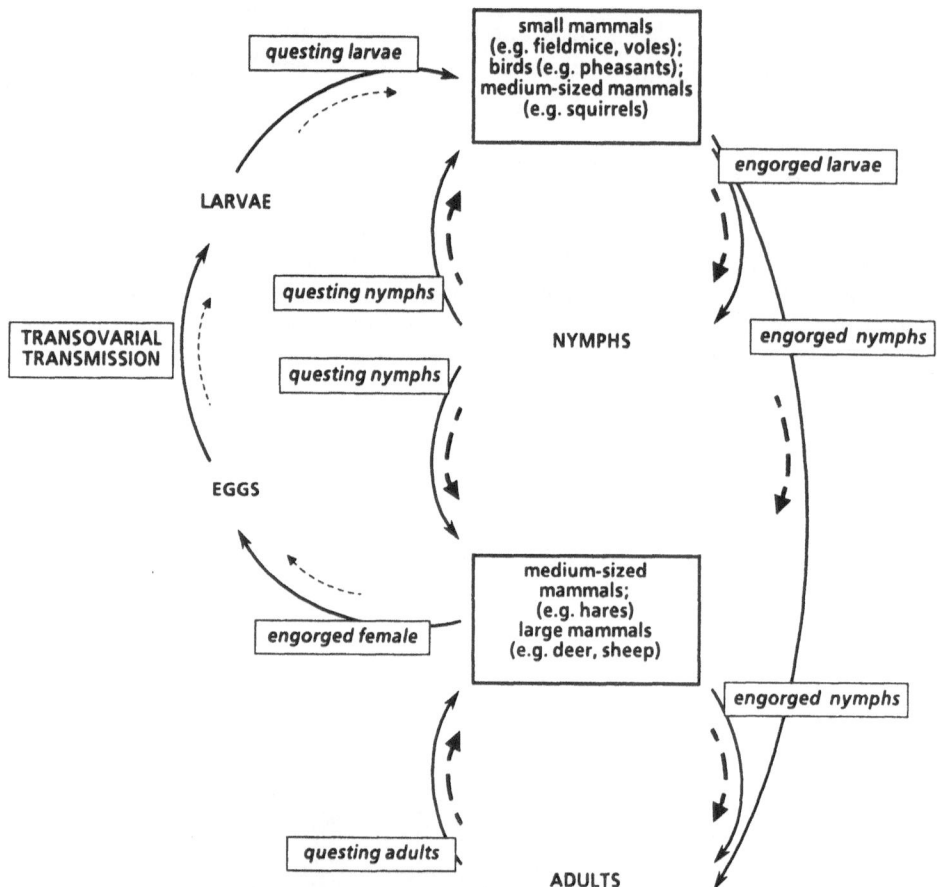

Figure 3. Model of *Borrelia burgdorferi* transmission cycle in the UK.
Solid arrows=life cycle of Ixodes ricinus; broken arrows=direction and intensity of
B. burgdorferi transmission.

TRANSMISSION CYCLE

The ecology of Lyme borreliosis is defined by the transmission cycle of *B. burgdorferi.* A model is shown in Fig. 3 based on our preliminary results. In depth studies in a woodland ecosystem (Thetford Forest) and a sheep-upland ecosystem (Cumbria) are being undertaken to test key points in the transmission cycle, e.g. the contribution of transovarial transmission to the transmission dynamics. Another important question is the identity of the vertebrate host species that ensures transmission from nymphal to larval populations. This requires that larvae and nymphs feed on the same individual hosts since nymphs are the principal source of infection for vertebrates on which uninfected larvae feed. The larva:nymph ratio for fieldmice trapped in Thetford Forest during 1991-92 was 274:1 which is consistent with observations of other workers.[10] In contrast, the autumn ratio for squirrels was 3:1 and on pheasants, 1:15. The comparative roles of rodents, squirrels and pheasants in maintaining *B. burgdorferi* in woodland habitats in the UK is under investigation, as is the role of sheep in faunistically impoverished areas of northern England.

Acknowledgements

The research was funded in part by the Health and Safety Executive and the British Council/Swiss National Science Foundation Joint Research Programme.

REFERENCES

1. M.A. Livesley, D. Carey, L. Gern, and P.A. Nuttall. Problems of isolating *Borrelia burgdorferi* from ticks collected in UK foci of Lyme disease. *Vet. Med. Entomol.* in press (1993).
2. K. Pelz, W. Wagner, and A. Vogt. *Ixodes ricinus* ticks as vectors of *Borrelia burgdorferi* in the Freiburg area. *Zbl. Bakt. Suppl* 18:35 (1989).
3. E. C. Guy and G. Stanek. Detection of *Borrelia burgdorferi* in patients with Lyme disease by the polymerase chain reaction. *J. Clin. Pathol.* 44:610 (1991).
4. M.A. Livesley, I.P. Thompson, M.J. Bailey, and P.A. Nuttall. Comparison of the fatty acid profiles of *Borrelia, Serpulina* and *Leptospira* species. *J. Gen. Microbiol.,* 139:889 (1993).
5. M. A. Livesley, P. A. Nuttall, I. P. Thompson, and L. Gern. Analysis of intra-specific variation in the fatty acid profiles of *Borrelia burgdorferi.* *J. Gen. Microbiol.,* 139:2197 (1993).
6. K. P. Martyn. "Provisional Atlas of the Ticks (Ixodoidea) of the British Isles," Biological Records Centre, Natural Environment Research Council, Institute of Terrestrial Ecology, Monks Wood, Huntingdon, UK (1988).
7. L. Gern, F. de Marval, and A. Aeschlimann. *Ixodes (Pholeoixodes) hexagonus,* an efficient vector of *Borrelia burgdorferi* in the laboratory. *Med. Vet. Entomol.* 5:431 (1991).
8. M. Panaccio and A. Lew. PCR based diagnosis in the presence of 8% (v/v) blood. *Nuc. Acids Res.* 19:1151 (1991).
9. O. Kahl. Lyme borreliosis - an ecological perspective of a tick-borne human disease. *Anz. Schädlingskde., Pflanzenschutz, Umweltschutz* 64:45 (1991).
10. J. S. Gray, O. Kahl, C. Janetzki, and J. Stein. Studies on the ecology of Lyme disease in a deer forest in county Galway, Ireland. *J. Med. Entomol.* 29:915 (1992).

THE RELATIVE CONTRIBUTIONS OF TRANSOVARIAL AND TRANSSTADIAL TRANSMISSION TO THE MAINTENANCE OF TICK-BORNE DISEASES

Sarah E. Randolph

Department of Zoology
University of Oxford
South Parks Road
Oxford OX1 3PS

INTRODUCTION

It is a common misconception that, because of the frequently observed low coefficient of transovarial transmission of many tick-borne diseases, this transmission route contributes very little to the maintenance of the diseases (*inter alia* Porco, 1991). A simple model, that includes the important epidemiological parameters of tick reproduction and mortality rates, shows that the potential contribution of transovarial transmission is greater than that of transstadial transmission, even when the transmission coefficient of the former is less than 10% of that of the latter. The actual contribution will depend on the specific vector/host relationships for each tick-borne disease, and the susceptibility to the disease of the host species for each tick stage. The example presented here concerns the maintenance of Lyme disease in wild vertebrate populations.

THE TICK'S LIFE CYCLE AND HOST RELATIONSHIPS

As ticks feed only once per life stage, with different host preferences at each stage, the basic reproductive number (defined as the average number of secondary infections produced when one infected individual is introduced into a host population where everyone is susceptible (Anderson & May, 1991)) of a tick-borne disease depends on: a) for transovarial transmission, the multiplication of the vector itself - the tick's reproduction rate; b) for transstadial transmission, the differential distribution of successive stages of ticks amongst infected and susceptible hosts. The mortality rate of the ticks at each stage of the life cycle has a profound effect on the basic reproductive number of any tick-borne disease. Approximate values for the reproduction and mortality rates at each stage of the life cycle for *Ixodes ricinus*, the major vector of *Borrelia burgdorferi* in Europe, are shown in Fig. 1, with typical patterns of host relationships.

THE BASIC REPRODUCTIVE NUMBER FOR LYME DISEASE

The parameters that determine the basic reproductive number for Lyme disease are shown schematically for the transstadial and transovarial transmission routes in Figs. 2 and 3 respectively. The examples shown are based on the frequently observed bimodal pattern of seasonal tick activity in the Spring and Autumn, starting with one infected host in the Autumn, but the scheme can be modified easily to fit the unimodal pattern also commonly seen (Steele & Randolph, 1985).

Lyme Borreliosis, Edited by J.S. Axford and
D.H.E. Rees, Plenum Press, New York, 1994

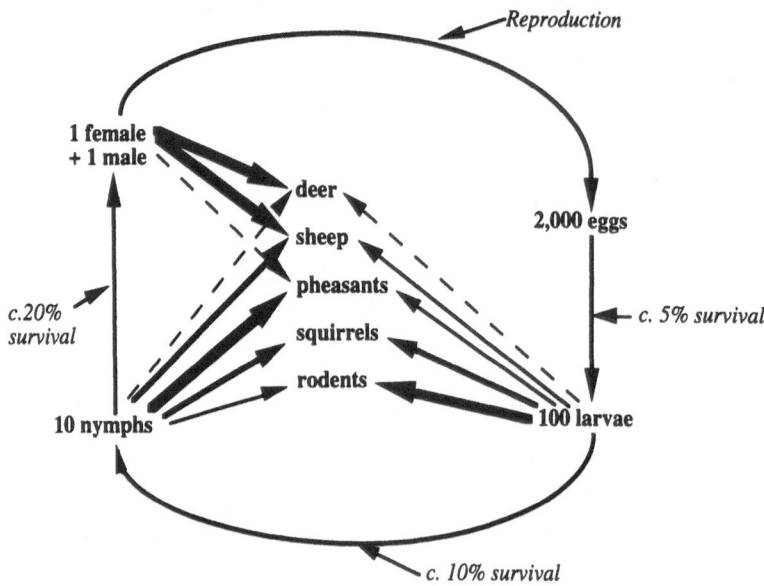

Figure 1. The life cycle of *Ixodes ricinus* with host preferences and approximate reproduction and mortality rates shown.

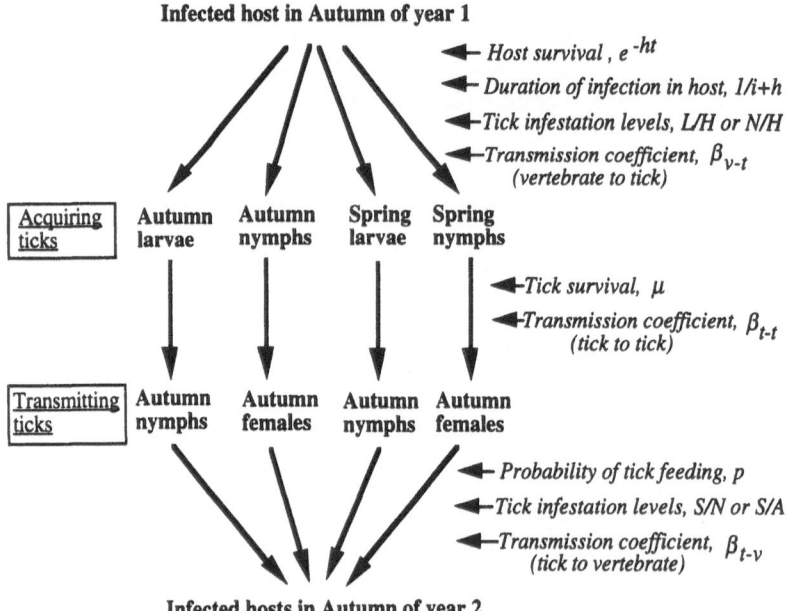

Figure 2. A scheme to illustrate the parameters relevant to the dynamics of transstadial transmission of Lyme disease by *Ixodes ricinus*

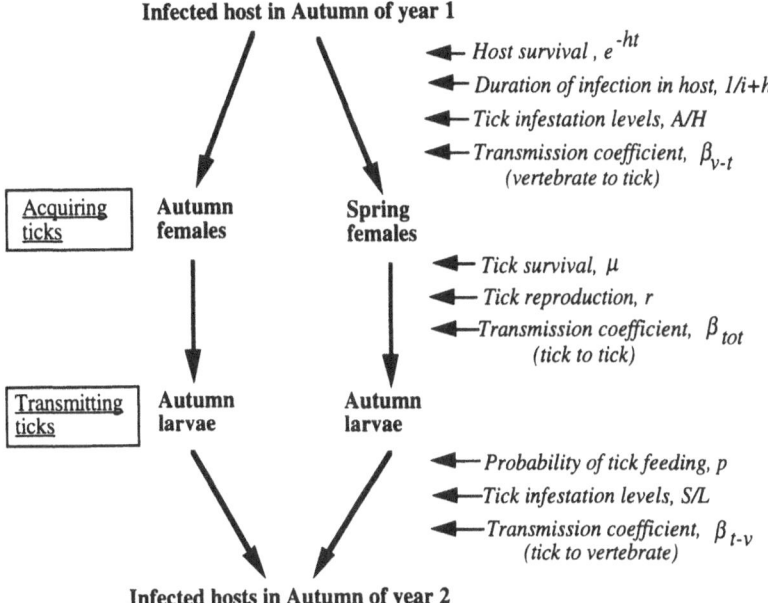

Infected host in Autumn of year 1

Host survival, e^{-ht}
Duration of infection in host, $1/i+h$
Tick infestation levels, A/H
Transmission coefficient, $\beta_{v\text{-}t}$
(vertebrate to tick)

Acquiring ticks   **Autumn females**   **Spring females**

Tick survival, μ
Tick reproduction, r
Transmission coefficient, β_{tot}
(tick to tick)

Transmitting ticks   **Autumn larvae**   **Autumn larvae**

Probability of tick feeding, p
Tick infestation levels, S/L
Transmission coefficient, $\beta_{t\text{-}v}$
(tick to vertebrate)

Infected hosts in Autumn of year 2

Figure 3. A scheme to illustrate the parameters relevant to the dynamics of transovarial transmission of Lyme disease by *Ixodes ricinus*.

The basic reproductive number of the disease, R_0, will have components from each route of transmission, which must be estimated for both autumn and spring ticks:

Partial R_0 due to transstadial transmission

$$= \sum_{sp_j}^{sp_1} L/H\, e^{-ht}\, 1/(i+h)\, \mu_{l\text{-}n}\, p_n\, \beta_{v\text{-}t}\, \beta_{t\text{-}t}\, \beta_{t\text{-}v}\, S/N \quad \text{(larvae-nymphs)}$$

$$+ \sum_{sp_j}^{sp_1} N/H\, e^{-ht}\, 1/(i+h)\, \mu_{n\text{-}a}\, p_a\, \beta_{v\text{-}t}\, \beta_{t\text{-}t}\, \beta_{t\text{-}v}\, S/A \quad \text{(nymphs-adult)}$$

Partial R_0 due to transovarial transmission

$$= \sum_{sp_j}^{sp_1} A/H\, e^{-ht}\, 1/(i+h)\, \mu_{a\text{-}l}\, r\, p_l\, \beta_{v\text{-}t}\, \beta_{tot}\, \beta_{t\text{-}v}\, S/L$$

N = total no. nymphs biting per day on total host population; L = total no. larvae biting per day on total host population; A = total no. female ticks biting per day on total host population; H = no. infected hosts; S = no. susceptible hosts; h = host mortality rate; i = rate of recovery from infection in host; μ = tick survival rate; r = tick reproduction rate; p = probability of tick feeding; β = transmission coefficient; t = period (days) between host being infected and feeding infectible ticks; $sp_{1\text{-}j}$ = species of hosts.

CONCLUSIONS

In the absence of precise quantitative data as yet, the relative contributions of transstadial and transovarial transmission to the maintenance of Lyme disease can only be assessed crudely. Table 1 shows the approximate relative values of each parameter, with the lower value set to 1 in each case. As the numbers of ticks decrease with each successive life stage, the size and life expectancy of the preferred host species increases, and host population density decreases; therefore the relevant tick infestation levels for each transmission route may not be very different.

Table 1. Approximate relative values of each parameter relevant to the dynamics of transmission of Lyme disease by *Ixodes ricinus* via the transstadial or transovarial route.

Parameter	Transstadial	Transovarial	Comments
e^{-ht}	1	10	larger hosts for adult ticks live longer
i	1	1	? infection life-long in all susceptible hosts
$1/(i+h)$	1	10	
μ	2	1	eggs and unfed larvae suffer highest mortality
P	10	1	nymphs (and females) have higher contact rate than larvae (Randolph & Steele, 1985)
r	1	1000	only females reproduce
$\beta_{\text{tick-tick}}$	10-80	1	empirical observations
$\beta_{\text{tick-vert}}$	-	-	depends on the susceptibility of each host species to *Borrelia burgdorferi*
?? 1	:	c.60	

Although these are extremely crude estimates, they do indicate that transovarial transmission should not be dismissed, even if the percent prevalence of infection in questing larvae is observed to be of the order of 1% (Piesman *et al*, 1986; Magnarelli *et al*, 1987). The huge reproduction rate of ticks more than compensates for this. If the vertebrate hosts to adult ticks do not develop infections to *B. burgdorferi*, the observed infection in questing larvae must arise from vertical transmission of the pathogen from nymphs to larvae without an infected blood meal at the adult stage, and this transmission route would also be amplified by the tick's reproduction. Estimation of more precise parameter values require much more detailed quantitative information from the field than is at present available.

REFERENCES

Anderson, R.M. & May, R.M., 1991, "Infectious Diseases of Humans", Oxford University Press.

Magnarelli, I.A., Anderson, J.F., & Fish, D., 1987, Transovarial transmission of *Borrelia burgdorferi* in *Ixodes dammini* (Acari: Ixodidae), *J. Infect. Dis.* 156:234.

Piesman, J, Donahue, T.N., Mather, T.N. & Spielman, A., 1986, Transovarially acquired Lyme disease spirochaetes (*Borrelia burgdorferi*) in field-collected larval *Ixodes dammini* (Acari: Ixodidae), *J. med. Ent.* 23:219.

Porco, T.C., 1991, A model of the enzootiology of Lyme disease in the Atlantic northeast of the United States. *Natural Resiources Modelling* 5:469.

Randolph, S.E. & Steele, G.M., 1985, An experimental evaluation of conventional control measures against the sheep tick, *Ixodes ricinus* (L.) (Acari: Ixodidae). II. The dynamics of the tick-host interaction, *Bull. ent. Res.* 75: 501.

Steele, G.M. & Randolph, S.E., 1985, An experimental evaluation of conventional control measures against the sheep tick, *Ixodes ricinus* (L.) (Acari: Ixodidae). I. A unimodal seasonal activity pattern, *Bull. ent. Res.* 75:489.

EPIDEMIOLOGY OF LYME DISEASE IN ITALY

Giusto Trevisan[1], Josef Simeoni[2], Paolo Conci[2], Anna Maria Bassot[2],
Carla Nobile[1], Giusseppe Stinco[1], Gerolamo Bianchi[3], and Guido Rovetta[3]

Italian Group for the Study of Lyme Borreliosis (IGSLB)
[1]Institute of Dermatology- University of Trieste
 Trieste, Italy
[2]Public Health Service
 Bolzano, Italy
[3]Institute E. Bruzzone- Rheumatological Center- University of Genova
 Genova, Italy

INTRODUCTION

Lyme borreliosis (LB) is a multisystemic disorder caused by a tick-borne spirochete, *Borrelia burgdorferi*. It is endemic in Italy, particularly in three areas: Liguria, Friuli-Venezia Giulia and Trentino-Alto Adige. In fact almost all the patients come from North Italy, while only sporadic cases of this disease were reported in the centre and in the south of the country. In 1983 Lyme disease was first described in Italy by Crovato and Trevisan. In 1989 in Trieste the firt Italian species of *Borrelia burgdorferi* was isolated in a skin biopsy specimen taken from Erythema Chronicum Migrans. It was called BITS1.

AIM OF THE STUDY

LB is a relatively new disease, which injures many organs and apparatuses; it is not subject to compulsory notification and it is very frequent to observe false positive and false negative serologic tests. For these reasons the correct data about its incidence and prevalence in Italy are incomplete or fragmentary. Therefore in 1990 Italian researchers interested in LB founded the Italian Group for the Study of Lyme Borreliosis (IGSLB) to gather epidemiologic data about LB in Italy and to co-ordinate all the researches on this disease.

PATIENTS AND METHODS

IGSLB proposes to evaluate patients with significant signs and symptoms of Lyme disease. In areas where the disease is endemic, tick bit patients, people at risk, healthy individuals and animals must be studied. Serum samples are examined by both immunofluorescence (IFI/ABS) and enzyme-linked immunosorbent assay (*flagellum* ELISA) with western blot as a confirmatory test. Biopsies are taken from skin lesions for histological examination (Warthin-Starry staining) and for culture in Barbour-Stoenner-Kelly's (BSK) medium. *Borrelia burgdorferi* may also be studied in the tissue by polymerase chain reaction (PCR). Spirochetes are also detected in ticks and animals.

In other areas where only sporadic cases are observed determination of borrelial antibody titers must be done in patients with supposed Lyme disease and in people at risk. Besides the aim of IGSLB is to study the prevalence of antibodies anti *Borrelia burgdorferi* in patients affected by flogistic rheumatisms, in blood donors coming from endemic and non endemic areas, and in forestry workers.

RESULTS

Preliminary epidemiologic data are available and they were presented at the first National Congress on LB that was organized by IGSLB in Trieste on November 27-28, 1992.

In Italy LB is frequent chiefly in the North, with endemic centres in Friuli-Venezia Giulia where in the last seven years over 600 cases were diagnosed, in Liguria with over 160 cases and in Trentino-Alto Adige with over 30 cases.

Prevalence rates presented by various authors at the first Italian Congress on LB are shown in table 1.

Serologic data are summarized in table 2 and 3.

Table 1. Prevalence of clinical manifestations of LB in endemic areas in Italy.

Erythema chronicum migrans	21 - 58%
Erythema chronicum migrans + other symptoms	7 - 26%
Lymphadenosis benigna cutis	0.6 - 8%
Acrodermatitis chronica atrophicans	1 - 3%
Neurological manifestations	24 - 27%
Arthritis	21 - 40%
Cardiovascular manifestations	1 - 2%
Other cutaneous manifestations	11- 14%

Table 2. Seropositivity rates in different clinical manifestations of Lyme disease.

Erythema chronicum migrans	39%
Neurological manifestations	85%
Chronic arthritis	97%

Table 3. IFI and ELISA seropositivity rates in endemic and non endemic areas.

	Endemic areas	Non endemic areas
IFI	35 - 40%	2.2 - 3.6%
ELISA	16 - 19%	8.9 - 12.7%

DISCUSSION

The numerous publications (more than 2500) about Lyme disease show that the number of case reports is costantly increasing. Today LB is considered the most common and diffuse tick-borne disease in Europe and in North America. In Europe some thousands of cases occurred and the disease was reported at least in 19 countries. The distribution of cases of LB mostly reflects the geographic areas where the principal vectors are present. The most important vector is a tick, *Ixodes ricinus* particularly in Europe, but also other arthropods can have this role. European and North American studies demonstrated that about 15% of *Ixodes ricinus* and *Ixodes dammini* ticks can bear *Borrelia burgdorferi*. In Austria, where Lyme disease has been carefully studied, about 15% of ticks are infected by *Borrelia burgdorferi*. In some areas endemic for Lyme disease, high percentage (50% and over) *Borrelia burgdorferi* borne ticks were observed. The Alps and neighbouring countries, in particular Austria, North Italy, Slovenia, Czechoslovakia, Bavaria, Switzerland are the areas where Lyme disease is most frequent. In fact also in Italy the endemic areas are close to the Alps as Liguria, Friuli-Venezia Giulia and Trentino Alto Adige, but also some Appenninic areas are endemic (North West of Tuscany and Emilia Romagna). The other regions of North Italy (Piedmont, Lombardy, Venetia) present a lower incidence while only sporadic cases are reported from the Southern and Insular regions. The epidemiologic findings of Lyme borreliosis drawn by preliminary data analysis is still very incomplete, because each author studied the disease alone. However our data give a tempestive idea of the distribution of LB in Italy. Therefore, we belive that it is important to standardize researches and data collection as IGSLB suggests. Only when we will have all the data collected with a standard method, we might know the frequency of clinical manifestations, the incidence and prevalence of LB in Italy and we shall compare them with other European countries and with the United States of America.

REFERENCES

Bianchi G, Buffrini L, Monteforte P, Chioni F, Murruzzu M, Mancardi G L, Parodi A, Crovato F, Rovetta G. Epidemiology and clinical manifestation of Lyme disease in Liguria. Abstract 1st National Congress of GISML, 27-28 November 1992 - Trieste.

Buffrini L, Brites Louro CA, Garzia V, Barberis A, Grignolo MC, Monteforte P, Rovetta G, Bianchi G. Prevalence of antibodies to *Borrelia burgdorferi* in blood donors: comparison between endemic area and non endemic area. Abstract 1st National Congress of GISML, 27-28 November 1992 - Trieste

Cinco M, Banfi E, Trevisan G, Stanek G. Characterization of the first tick isolate of *Borrelia burgdorferi* from Italy. APMIS (1989) 97: 381-382.

Cinco M, Balanzin D, Benussi P, Trevisan G. Seroprevalence and incidence of Lyme Borreliosis in forestry worker in Friuli-Venezia Giulia (Northern Italy). Alpe Adria Microbiology Journal 1993, in press.

Crovato F, Nazzari G, Fumarola D, Rovetta G, Cimmino MA, Bianchi G. Lyme disease in Italy: first reported case. Ann Rheum Dis (1985) 44: 570-571.

Dettori G, Grillo R, Speziale D, Zanetti S, Calderaro A, Chezzi C, Fadda G. Prevalence of antibodies to *Borrelia burgdorferi* in adults in Sassari province. Abstract 1st National Congress of GISML, 27-28 November 1992 - Trieste.

Gaddoni G, Baldassari L, Selvi M, Pavan WO. Clinical, anamnestic, serological aspects of Lyme Borreliosis in Faenza. Abstract 1st National Congress of GISML, 27-28 November 1992 - Trieste.

Marini M, Dettori G, Viani I, Calderaro A, Chezzi C. Serologic study about the presence of Lyme disease patients in province of Parma. Abstract 1st National Congress of GISML, 27-28 November 1992 - Trieste.

Nocerino A, Dettori G, Grillo R, Strano CG, Manini M, Di Martino L. Lyme Arthritis in Campania. Abstract 1st National Congress of GISML, 27-28 November 1992 - Trieste.

Proietti A, Crucil C, Nocerino A, Di Lenardo C, Pitzus E. Antibodies against *Borrelia burgdorferi* in province of Udine: 30 months experience. Abstract 1st National Congress of GISML, 27-28 November 1992 - Trieste.

Pugina P, Benzi G, Rassu M, Pagani W, Grosso F, Chiacchiarini, Spoladori S, Lorenzi G. Lyme disease prevalence in Polesine: diagnostic possibility. Abstract 1st National Congress of GISML, 27-28 November 1992 - Trieste.

Ruscio M, Salvatorelli E, Venchiarutti A, Pesente F. Epidemiology of Lyme disease in North Friuli. Abstract First National Congress of GISML, 27-28 November 1992 - Trieste.

Trevisan G. Lyme disease: about a case. Ann It Derm Clin Sperim 40: 91-95 (1986).

Trevisan G, Crovato F, Marcuccio C, Fumarola D, Scarpa C. Lyme disease in Italy. Zbl Bakt Hyg (1986) 263: 459-463.

Trevisan G. Lyme Borreliosis epidemiology in Friuli-Venezia Giulia. Ann It Derm Clin Sperim (1990) 44: 127-137.

Trevisan G, Pavan W, Rorai E. Lyme disease: a case report in Cavarzere - Venice. G It Derm Venereol (1991) 126: 435-437.

ROLE OF HOST DENSITY IN THE ECOLOGY OF LYME DISEASE

Tamara E. Awerbuch[1] and Andrew Spielman[2]

[1]Department of Biostatistics
Harvard School of Public Health
677 Huntington Avenue
Boston, MA 02115

[2]Department of Tropical Public Health
Harvard School of Public Health
665 Huntington Avenue
Boston, MA 02115

ABSTRACT

To explore the effects the abundance of deer and mice, the main hosts of the vector of Lyme disease in the Northeastern part of the USA, on that of deer ticks, we implemented a model representing the life cycle of these vector ticks using realistic parameter estimates taken from 2 field sites in coastal Massachusetts. The main inputs of the model were monthly scanning capacities and host abundance. The equation for scanning capacity has 4 components: tick density on hosts in a particular month, duration of activity that month, duration of tick attachment and density of questing ticks.

In a site in which the abundance of deer remained constant and the abundance of mice varied from year to year, that of recently emerged larvae varied as well. Our studies showed that had the density of mice been held at a level corresponding to that in a year of exceptional mouse abundance, the ticks would have thrived. Had such hosts remained scarce, tick abundance would have waned. A stable density of ticks accompanied an "ordinary" density of mice.

In the second site where deer abundance was systematically reduced and mice abundance fluctuated, the tick population continued to grow. Our simulations suggested that the critical threshold of deer abundance is 8 animals. We conclude that the abundance of deer ticks is sensitive both to the abundance of mice and that of deer.

INTRODUCTION

White-tailed deer (Odocoileus virginianus) serve as the definitive host of the deer tick (Ixodes dammini), the vector that transmits the agent of Lyme disease in eastern North America. The adult stage of this tick, generally feeds on deer and its subadult stages on white-footed mice (Peromyscus leucopus) (Spielman 1988). Deer are considered crucial in the life cycle of these ticks and are an important vehicle for their geographical spread (Spielman 1988, Wilson et al. 1985, Wilson et al. 1988). At times, subadult deer ticks will also feed on deer, although they generally feed on

Lyme Borreliosis, Edited by J.S. Axford and
D.H.E. Rees, Plenum Press, New York, 1994

mice or other small animals (Piesman et al. 1979). Another level of complexity is added by the fact that immature stages of this tick appear to feed mainly on mice while adult deer ticks never parasitize these host; thus making it difficult to define the density of hosts required for perpetuation of the deer tick. It may be that the density of both kinds of hosts determines the abundance of the tick.

To explore the role of the various kinds of hosts in determining the density of I. dammini in a particular geographic location, we applied a mathematical model formulated to represent the life cycle of deer ticks (Sandberg et al. 1992) and implemented it with actual parameter estimates based on field observations in 2 sites in coastal Massachusetts. In one site, the deer population remained constant while the mouse population fluctuated during the 8-year period of the study. In the other site, the mouse population mildly fluctuated throughout the year while deer were removed by limited hunting. We used computer simulation to explore the effect of realistic changes in mouse and deer abundance on the population biology of these ticks.

DESCRIPTION OF STUDY SITE AND FIELD METHODOLOGY

The data in our studies was derived from the Nantucket Field Station of the University of Massachusetts, on Nantucket Island, Massachusetts. This 32 ha shoreline site was studied regularly during 1985 through 1991. The vegetation was comprised equally of grassy meadows separating stands of dense coppice. Voles were most abundant in the meadows and white-footed mice in ecotonal situations. Although shrews, cottontail rabbits and domestic animals were abundant there, few animals of other kinds inhabit this restricted site.

Observations were made monthly between April and October. Deer density was estimated by direct observation and that of mice by a minimum-number-alive estimate (Wilson et al. 1988). The abundance of rodents was estimated by placing oat-baited live traps (Longworth Co, Abingdon, England, UK) in a permanent 7 X 7 grid in the field site. Each captured animal was marked by means of a numbered ear-tag. The minimum number alive method of estimating abundance was employed (Hilborn et al., 1976). The abundance of feeding ticks was determined by visually inspecting each trapped animal and animals promptly released. Ticks were removed for subsequent identification.

The study on the effect of incremental removal of deer on the density of the deer tick was undertaken in Ipswich, Massachusetts, on a 567 ha coastal site maintained by the Trustees of Reservations. Much of the property is comprised of a 9 km long barrier island characterized by beach, dunes, salt marsh and woodland. Lyme disease affects numerous residents of nearby sites (Lastavica et al. 1989). Mouse abundance an density of ticks feeding on mice was monitored as described. Deer abundance was reduced from an estimated 430 in 1983 to about 150 in 1991.

THE MODEL

Seasonal Transitions

The seasonally punctuated life cycle of the deer tick can be represented by a mathematical vector consisting of 11 stages, and the monthly transitions by 12 matrices, each capturing the changes occurring in a particular month of the year (Sandberg et al. 1992). The main input variables were considered to be host density and efficiency of host-finding (scanning capacity) for each developmental stage of the tick. The present effort is designed to represent conditions actually observed during 1984-1991 in the Nantucket field site and in Ipswich, Massachusetts. The density of ticks present during each year is calculated as the product of the mathematical vector for the previous year and of the 12 transition matrices representing each month of the year in question:

$$X(t+1) = \prod_{i=1}^{12} A_i \; X(t)$$

where A represents an 11X11 matrix containing the monthly transition parameters and $X(t)$ is the vector representing tick density during each of the 11 designated activity stages of the tick in year (t) (Sandberg et al. 1992).

<u>Parameter Estimation</u>

The 2 main input parameters that are used in the matrices are derived from the sequence of field observations actually conducted in the study sites as well as certain supplementary observations conducted in the laboratory. One of these variables, host density, is determined by direct observation (of deer) and by mark-release-recapture (of rodents).

The other main input parameter is estimated indirectly. Scanning capacity of a particular stage of the deer tick, questing in a particular month after a particular kind of host, is comprised of the following elements:

$$\frac{\begin{array}{cc} \text{density of ticks} & \text{duration of feeding} \\ \text{per host that month} & \times \quad \text{activity that month} \end{array}}{\begin{array}{cc} \text{duration of attachment} & \text{density of questing} \\ \text{of that stage of tick} & \times \quad \text{ticks that month} \end{array}}$$

The first two components, tick density on hosts and duration of activity that month, are derived from field observations. The third, duration of tick attachment, is determined in the laboratory (Piesman et al., 1979). Biological assumptions are required to estimate the fourth component in this formula, density of questing ticks. These assumptions are based on certain features of the seasonally discrete life cycle of this tick (Yuval and Spielman, 1990).

The developmental cycle of the stages of the deer tick proceeds as follows:

Egg -> Larva -> Nymph -> Adult -> Egg

Ticks feed once during each trophic stage of development before they molt to the next stage. The rigid seasonality of this feeding behavior in the case of the deer tick punctuates its developmental cycle as follows: Eggs deposited in June and July hatch in August. Larvae abundantly quest for hosts in August through September and again in April through July. Nymphs abundantly quest in May through July, adults from mid-October through December and again in February through mid-April.

Density of feeding ticks is calculated, month by month, on the basis of the observed density of ticks per host and the observed abundance of that host. We assume that 2000 eggs are laid by each fertilized female, of which half are potential female which will start the cycle anew. We further assume that all eggs hatch to produce an equivalent number of questing larvae. The cohort of ticks that hatches in the fall of 1984 provides the basis of our analysis. Some of these larval ticks will attach to available hosts, and all of these will feed to repletion in 3 days. Those that fail to find a host during the fall of that year, will resume questing during the

141

subsequent spring. We assume that all such engorged larvae ultimately molt to the nymphal stage and that all seek hosts beginning in the spring of 1985. Some of these nymphs will attach to available hosts as the opportunity presents itself during the spring and summer of 1985 and of 1986, and all of these will feed to repletion in 5 days. We assume that all nymphs that engorge in 1985 molt to the adult stage and that all will seek hosts beginning in the fall of the same year. Some of these adults will attach to available hosts, and all of these will feed to repletion in 10 days. All engorged adults ultimately oviposit beginning in the summer of 1986. In this manner, the population of deer ticks completes its developmental cycle.

The density of questing deer ticks is estimated by combining the results of the monthly field observations with derivations based on these assumptions. The calculation begins with the assumption that all the newly hatched larvae began to quest early in August of 1984. The activity of this cohort of ticks is followed until they mature to the adult stage and ultimately oviposit. We calculate the number of larvae questing at the beginning of each month by subtracting the calculated number of larvae that fed during the previous month from the number estimated to be questing during that month. In the event that no larvae were found on hosts during a particular month, we assume that diapause prevented questing activity. The number of questing nymphs and adults is calculated similarly. Larvae that engorge during the summer season are assumed to molt and to recommence questing as nymphs during the following season. These rules dictate the manner of calculating scanning capacity.

Scanning capacity of these ticks is calculated for a particular stage of the tick seeking a host in a particular month. That of an adult in October, for example, is:

$$(86 \times 15) / (10 \times 21{,}996) = 0.005865$$

which is the probability that an adult tick may attach to a deer. Thus, only 1 out of 171 adult deer ticks that are present in the site in October would find a host. A similar calculation is repeated for each of the other stages of the tick, for each month and for each kind of host. Scanning capacity is calculated and averaged over each of the years for which data are available.

Implementation of the model

To implement the model, 2 programs are constructed; one for calculating scanning capacity and the other for vector density. For convenience, both employ computerized spreadsheets (Lotus 1-2-3).

Because scanning capacity is calculated for each month of the year, a column in the spreadsheet is designated for each month. The rows receive the observed data on host and tick density as well as the equations for calculating scanning capacity. The output of each columnar calculation is placed in designated positions in the appropriate matrix. numerical values when matrices are multiplied.

In constructing the program for tick density at each developmental stage, difference equations represent matrix multiplication. The columns represent density of the various stages of the vector for each month of the year. Vector density during each month depends on density during the previous month and on a transition parameter that represents feeding, molting or death of the tick. In this manner stage-specific vector density is calculated for each of the designated months of the year.

RESULTS

The pattern of annual variation in mouse abundance in the Nantucket study site is presented in Figure 1. Mouse abundance fluctuated annually, with a 6-year rising trend between the years 1984-1989, followed by a 2-year pattern of decrease.

We determined whether the abundance of ticks vary in parallel with the observed variations in the abundance of mice. Abundance of larvae during midwinter was chosen to represent the dynamics of the tick population because this phase of tick development directly reflects reproductive activity and is subject to few environmental variables. Abundance of questing larvae closely follows that of mice, generally lagging by a 1 or 2-year interval (Fig 1). The abundance of newly emergent larvae varies with the abundance of mice.

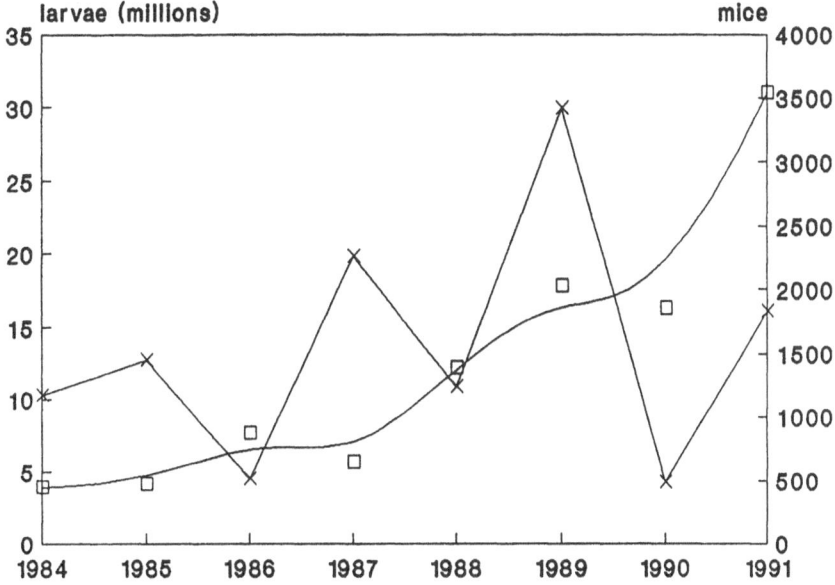

Fig.1 Changes in the observed abundance of mice present in the field site and simulataneous changes in the abundance of larval deer ticks predicted for the first of December for each year of the study.

In order to examine the effect of superabundant hosts on the abundance of newly emerged larval deer ticks, we determined how many would develop if their rodent hosts remained as abundant as they were in an exceptional year. A simulation was ran starting with one female that deposited a clutch of 1,000 fertile eggs in a site that had previously been devoid of ticks and mouse abundance was as high as in the year 1989. Tick density increased exponentially, with a doubling time of 1.06 years (Figure 2). When hosts for the subadult stages of this tick are exceptionally abundant, the tick population increases rapidly and continuously.

We then determined how an exceptionally scarce population of hosts may affect the abundance of newly emergent larvae. Because mice were particularly scarce in the study site in 1990, that year was chosen as the standard of host scarcity (Fig. 1). The simulation began, as in the previous simulation, with the deposition of a clutch of 1,000 fertile eggs in a site that had previously been devoid of ticks. Tick density waned and ultimately disappeared (Fig 3). Under these conditions of exceptional mouse scarcity and a continuing presence of 10 deer, deer ticks might become endangered.

To determine whether a particular level of mouse density may result in a stable density of ticks, we applied a similar process using the data of 1986, a year in which the density of mice noted in the study site somewhat exceeded that in the designated year of paucity, 1990 (Fig. 1). After 1,000 eggs were "deposited in the empty field" of the study site, the tick population oscillated (Fig. 3) and ultimately stabilized at a density that was about half that of the first year of the simulation.

This study indicates that there is a threshold of host density below which the tick population will fail to maintain itself.

In the second study site were deer abunadance was systematically reduced from 400 to 100, the tick population continued to grow (Fig. 4). Our simulations show that in order to limit the tick population on the site, deer would have to be reduced to a very low number. In our studies the critical abundance is 8 animals (Fig. 5).

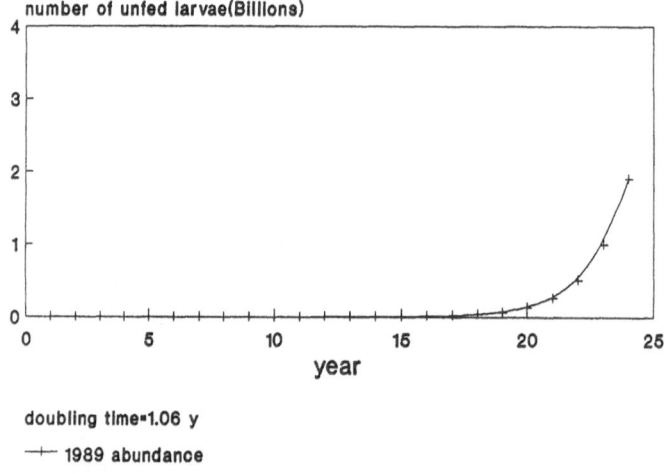

Fig 2. Increase in the simulated abundance of larval deer ticks assuming that mice continuously remain exceptionally dense (as observed in 1989).

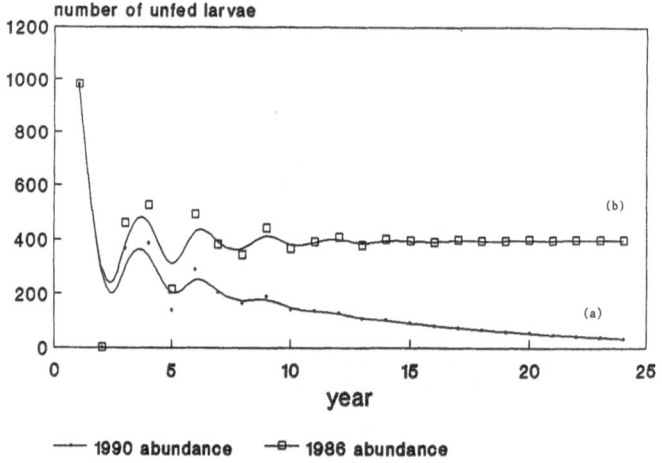

Fig 3. (a) Decrease in the simulated abundance of larval deer ticks assuming that mice continuously remained exceptionally scarce (as observed in 1990). (b) Stabilization in the simulated abundance of larval deer ticks assuming that mice continuously remained moderately scarce (as observed in 1986).

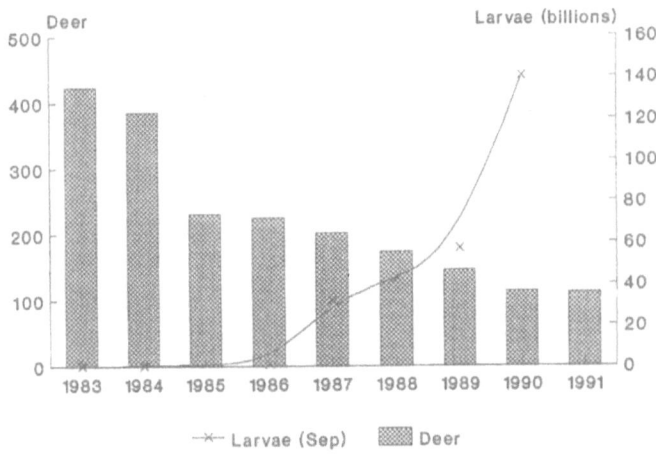

Fig 4. Abundance of deer ticks following removal of deer in the Ipswich study site.

DISCUSSION

Host availability is a critical determinant of tick dynamics. Only a minute portion of the tick population proceeds from stage to stage, mainly due to a failure to find hosts (Sandberg et al. 1992). Longevity of these ticks seems to be closely programmed in nature; a questing larva does not live longer than 11 months, a nymph 14 months and an adult 8 months (Yuval and Spielman 1990). Successful development of a tick depends upon its scanning capacity. Host abundance similarly affects the ability of a tick to find a host. Mouse density varies in a complex manner (Adler et al. 1984). They begin to reproduce in the spring and continue to increase until fall. This temporal coincidence facilitates feeding activity of deer ticks by matching maximum questing activity of larvae, which eclode in August (Yuval and Spielman 1990), with maximum availability of their hosts. Years of relative mouse abundance tend to follow years of scarcity. This alternation buffers the abundance of deer ticks because the same cohort of ticks generally experiences both extremes in their different subadult stages of development which spans 2 years or more. Multi-year cycles also occur. The 5-year cycle frequently observed in populations of these white-footed mice (Adler et al. 1984; Krohne et al. 1982) is consistent with the pattern observed in our study. All of these crucial and temporally variable relationships are incorporated as elements in the transition matrices. The derivation of scanning capacity combines estimates of potential turn-over of ticks on hosts and on efficiency of host-finding (Sandberg et al. 1992). Although separate calculations are made for each stage of the tick questing during each month of the year, our simulations are based on average estimates spanning the 8-year period of observation. Estimates of the abundance of questing ticks are required for directly calculating the scanning capacity of ticks in a particular site. Because field observations are not available for direct estimation of this parameter, our estimates are based on an algorithm that uses the density of feeding ticks on hosts as one of its major inputs.

The density of feeding ticks is more easily determined in the field than is that of questing ticks. Questing deer ticks generally are sampled by sweeping vegetation with a section of fabric intended to represent a surrogate host. Such an operation, however, is fraught with uncertainty; only a tiny fraction of the population can be sampled.

The presence of numerous deer generally is regarded as prerequisite to an abundant infestation of deer ticks (Spielman 1988). Qualitative observations have established that stable infestations of the tick are evident solely where deer are resident. Although correlative observations demonstrate a relationship between the density of larval deer ticks feeding on mice and the density of deer (Wilson et al. 1985), reduction of an established infestation of deer may not inhibit population growth of the deer tick. This result is consistant with experimental evidence that demonstrated that only a major destruction of resident deer is followed by the virtual elimination of these ticks (Wilson et al. 1988).

LITERATURE CITED

Adler, G. H. and Tamarin, R. H. (1984). Demography and reproduction in island and mainland white-footed mice (Peromyscus leucopus) in southeastern Massachusetts. Canadian Journal of Zoology 62: 58-64.

Caswell, H. 1989. Matrix population models. Sinauer, Sunderland, MA.

Daniels T. J., D. Fish and R. C. Falco. 1989. Seasonal activity and survival of adult Ixodes dammini (Acari: Ixodidae) in southern New York State. Journal of Medical Entomology 26:610-614.

Hilborn, R., J. A. Redfield and C. J. Krebs. 1976. On the reliability of enumeration for mark and recapture census of voles. Canadian Journal of Zoology 54:1019-1024.

Krohne, D. T., J. F. Merritt, S. H. Vessey and J. O. Wolfe. 1988. Comparative demography of forest Peromyscus leucopus. Canadian Journal of Zoology 66:2170-2176.

Leslie, P. H. 1945. On the use of matrices in certain population mathematics. Biometrika 35:183-212.

Mather, T. N. and A. Spielman. 1986. Diurnal detachment of immature deer ticks (Ixodes dammini) from nocturnal hosts. American Jounal of Tropical Medicine and Hygiene 35:182-186.

Piesman J., A. Spielman, P. Etkind, T. K. Reubush Jr. and D. Juranek. 1979. Role of deer in the epizootiology of Babesia microti in Massachusetts, U. S. A. Journal of Medical Entomology 15:537-540.

Sandberg, S., T. E. Awerbuch and A. Spielman. 1992. A comprehensive multiple matrix model representing the life cycle of the tick that transmits the agent of Lyme disease. Journal of Theoretical Biology in press.

Semtner, P. J. and J. A. Hair. 1975. Evaluation of CO_2-baited traps for survey of Amblyomma americanum Koch and Dermacentor variabilis Say (Acarina: Ixodidae). Journal of Medical Entomology 12:137-138.

Spielman, A. 1988. Lyme disease and human babesiosis: evidence incriminating vector and reservoir hosts. Pages 147-165. in P. T. Englund and A. Sher, eds. The Biology of Parasitism , New York: Alan R. Liss, New York.

Wilson, M. L., S. R. Telford III, J. Piesman and A. Spielman. 1988. Reduced abundance of immature Ixodes dammini (Acari: Ixodidae) following elimination of deer. Journal of Medical Entomology 25:224-228.

Wilson, M. L., G. H. Adler and A. Spielman. 1985. Correlation between abundance of deer and that of the deer tick, Ixodes dammini (Acari: Ixodidae). Annals of the Entomological Society of America 78:172-176.

Yuval B. and A. Spielman. 1990. Duration and regulation of the developmental cycle of Ixodes dammini (Acari: Ixodidae). Journal of Medical Entomology 27:196-201.

BORRELIA BURGDORFERI STUDIES IN MAN AND TICKS IN SCOTLAND

S.M. Curtin and T.H. Pennington

Department of Medical Microbiology, Medical School
Buildings, Foresterhill, Aberdeen AB9 2ZD
0224 681818 ext 52864   Fax 0224 685604

ABSTRACT

Objectives To determine the seroprevalence of *B. burgdorferi* antibodies in a group of high risk workers in Scotland, the carriage rate of *B. burgdorferi* in *I. ricinus* ticks and to study the relationship between the two.
Methods Seroprevalence was tested by ELISA with workers answering a questionnaire to determine frequency of tick exposure and tick bites. Carriage rate was determined by PCR analysis.
Results 19.0% of high risk workers were found to have IgG anti-*B. burgdorferi* antibodies. A higher seroprevalence was found in the West of Scotland, 26.0% as opposed to the East (14.0%). This was related to the carriage rate of *B. burgdorferi* in nymphal *I. ricinus* ticks with a carriage rate of 34.8% on the West coast and 25.6% on the East. There seemed little correlation between the number of tick bites and frequency of exposure to ticks.
Conclusion There is a higher chance of becoming infected with *B. burgdorferi*, when bitten by *I. ricinus* nymphs on the West coast of Scotland than on the East coast, if the risk of tick bites are unknown.

INTRODUCTION

Since the recognition of Lyme disease in 1975 (1), much controversy has centred around the importance of it's causative agent *Borrelia burgdorferi* as a vector borne pathogen. Isolated studies (2,3) have measured the seroprevalence of *B. burgdorferi* in patients in Scotland, but until now no-one has investigated the risk of infection in a high risk group, or the relationship of seroprevalence to carriage rate of *B. burgdorferi* in ticks. Ixodes ticks are widely distributed in Scotland . *I. hexagonus* and *I. canisuga* can be excluded as vectors for human *B. burgdorferi* infections, as they bite only hedgehogs and dogs respectively. *I. uriae* only bites birds, but has proved to be a competent vector (4) and should be taken into account when determining the enzoootic cycle of *B. burgdorferi* in an area. The other Ixodes species found in Scotland are *I. ricinus* and *I triangulceps*, Campbell (5) has shown that *I. ricinus* is the predominant species of tick found on small mammals . *I.triangulceps* is found only in burrows and so is unlikely to be an important vector of human infections. It seems probable therefore that *I. ricinus*, while possibly not solely responsible for maintenance of *B. burgdorferi* in the wild, is responsible for transmission to humans.

Lyme Borreliosis, Edited by J.S. Axford and
D.H.E. Rees, Plenum Press, New York, 1994

We have examined two sites in Scotland for the presence of *B. burgdorferi* in ticks. Tentsmuir forest (Site 1) is situated on the East coast of Scotland, and consists of a mixed coniferous and Beech wood, with a predominantly bracken undergrowth. The Isle of Rum (Site 2) is situated off on the West coast and is covered mainly with rough grassland.

Seroprevalence was examined in Scottish National Heritage workers, a high risk group of outdoor workers, based throughout Scotland as shown in Figure 1.

Fig. 1. Distribution of Scottish National Heritage Bases in Scotland.

METHODS

Tick Collection

Ticks used in this study were collected by flagging, which whenever possible, took place in the evening, when questing ticks were most abundant. Ticks were stored at 4°C for a maximum of 2 months in humid conditions, separated into batches of nymphs (50-70), adult males (approx 10) and adult females (approx 10).

Immunofluorescence Assay

Live adult ticks were washed in 100% ethanol, placed on an 8-well immunofluorescence slide and cut in half with a fresh blade. The gut and salivary region from each half tick were smeared into separate wells. Each sample was then mixed with $20\mu l$ Phosphate Buffer Saline, pH 7.4 (PBS), air dried and fixed with $10\mu l$ acetone. Slides were incubated with $20\mu l$ of sheep anti-*B. burgdorferi* fluorescence conjugate antibody, incubated for 30mins at 37°C and then given 2 x 15 minute washes with PBS, followed by Evans Blue counterstain. After air drying, samples were examined under a fluorescence microscope.

Polymerase Chain Assay

PCR primers used were based on the sequence of the American B31 strain (6). These primers amplify a 550 base pair (bp) segment of the *Osp* A gene. $10\mu l$ of target DNA was assayed in a total reaction volume of $100\mu l$, containing 2U Taq Polymerase, $200\mu M$ oligonucleotides, 125nm each primer, manufacture's reaction buffer and distiled water overlaid with mineral oil. Samples were subjected to an initial 3 minute cycle at 95°C, followed by 30 cycles of 1 minute at 93°C, 1.5 minutes at 55°C and 1 minute at 72°C. A final extension step of 7 minutes at 72°C was included. PCR products were analysed on a 2% agarose gel and visualised by ethidium bromide staining.

Polymerase Chain Reaction - Sample Preparation

Two methods of DNA extraction were used to prepare ticks for PCR analysis. The longer, standard DNA extraction was used for pooled ticks, whereas a more simple rapid method was used for single ticks.

Ticks - Standard Preparation Whole ticks were quick frozen at -70°C for 30 minutes, which allowed simple homogenization with a pipette tip and the use of clean sterile materials for each tick preparation. Homogenised ticks were suspended in (1% (wt vol^{-1}) sodium dodecyl sulphate (SDS), 0.8% (wt vol^{-1}) sodium hydroxide in water), samples were centrifuged at 12,000g for 1 minute and the supernatant discarded. Each sample was washed in DNA buffer (0.44% (wt vol^{-1}) sodium chloride, 0.89% (wt vol^{-1}) ethylene diamine tetracetic acid (EDTA), pH 8.0), resuspended on $400\mu l$ DNA buffer containing $50\mu l$ Proteinase K (10mg ml^{-1}) and $25\mu l$ 10% SDS, and incubated overnight at 37°C. $5\mu l$ of 0.1M Phenylemethyl Sulfanyl Fluoride (PMSF) was added and incubated for 1 hour at 25°C to inactivate Proteinase K. A phenol extraction using water saturated phenol, followed by a back extraction using TE buffer (10mM Tris/HCL and 1mM EDTA, pH 7.6), was performed on each sample. After a further wash in chloroform:isoamyl alcohol (24:1 by volume), $400\mu l$ of the aqueous layer was removed and 1/10th volume 3M sodium acetate, 2 volumes 100% ice cold ethanol were added. Pelleted samples were air dried and then resuspended in $20\mu l$ water.

Ticks - Rapid Method Single adult ticks were homogenised and suspended in $200\mu l$ TP buffer (0.5% Nonidet P-40 (by volume) 5% Tween 20 (by volume) and 10mgs/ml Proteinase K in TE buffer). Samples were then heated for 1 hour at 55°C, followed by 10 minutes at 95°C to deactivate Proteinase K. After centrifugation at 12,000g for 1 minute, the supernatant was removed and stored at 4°C until assayed.

Double PCR

Double PCR was initially used to increase our stock of interesting PCR positive samples. It also proved useful to increase the amount of DNA in samples showing positive by Southern blot but not by gel analysis. The PCR product was purified, resuspended in 20µl distiled water and run on a 2% agarose gel. Bands were excised and stored overnight at 4°C in TE buffer. DNA eluting from the agarose into the TE buffer provided an ideal target for PCR analysis.

Southern Blot Analysis

To confirm positive bands as seen by gel electrophoresis all gels were probed using the Boehringer Mannheim digoxigenin (DIG) labelling system. The probe was manufactured by PCR amplification of the primer reference strain B31 including 5µl of the DIG reaction mixture. Successful labelling was confirmed when our probe was seen to be heavier than standard B31 PCR product by gel analysis.

Tick Culture

Live ticks were washed in 100% ethanol, roughly dissected into four using a fresh sterile blade and inoculated into 9mls of Barbour Stonner Kelly medium (BSK 11) containing 4µg/ml Trimethoprim, 256µg/ml Sulfathoxazole and 100mg/ml Neomycin. Nymphs were inoculated in pools of 5; adult ticks were inoculated singly. Cultures were incubated at 34°C and examined weekly, or when cultures turned yellow, by dark-field microscopy. Contaminated cultures were discarded.

Blood Collection

Blood samples were collected at four points between October and November at the Annual General Meetings of Scottish National Heritage workers (SNH). While donating blood each volunteer filled out a questionnaire giving details of age, sex, base in Scotland and frequency of tick bites. 5mls of blood was collected from each volunteer and the sera frozen at -20°C until assayed. Both IgG and IgM ELISA tests were carried out using the DAKO 2nd Generation anti-flagellin Lyme Borreliosis kit, according to the manufacturers protocol. Samples were tested in duplicate in batches of 50, with borderline positives being tested a third time. IgG positive samples were tested in duplicate for anti-*B. burgdorferi* IgM. Control sera (100), from Aberdeen Royal Infirmary Blood Transfusion were tested for IgG antibodies.

Statistical analysis

Significance values for the differences in carriage rates/ seroprevalence were estimated using measures of Students t distribution (8). Product moment correlation coefficients were estimated and compared by standard means (Clarke & Cooke, 1983) (9). The linear regression for reported tick bites / exposure levels was calculated and plotted using facilities within the Harvard Graphics computer program.

RESULTS

4,000 *I. ricinus* ticks were collected, of which 80% (3,200) were nymphs and 20% (800) were adults. In spring and early summer, nymphs were far more common than adults, whereas in autumn the differences were less marked. Larvae were found in abundance at one site in August, but otherwise they were extremely uncommon and have not been included in this study. Dragging took place next to small tracks and paths where the highest number of ticks were found.

Immunofluorescence Assay

100 *I.ricinus* adults were examined by immunofluorescence for **B.burgdorferi**. 20% (20) were found to be positive.

Polymerase Chain Reaction and Southern blot hybridization

Ticks were separated into "nymph" and "adult" groups before being screened for *B.burgdorferi* by PCR using primers based upon the sequence of *Osp* A. From 25 pools of 10 nymphs, 9 positive results for *B.burgdorferi* were obtained by single step PCR. Given this relatively low rate, it might be presumed that only 1 tick in each positive pool was carrying *B. burgdorferi*, giving an overall prevalence of 3.6%. However, when 250 nymphs were examined individually, 87 were found to be positive, giving a higher overall carriage rate of 34.8%. Therefore subsequent PCR analysis was always performed on individual ticks.

Southern blot hybridization of PCR product run in agarose gels was used to confirm all samples judged to be positive by ethidium bromide-staining. A few samples produced multiple banding patterns which were probe-positive, and some samples from Scottish ticks produced Southern blot-positive PCR-amplification bands which differed in size (390 bp) compared with those from the American control strain, *B.burgdorferi* B31. In addition, Southern blot hybridization was also useful in detecting probe positives when bands could not be not seen by ethidium bromide-stained gel analysis. These PCR products could subsequently be subjected to double PCR assay to produce PCR product suitable for further analysis.

Table 1: Carriage rate of *B.burgdorferi* in *I.ricinus*

| | PCR analysis | | | Immunofluorescence |
	East	West	Total	
Nymphs	64/250 (25.6%)	87/250 (34.8%)	151/500 (30.2%)	
Adults	13/100 (13%)	18/100 (18%)	31/200 (15.5%)	20/100 (20%)

As shown in Table 1, significantly higher carriage rates were found in nymphs as compared to adult ticks ($p < 0.0001$). When carriage rates were compared for ticks according to their site of isolation, a significantly higher carriage rate of *B. burgdorferi* was found in nymphs from the west coast of Scotland compared to those from the east ($p < 0.02$). Although there was also a lower carriage rate of *B.burgdorferi* in adult ticks from the east coast compared to those isolated from the west coast, this difference was not statistically significant.

Culture for *B.burgdorferi*

1,600 *I.ricinus* nymphs and 400 adults were cultured using two methods. Given that the sole available British isolate of *B.burgdorferi*, strain TK100, can only survive in culture in symbiosis with a *Pseudomonas aeruginosa* contaminant, minced ticks were inoculated into standard BSK II, and also BSK II previously inoculated with the contaminant *P.aeruginosa* in parallel. All cultures proved to be negative.

B.burgdorferi seroprevalence in a human population

Sera from 153 Scottish National Heritage Workers were tested for IgG antibody against **B. burgdorferi**. 29 (19.0%) were found to be positive or strong positive by Dako ELISA. 88 were found to be borderline positive. A significantly higher seroprevalence (positive/strong positive) was found in workers based in the West of Scotland (16/61: 26%) as compared with those from the East (8/57: 14%), $p < 0.0001$. Control sera were found to be 1.0% (1/100) positive, and 3.0% (3/100) borderline positive. Of the 29 subjects found to be IgG positive/strong positive, 6 (20.7%) also proved to be positive for anti-**B.burgdorferi** IgM, giving an overall rate of 3.9% (6/153) IgM seroprevalence.

Table 2: Antibody levels according to exposure level

Exposure level	Negative	Borderline positive	Positive	Strong positive	(Total positive
Daily	35 (66.1%)	9 (17.0%)	8 (15.1%)	1 (1.8%)	18 (33.9%)
Weekly	45 (61.6%)	13 (17.8%)	14 (19.2%)	1 (1.4%)	28 (38.4%)
Monthly	23 (57.5%)	6 (15.0%)	7 (17.5%)	4 (10%)	17 (42.5%)
Never	3 (75.0%)	1 (25.0%)	0 (0%)	0 (0%)	1 (25.0%)

Table 2 shows the results for IgG levels stratified according to the degree of tick exposure as measured by questionnaire at the time of blood sampling. Contrary to what might be expected, seronegativity does not seem to decrease with increasing exposure as those with daily exposure have a lower seronegativity than those with weekly exposure who, in turn, have lower seronegativity than those with monthly exposure. In addition, in those who are seropositive, the level of antibody appears to be inversely related to exposure level. We determined the relationship between exposure level and the number of tick bites reported per year. Although the number of reported tick bites does increase with increasing tick bites, the correlation appears to be a weak one, $r=0.288$. This correlation coefficient was then tested for use as a surrogate marker of individuals' risk assessment. The correlation between exposure level and reported tick bites was calculated for each antibody level category as follows: antibody negative, $r=0.371$; borderline positive, $r=0.327$; positive, $r= 0.201$; and strong positive, $r=-0.330$.

DISCUSSION

Initially we intended to use IFA to determine the carriage rate of **B.burgdorferi** in Scottish ticks, but the development of PCR-based assays allowed the rapid processing of a larger number of ticks, while also giving at the same time a product suitable for further analysis. However, in our hands, IFA and PCR on individual ticks appear to be equally sensitive, although IFA was useful visually to confirm the presence of the spirochaete in local ticks. False positives are a well known complication of PCR work; to reduce this risk clean rooms were used throughout our study. If a false positive was suspected all reagents were discarded and fresh stock solutions and a new DNA extraction prepared.

The reason for the lower estimate of prevalence of *B. burgdorferi* in pooled ticks compared with the results obtained for single isolates is unclear. The most likely explanation is that by pooling ticks when the carriage level is relatively low, the ratio of PCR inhibitory substances to the number of DNA copies is high, such that only relatively inefficient amplification takes place. By contrast, for single ticks, if DNA is present, it is present in levels sufficient to overcome the concentration of inhibitors. This has serious consequences for the screening of tick populations by PCR, where traditionally pooling has been regarded as the most efficient method of screening.

The regular use of Southern blotting in this study confirmed its value as reliable confirmatory tool; it showed increased sensitivity compared with conventional ethidium bromide staining of agarose gels and had the ability to detect authentic PCR product that was not of the size predicte from the control strain results. The digoxigenin-based non-radioactive labelling system used for the probe was rapid, cheap and efficient.

The difference in *B.burgdorferi* carriage rates between ticks collected at the two sites is unexplained. As ticks were not collected throughout the whole year, our carriage rates may be higher than the mean, as studies in Sweden have shown a marked variation in carriage rate of *B. burgdorferi* in Ixodes ticks throughout the year (T.G Jaenson, personal correspondence). More ticks were found per 100 yards on the west coast of Scotland compared with the east, and this increased tick density may be important. In addition, weather conditions and vegetation differ between the two sites, but the effect of this is unknown. Our finding of a higher carriage rate in nymphs compared with adults is unusual and implies that transstadial transmission of *B.burgdorferi* between nymphs and adults is not universal or alternatively that adults are less competent as reservoirs than nymphs and that carriage, is not maintained. The high carriage rate in nymphs may be important in the dissemination of infection to humans as our questionnaire datum indicates that humans are mostly frequently bitten by nymphs.

The overall seroprevalence in this high risk group of workers demonstrates that there is a considerable exposure to *B.burgdorferi* in outdoor workers in Scotland. This compares with studies among Dutch Forester workers (10), but is slightly lower than studies of English Forester Workers (11) and Swiss Orienteers (12). There were no differences in prevalence of antibodies between males and females. The significantly higher prevalence of IgG antibody found in west coast workers compared with those from the east coast is probably explained by the higher carriage rate for *B.burgdorferi* in west coast ticks as well as by their higher density. It is hard to explain to the pattern of antibody response seen when the results are stratified by degree of exposure. The strong antibody in those with occasional exposure may represent acute exposure to *B.burgdorferi* antigens; the weak antibody response more typical of the high exposure group may reflect the effect of chronic exposure to *B.burgdorferi*.

The low level correlation between the level of exposure (daily/weekly/monthly/never - reliable) and the reported number of tick bites is a surprising one (see Fig 1). There is considerable spread in results which suggests that either it is extremely difficult to notice ticks biting when working outdoors or that these high risk workers underestimate the degree of risk from ticks and pay them too little attention. If this correlation is a surrogate marker of risk then might this explain its relationship to antibody levels.

The failure of attempts to culture *B. burgdorferi* from more than 2,000 ticks, many positive by PCR and IFA, requires explanation. One possibility is that the Scottish strain differs fundamentally from American and European laboratory adapted strains in that it does not grow in BSK 11 media. The fact that some PCR positive samples gave products with a smaller molecular weight than the control B31 strain could reflect a deletion in the Osp A-B linear plasmid.

ACKNOWLEDGEMENTS

We would like to thank Mr. John Campbell of the Scottish National Heritage group for his help in organising the collection of blood samples for our study. Thanks also to the Aberdeen branch of SNH for supplying the map shown in figure 1. This work was supported in part by the Scottish Home and Health department. The Isle of Rum Wilf Nelson fund and the Scottish Red Deer Commission.

REFERENCES

1. Steere AC., S.E. Malawista, J.A. Hardin, S. Ruddy, P.W. Askenase and W.A Andiman. Erythema Chronicum Migrans and Lyme Arthritis. The Enlarging Clinical Spectrum. *Ann Intern Med.* 1977; **86**: 685-98

2. Baird A.G., J.C.M. Gillies, F.J. Bone, B.A.S. Dale and N.T. Miscampbell. Prevalence of Antibody Indicating Lyme Disease in Farmers in Wigtownshire. *BMJ* 1989; **299**: 836-37

3. Ho-Yen D. and A.J. Bennet. 1990, Lyme Disease in the Highlands. *Scot Med J* 1990; **35**: 168-70

4. Olsen B., T.G. Jaenson, L. Noppa, J. Bunikis and S. Bergstrom. A Lyme Borreliosis Cycle in Seabirds and *Ixodes Uriae* Ticks. *Nature* 1993; **362**(6418): 340-2

5. J. Campbell Studies of the Sheep Tick, *Ixodes ricinus*, in S.W. Scotland. *Msc Thesis* 1988; University Glasgow, Britain.

6. Guy E.C. and G. Stanek. Detection of *Borrelia burgdorferi* in patients with Lyme Disease by the Polymerase Chain Reaction. *J Clin Path* 1991; **44**(7): 610-1

7. Barbour A.G. Isolation and Cultivation of Lyme Disease Spirochaetes. *Yale J Biol Med* 1984; **57**: 71-75

8. Bland M. An Introduction to Medical Statistics. *Oxford Medical Publications* 1987;

9. Clarke G.M. and Cooke D. A Basic Course in Statistics. Second Edition. *Hodder and Stoughton* 1983

10. Kuiper H., B.M. de Jongh, A.P. Nuata, H. Houweling, L.G. Wiessing, A.W. van Charante and L. Spanjaard. Lyme Borreliosis in Dutch Forestery Workers. *J Infect* 1991; **23**(3): 279-86

11. Guy E.C., D.E. Bateman, C.N. Martyn, J.E. Heckels and N.F. Lawton. Lyme Disease: Prevalence and Clinical Importance of *Borrelia burgdorferi* Specific IgG in Forestry Workers. *Lancet* 1989; **Mar 4**: 484-486

12. Fahrer H., S.M. van der Linden, M.J. Sauvin, L. Gern. E. Zhioua and A. Aeschlimann. The Prevalence and Incidence of Clinical and Asymptomatic Lyme Borreliosis in a Population at Risk. *J Infect Dis* 1991; **163**(2): 305-10

EXPRESSION OF PUBLIC IDIOTYPES IN PATIENTS WITH LYME ARTHRITIS

J S Axford[1], R A Watts[2], A A Long[3], D A Isenberg[2], A C Steere[3]

Academic Rheumatology Group, Division of Immunology, St George's Hospital Medical School, London, UK[1], Department of Rheumatology Research, University College and Middlesex School of Medicine, University of London, UK[2], Divisions of Rheumatology/ Immunology & Allergy, New England Medical Centre, Tufts University School of Medicine, Boston, Massachusetts, USA[3]

ABSTRACT

Objectives

Joints are often affected in Lyme disease and in some instances this may be due to immune auto-reactivity. To characterise further the immune response in this disease investigations were carried out to determine the expression of three public idiotypes on serum immunoglobulins in patients with Lyme disease during the development of varying degrees of arthritis.

Methods

The expression of idiotypes (Ids) 16/6, BEG2, and PR4, first identified on monoclonal antibodies to DNA, was determined by an enzyme linked immuno-sorbent assay (ELISA) in serial blood samples from 12 patients with Lyme disease over a mean period of six years during the development of a variety of arthritic symptoms, and in serum samples from healthy control subjects and control subjects with systemic lupus erythematosus.

Results

Expression of serum IgM or IgG public Ids 16/6 and BEG2 was significantly increased in patients with Lyme disease. IgA Id 16/6 expression, in contrast, was significantly increased only during episodes of arthritis and was also related to its severity. IgM and IgG Id 16/6 expression was related to their respective total immunoglobulin concentration and, in the case of IgM, to the level of IgM antibodies to Borrelia burgdorferi, whereas similar findings were not apparent with IgA antibodies.

This may indicate that the IgA response is related to the pathogenesis of arthritis, especially as total IgA and IgA Id 16/6 levels were found to increase over the duration of disease. Sequential analysis of antibodies also showed restriction in the expression of Id 16/6 as it was never found on all immunoglobulin isotypes at the same time, and Id 16/6 and PR4 was never expressed. Ids 16/6 and BEG2 expression, however, may be associated as seven patients expressed these idiotypes simultaneously.

Conclusions

These data indicate the use of public idiotypes in the immune response against B burgdorferi, which may be restricted in terms of idiotype class and isotype expression, and a possible association between IgA antibodies bearing Id 16/6 with arthritis.

INTRODUCTION

Joint symptoms in Lyme disease often begin with migratory arthralgia followed months later by brief attacks of oligoarthritis (1). In about 10% of patients with joint disease chronic arthritis develops during the second or third year of illness, sometimes leading to erosion of the cartilage and bone. The synovial lesion in Lyme arthritis is histologically similar to that found in rheumatoid arthritis, and chronic Lyme arthritis is thought to have an autoimmune basis because of the association with the HLA class II DR4 and DR2 genes. The cellular and humoral immune responses in Lyme disease develop gradually over a period of months to an increasing array of spirochaetal polypeptides (2), which may be associated with the production of a population of anti-cardiolipin (aCL) antibodies, possibly as a direct response to B burgdorferi phospholipid antigen (3). Interestingly, in most patients Lyme arthritis can be successfully treated with antibiotics, but a small percentage of patients, primarily those with HLA-DR4, do not respond to this treatment.

Idiotypes (Ids) are variable region structures on immunoglobulins which are located at or adjacent to the binding site. Anti-idiotypic antibodies can be used to identify public (cross reactive) and private idiotypic determinants, and their analysis allows the study of the genetic relation between antibodies. Ids 16/6 (4), PR4 (5),and BEG2 (6) are public Ids that were first identified on human monoclonal antibodies to DNA derived from the peripheral blood lymphocytes of patients with haemolytic anaemia (16/6) and leprosy (PR4), and from normal fetal hepatocytes (BEG2). They have subsequently been found to be expressed on serum and tissue bound immunoglobulins of patients with a variety of autoimmune rheumatic diseases. Id 16/6 has also been identified on antibodies to K30 in patients with Klebsiella infections and on the serum immunoglobulins of 60% of patients with Mycobacterium tuberculosis infection. It has been shown by mRNA sequencing that the heavy chain of the monoclonal antibody bearing 16/6 is identical to that of the germline gene V_H26 (7). It is thought that Id 16/6 is on the heavy chain. Thus the public expression of Id 16/6 may well be the result of a conserved germline V gene, primarily involved in antibody production against environmental pathogens. It may also be associated with pathological autoantibodies, however, as it has been detected on immunoglobulins deposited in the kidneys of patients with systemic lupus erythematosus (SLE) (8).

We report here the longitudinal expression of these three public idiotypes in serial serum samples from 12 patients with erythema migrans followed by brief or prolonged episodes of Lyme arthritis.

PATIENTS AND METHODS

Patients and control subjects

We studied 68 serial serum samples in 12 patients from New England, USA, who had erythema migrans followed by arthralgia or one brief attack of arthritis lasting no longer than six weeks (brief arthritis, four subjects), or by intermittent or chronic arthritis for three to six years (prolonged arthritis, eight subjects). Five patients had neurological symptoms early in their illness (four meningitis, one Bell's palsy). The 12 patients had a mean age of 45 years (range 9-71 years), 10 were men and 2 women. They were followed up for a mean duration of six years (range one to nine). Lyme disease was initially diagnosed in these patients by the presence of erythema migrans before the causative agent or efficacy of antibiotic treatment was known. It was later shown that all 12 patients had increased IgG antibody titres to B burgdorferi. Patients were scored for the severity of joint swelling as follows: 0=no swelling, 1=mild, 2=moderate, and 3-severe swelling.

For the investigation of Ids 16/6 and PR4, serum samples from 23 healthy subjects with no personal or family history of autoimmune disease were used as healthy controls and serum samples from 15 patients with active SLE were used as disease controls. In the Id BEG2 assay a different panel of serum samples from 20 healthy subjects was used, again with no personal or family history of autoimmune disease.

Idiotype assays

Antibodies against idiotypes 16/6, BEG2 and PR4 were prepared in rabbits, extensively adsorbed against human IgG/IgM, and used at a concentration at which there is negligible background binding of irrelevant rabbit polyclonal antibody, as detailed elsewhere (4-6). The identification of Ids 16/6 and PR4 on serum immunoglobulins (Ig) was carried out by an enzyme linked immunosorbent assay (ELISA) using the rabbit polyclonal anti-idiotype reagent. Negative and positive control serum samples were used on each plate together with test serum samples. Serum dilution and capture antibody concentration were chosen to ensure saturation by serum immunoglobulins. The following techniques were used:

Immulon 1 (Dynatech) microtitre plates were coated with goat anti-human IgG Fc (Sigma Chemicals, 1µg/ml in borate buffer), which had been previously adsorbed against normal rabbit serum for 18 hours at 4°C. Immulon 2 and Nunc plates were similarly coated and incubated with goat anti-human IgM (Capell, 1µg/ml in borate buffer) and goat anti-human IgA (Sigma Chemicals, 0.5 µg/ml in borate buffer) respectively. Each well was washed and then blocked with phosphate buffered saline (PBS)-3% bovine serum albumin (BSA) and then washed five times with PBS-0.05% Tween (PBS-T). A 75µl volume of serum, diluted 1:200 in PBS, 3% goat serum, 1% BSA and 0.05% Tween (Sigma Chemicals), was added in duplicate to the wells and incubated for two hours at 37°C. Each plate was washed five times with PBS-T. Aliquots (75µl) of either rabbit anti-16/6 (1/20 000) or rabbit anti-PR4 (1/2000), diluted in 3% goat serum, 1% BSA, and PBS-T were then added to the wells and incubated for two hours at 37°C. Each plate was again washed five times with PBS-T before incubating overnight with 75µl aliquots of goat (F(ab')2 anti-rabbit Ig alkaline phosphatase conjugate (Sigma Chemicals), which had previously been adsorbed against human serum and diluted in PBS-3% goat serum, 1% BSA, and 0.05% Tween. Each plate was developed using Sigma 104 alkaline phosphatase substrate after washing five times with PBS-T, and absorbances were subsequently measured at 405nm after six hours at room temperature.

Reference to healthy control serum samples on each ELISA plate was made to enable accurate comparison of absorbance readings obtained from different experiments and the presence of Ids 16/6 and PR4 was expressed at ratios of the average absorbance obtained for the duplicate Lyme disease and SLE samples against the average absorbance obtained from the healthy subjects on each plate.

A capture of ELISA was used to detect the Id BEG2 (ß) on Ig in human serum samples (6). Id BEG2 levels were expressed in arbitrary units (100 BEG2 units = 100% binding of standard serum) and this assay does not distinguish between IgG or IgM isotypes.

Quantification of the level of expression was made by comparison to the number of standard deviations (SD) greater than the healthy control mean value. Antibody levels are reported as positive if levels were greater than or equal to two SD above the mean.

Measurement of serum antibodies to Borrelia burgdorferi and total immunoglobulin levels

Antibodies to B burgdorferi antigen were measured by ELISA as described previously (9). Briefly, IgM and IgA antibodies to B burgdorferi were detected using an immunoglobulin capture technique followed by exposure of the immunoglobulin to B burgdorferi antigen and then detection of antibodies specific to B burgdorferi with a rabbit antibody against the Borrelia antigen. IgG antibodies to B burgdorferi were detected by their ability to directly bind sonicated B burgdorferi antigen. Antibody to B burgdorferi values are expressed either as direct absorbances or as titres, or, for IgA, as an absorbance ratio of a positive standard. IgM titres $\geq$ 1/200, IgG titres $\geq$ 1/400, and IgA ratios $\geq$ 2.0 were considered increased. Total immunoglobulins were measured by standard nephelometry techniques (Beckman assay immunochemistry system). Normal ranges were IgM 0.45-1.50, IgG 8.0-15.0, and IgA 0.98-3.25 g/l.

Measurement of anti-cardiolipin antibodies

Anti-cardiolipin antibodies were detected by ELISA as reported previously (10). Essentially, serum samples were diluted 1/50 in PBS-10% adult bovine serum and assayed in triplicate for IgM aCL and IgG aCL antibodies. Each plate included two positive and eight negative control samples and positive results were those with absorbances above the mean +2 SD of the negative controls.

Measurement of antibodies to B burgdorferi expressing Id 16/6

To determine whether serum antibodies with B burgdorferi antigen binding specificity from patients with Lyme arthritis expressed Id 16/6, a modification of the Id 16/6 ELISA assay was used. Serum samples from three patients with prolonged arthritis who expressed Id 16/6 on their total serum Ig were selected, together with serum samples from one patient with active SLE and from one healthy subject. Nunc plates were coated with sonicated B burgdorferi (strain B31, 5µg/ml in carbonate buffer), antibodies captured from the serum samples were doubly diluted four times (1/300-1/2400) and Id 16/6 expression determined as described previously.

Statistics

Statistics used in analysing the data were Student's t test, the Pearson co-efficient of product-moment correlation, the standard error of the mean (SEM), and the standard normal deviant test was used to compare regression analyses.

RESULTS

Total serum antibodies, antibodies to B burgdorferi and anti-cardiolipin antibodies

In patients with prolonged arthritis, the mean total IgA concentration decreased over the first two years of illness (r=0.996; p<0.03), but increased over the subsequent four years (r=0.947; p<0.02). The reverse was found with levels of IgA antibodies to B burgdorferi and a significant reduction in levels (mean (SEM)) was observed between years two (7.05(1.84)) and six (4.02(0.75)) of follow up (p<0.03). In contrast, the mean total serum IgG and IgM levels decreased over the period of investigation (IgG r=-0.939, p=0.002; IgM r=-0.831, p<0.02) and though a slight increase in levels of IgG antibodies to B burgdorferi was noted over the initial two years, levels of IgG and IgM antibodies to B burgdorferi also decreased (IgG r=-0.830, p<0.05; IgM r=-0.526, p<0.05).

In patients with brief arthritis, the mean total IgM levels decreased over the period of investigation (r=-0.987; p<0.05). Total IgG and A values were always within normal limits and levels of antibodies to B burgdorferi of all isotypes were increased and did not fluctuate over the period of investigation.

Anti-cardiolipin antibody levels were increased on one occasion in two asymptomatic patients with Lyme disease with no previous history of neurological disease - one with brief arthritis (IgG antibody greater than three SD of the negative control mean) and the other with prolonged arthritis (IgM) antibody greater than five SD of the negative control mean).

Id 16/6 expression on total serum antibodies

The table gives the total mean (range) (SEM) immunoglobulin (Ig) absorbance readings after six hours from all experiments.

Levels of IgM and IgG antibodies from patients with Lyme disease and SLE were significantly increased compared with those from healthy control subjects (Lyme disease IgM p<0.01, IgG p<0.02; SLE IgM p<0.05, IgG p<0.001).

Expressing the data as a ratio of the absorbance value obtained from healthy subjects, the levels of IgM and IgG isotypes (mean (SEM)) bearing Id 16/6 were significantly increased in patients with erythema migrans followed by brief or prolonged episodes of Lyme arthritis (IgM 2.0(0.2), p<0.01; IgG 1.3(0.1), p<0.003) compared with healthy control subjects, but the mean level of IgA immunoglobulins bearing this idiotype was not increased (Fig 1). Positive levels of IgM and IgG Id 16/6 antibodies were detected on 5/12 and 6/12 patients respectively, primarily during the first several years of illness. In contrast, significantly increased levels of IgA Id 16/6 antibodies (1.4(0.2); p=0.05) were found only during the period of joint disease when compared with healthy control subjects, and positive levels were detected in three of eight patients who developed prolonged arthritis.
In the 15 control patients with SLE, as in the patients with Lyme disease, the mean level of IgM and IgG Id 16/6 antibodies was significantly increased (IgM 2.6(0.9), p<0.03; IgG 2.2(0.7), p<0.005). These patients did not have increased Id 16/6 expression on IgA antibodies, however.
In the eight patients with erythema migrans followed by prolonged arthritis, IgA Ids 16/6 expression increased significantly during the first (from 1.1(0.1) to 1.7(0.3); p<0.05) and third (from 1.2(0.3) to 1.8(0.2); p<0.05) years of illness and remained high thereafter (Figure 2); levels of IgG Id 16/6 remained high throughout the initial years of illness; and expression of IgM Id 16/6 was highest during the first year of illness and generally decreased in each subsequent year (r=-0.792; p<0.05). In contrast, in the four patients with erythema migrans

Table 1

The total mean (range) ± SEM immunoglobulin (Ig) optical density readings after six hours from experiments conducted to detect idiotype (Id) expression on total serum antibodies. (* p≤ 0.05, ** p≤ 0.02, ~ p≤ 0.01, ~~ p≤ 0.001. ND = not done)

IDIOTYPE	ISOTYPE	LYME DISEASE	SLE	HEALTHY
Id16/6	IgM	162.8(0-722)±15.6 ~	210.2(23-1200)±74.0 *	92.2(0-325)±16.6
	IgG	44.8(0-136)±2.9 **	67.3(6-190)±18.9 ~~	33.9(0-63)±3.4
	IgA	19.3(3-47)±1.2	19.1(2-44)±3.0	14.9(0-45)±2.9
IdBEG2	IgG/M	243.9(31-611)±17.4	ND	229.6(25-398)±17.6
IdPR4	IgM	333.5(44-1500)±16.8	396.0(17-1500)±59.0	302.8(79-704)±26.6
	IgG	143.1(13-1100)±9.8	140.0(57-474)±13.9	131.6(17-263)±10.0
	IgA	109.8(2-305)±6.8	245.2(21-1000)±42.1 **	146.1(12-314)±16.1

followed by brief episodes of arthritis, Id 16/6 expression on all isotypes was highest when the skin lesion was present and decreased thereafter. No patient had Id 16/6 expression on all isotypes simultaneously, and only one patient with prolonged arthritis had increased expression of this idiotype on each isotype at different times.

IgA Id 16/6 expression correlated directly with the severity of arthritis (r=0.258; p<0.02), but not with the erythrocyte sedimentation rate (ESR); IgM and IgG Id 16/6 expression correlated with neither the severity of arthritis nor the ESR. The levels of IgM and IgG Id 16/6 correlated directly with their respective total immunoglobulin concentrations (IgM r=0.538; IgG r=0.308; in both instances p<0.001), whereas the levels of IgA Id 16/6 did not. Similarly, IgM Id 16/6 expression showed a positive correlation with levels of IgM antibodies to B burgdorferi (r=0.264; p<0.025), but the other antibody isotypes did not. There was a significant difference (p<0.05) between the overall positive correlation of IgA Id 16/6 expression (r=0.439; p=NS) and the negative correlation of IgM Id 16/6 expression with disease duration.

Id 16/6 expression on antibodies to B burgdorferi

The total mean (SEM) absorbance readings for the three patients with Lyme disease were as follows: 1/300, 338.3 (84.8; 1/600, 253.3(103.6); 1/1200, 173.3 (66.9), and 1/2400, 135.0(60.7). Id 16/6 expression was not detected when serum samples, from the patient with SLE nor from the healthy subject were used.

Id BEG2 expression on total serum antibodies

The table gives the total mean (range) (SEM) absorbance readings from all experiments.

When the data were expressed as the percentage binding relative to a standard serum sample, patients with Lyme disease had significantly increased levels of expression of IgM and IgG BEG2 when compared with healthy subjects (37(5) v 20(2); p< 0.02). Positive levels of IgM and IgG BEG2 expression were detected in 6/12 patients, which included some with brief or prolonged arthritis, and this was predominantly early in the illness. There was no association between Id BEG2 expression and arthritis score, ESR, total immunoglobulin, or levels of antibodies to B burgdorferi.

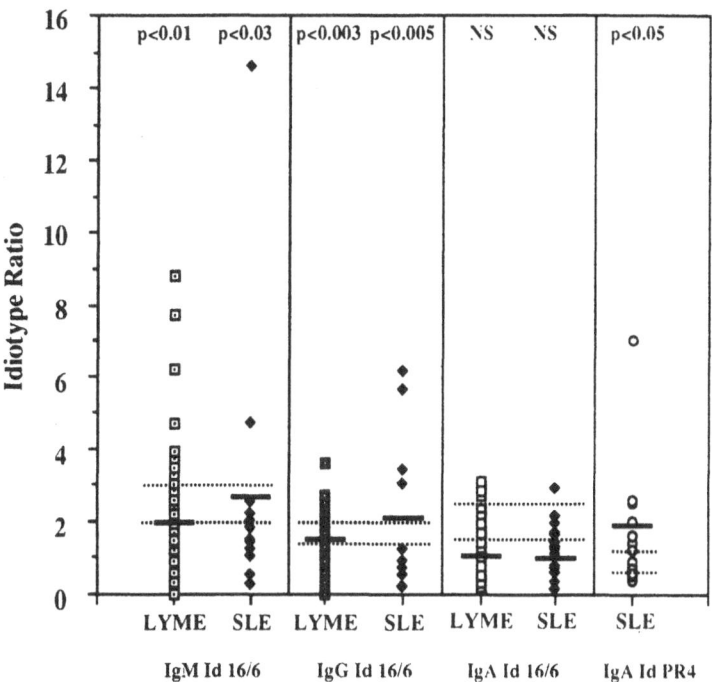

Figure 1

Expression of idiotypes 16/6 and PR4 on serum immunoglobulins in Lyme disease (n=68) and SLE (n=15). Idiotype ratio is the mean optical density ratio of test/control sera. The solid bars indicate mean values, the broken bars indicate standard deviations above the control mean (x2 = upper bar), x1 = lower bar), and the dots represent each serum sample tested. A significant increase in IgM and IgG Id 16/6 expression was found in Lyme disease, but IgA Id 16/6 expression was only increased in those patients developing Lyme arthritis.

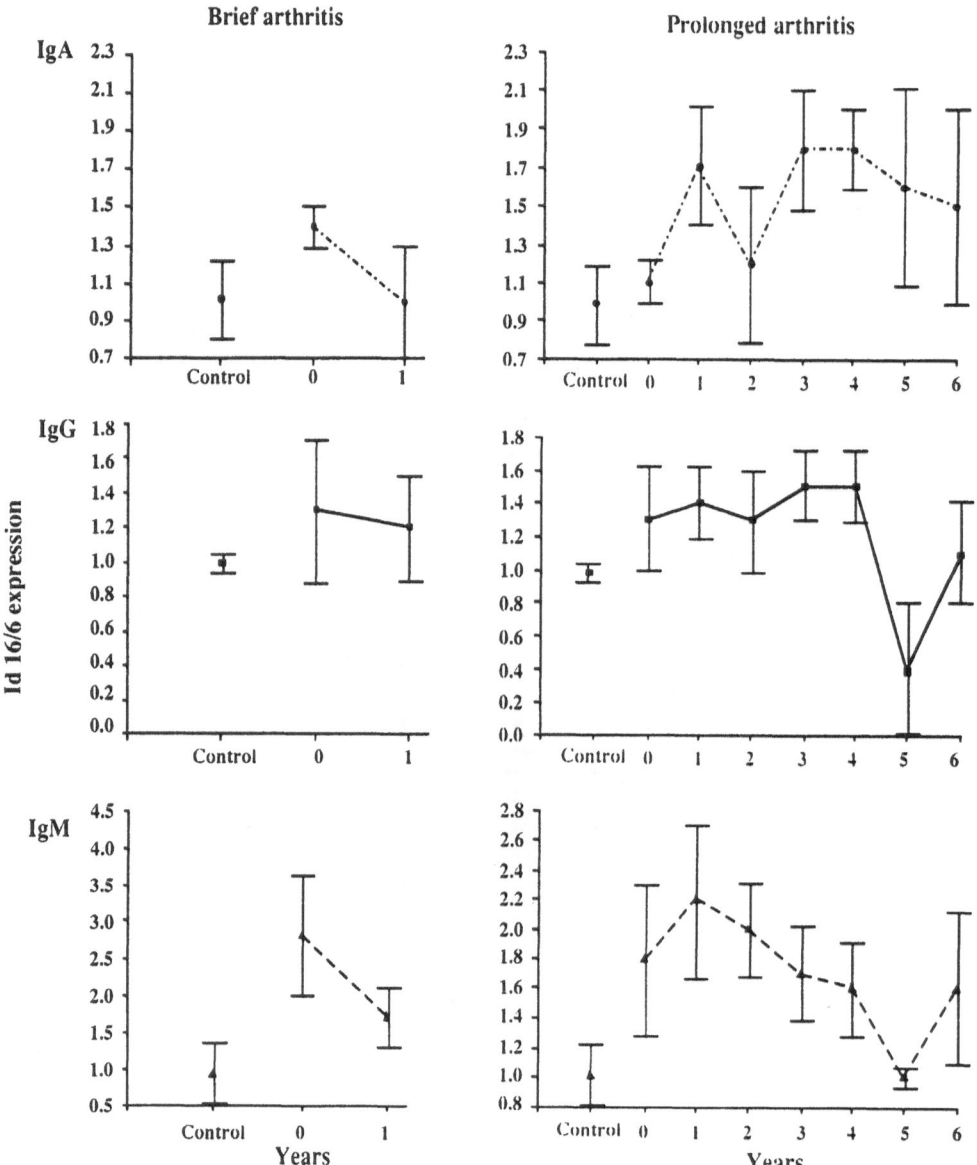

Figure 2

Serum immunoglobulin Id 16/6 expression (mean ratio ± SEM), in Lyme disease patients who developed brief arthritis (n=4) and prolonged arthritis (n=8) and in healthy controls (n=23). In patients with prolonged arthritis, IgA Id 16/6 expression rose significantly during the first and third year (p <0.05), whereas there was a significant reduction (p <0.05) in expression of IgM Id 16/6 with time. There was a significant difference between the increase and decrease in expression of IgA and IgM 16/6 respectively (p <0.05). In those with brief arthritis, Id 16/6 expression on all idotypes tended to reduce.

Id PR4 expression on total serum antibodies

The table gives the total mean (range) (SEM) absorbance readings for all experiments.

The IgA absorbance readings for subjects with SLE were significantly increase (p>0.02) compared with healthy control subjects.

When the data were expressed as a ratio of the absorbance value obtained from healthy subjects, Id PR4 expression was again not significantly different in the patients with Lyme disease compared with the healthy control subjects, but the control patients with SLE had increased expression of IgA PR4 (1.7(0.4); p<0.05) compared with healthy subjects.

Co-expression of Ids

Seven patients with Lyme disease had simultaneously increased expression of Ids 16/6 and BEG2. Only one patient with prolonged arthritis had no expression on any of the isotypes.

CASE REPORT

A 58 year old man had erythema migrans followed by several brief episodes of arthritis affecting his wrists, knees or interphalangeal joints. During the third year of illness he developed chronic arthritis of the right shoulder and hip, which was treated successfully with antibiotics during the fifth year of illness. IgM Id 16/6 expression was greatest during the first year of illness; it decreased rapidly and increased before a flare of chronic arthritis in the second year and again during a period of arthritis in the fifth year (Figure 3). IgG Id 16/6 expression was greatest during the initial three years of illness. In contrast, IgA Id 16/6 expression increased during the fifth year. No change in IgA Id 16/6 arthritis in the second year of illness. IgM ID 16/6 levels correlated with total and B burgdorferi specific IgM levels, but the amounts of IgG and IgA Id 16/6 did not correlate with their respective total or specific B burgdorferi immunoglobulin levels. Id 16/6 expression on all isotypes decreased simultaneously after treatment with antibiotics.

Similar findings were also found in the other seven patients investigated, although this patient was one of the more extensively investigated and had one of the longer follow up times.

DISCUSSION

Data are presented that characterise the expression of public Ids in a cohort of patients with Lyme disease who developed varying degrees of arthritis, and this offers a unique opportunity to observe the humoral response to B burgdorferi in patients not treated with antibiotics.

In Lyme disease a significant increase in the expression of serum IgM or IgG public Ids 16/6 BEG2 was shown and positive levels detected (greater than two SD above the mean for healthy control subjects) in half the patients investigated at some stage of illness. In contrast, IgA Id 16/6 expression was increased significantly only during episodes of arthritis and positive levels were detected in three of the eight patients with prolonged arthritis. IgA Id 16/6 expression was also related to the severity of arthritis, but not the acute phase response, whereas IgM and IgG expression were not. Interestingly, IgM and IgG Id 16/6

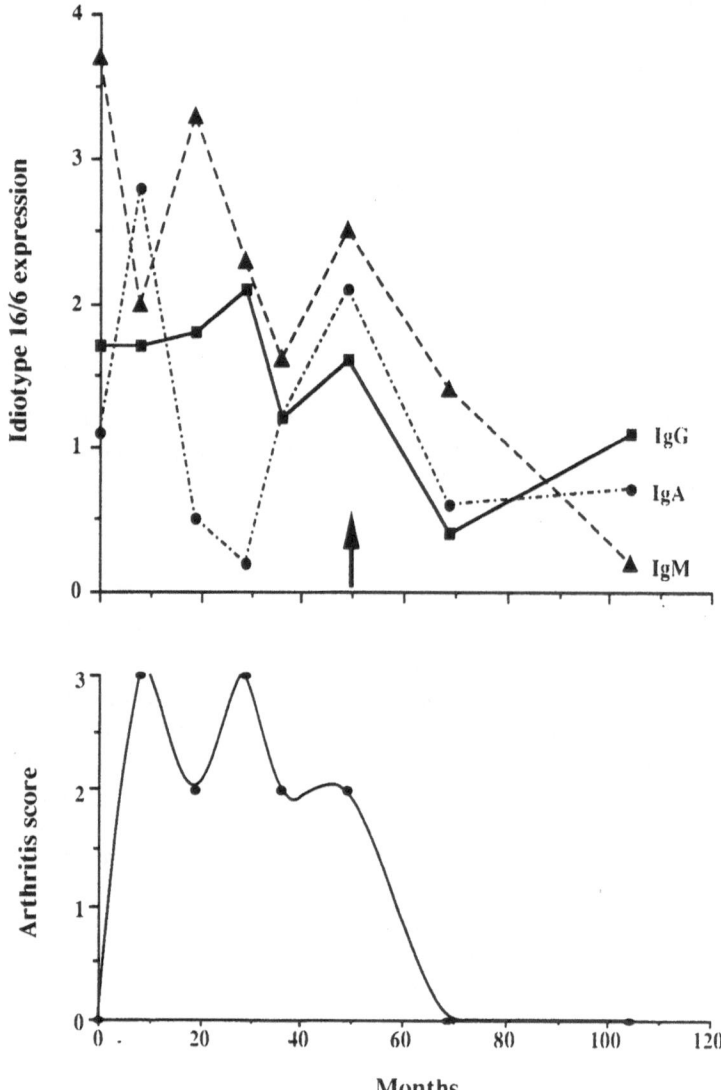

Figure 3

Serum idiotype 16/6 expression and arthritis score in a 58 year old male with chronic Lyme arthritis.

IgM idiotype 16/6 expression and arthritis (traingles and dashed line) reduced from initially raised values on presentation with EM. IgG idiotype 16/6 expression (squares and unbroken line) increased to a peak at thirty months of disease and then decreased to those of the healthy controls. IgA idiotype 16/6 expression (circles and dots/dashed line) peaked at ten and fifty months, before decreasing to those of the healthy controls. Antibiotic treatment, indicated by the arrow, was associated with a simultaneous decrease in Id 16/6 expression on all isotypes.

expression correlated with their respective total immunoglobulin concentrations and, for IgM, to the level of IgM antibodies to B burgdorferi, whereas similar findings were not found with IgA antibodies. Id 16/6 bearing antibodies bind to the B burgdorferi antigen, as was subsequently confirmed using B burgdorferi sonicate in a capture ELISA. These data indicate that the IgM and IgG production is probably the major response against B burgdorferi and perhaps IgA Id 16/6 expression is involved with more subtle factors relating to the pathogenesis of arthritis perhaps associated with specific antibodies to B burgdorferi, for example, against the outer surface protein A.

Sequential antibody studies showed that IgM and IgG Id 16/6 expression tended to decrease during prolonged illness, whereas IgA Id 16/6 expression tended to increase. Furthermore, the difference between IgM and IgA expression was significant, which may indicate that isotype switching is occurring. In this respect it is pertinent that the humoral response against outer surface protein A in Lyme arthritis occurs late in the disease (11) and in the patients reported here total IgA levels increased as the disease progressed. No evidence was found to suggest an association between the production of aCL antibodies and arthritis, indicating that the Id 16/6 bearing antibodies do not also have cardiolipin specificity. Comparable findings were observed in the patient reported, though the effects of antibiotic treatment seemed to correct these changes, giving further weight to the data indicating that Id 16/6 is expressed on antibodies to B burgdorferi. These changes were not apparent if the illness was brief, though the patients were not investigated for the same period of time.

Restriction in the expression of Id 16/6 was also observed. Simultaneous expression of Id 16/6 on all isotypes was never found and only one patient had increased expression of this idiotype on each isotype at different disease stages. Restriction was also found when Id PR4 was examined, as this was never expressed in Lyme disease; however, Ids 16/6 and BEG2 seem to be associated with a similar repertoire of antibodies to B burgdorferi as seven patients expressed Id 16/6 simultaneously with Id BEG2. These Ids are not associated with the same V gene family, however; Id 16/6 and Id BEG2 are VH3 and VH4 associated respectively. By using Ids 16/6 and BEG2 as markers of VH gene usage, an impression of the complexity of the humoral response to the B burgdorferi bacterium can be seen. It is evident that the response if far from static as these Ids were not expressed continuously, but as a seemingly well orchestrated response with Ids appearing on different isotypes at different times. It is interesting that there is a similar expression of Ids 16/6 and BEG2 in the patients with SLE studied, but with the important exception that Id PR4 rather than Id 16/6 was expressed on IgA. Perhaps these subtle Id fluctuations have clinical associations.

So does Id 16/6 play a part in the pathogenesis of the disease? These data suggest that it does. It may simply be a marker of infection related antibodies or it may be a marker of a component of an idiotype network in which a cascade of interrelating antibodies is generated, any one of which may be pathogenic (12). For example, IgA bearing 16/6 may be antibody 3 (anti-AB2) in a parallel set of Id 16/6 positive and DNA binding negative autoantibodies (13) which may have antigen binding specificity to B burgdorferi antigen due to the presence of an internal image of the B burgdorferi antigen occurring at the V region of antibody 2 (anti-AB1 - that is, the B burgdorferi antibody) (14). After processing the 16/6 idiotope may be seen by T cells, which, in the presence of certain HLA Class II gene products (DR4, for example), cause them to become auto-reactive and arthritogenic (15). Subsequent B cell activation may follow, which could result in perturbation of the antibody response.

In summary, these data indicate restriction in the expression of some public idiotypes in Lyme disease and a possible association between IgA antibodies bearing Id 16/6 and the development of arthritis. The associations with infection shown here further strengthen the suggestion that the main reason for the conservation of the public idiotypes is not to form part of an autoantibody response, but the help protect the species against environmental pathogens.

ACKNOWLEDGEMENTS

The authors thank Dr R S Schwartz for critical comment and for allowing us to use idiotype reagents prepared in his laboratory. We also thank Dr A Tabson for measuring immunoglobulin levels and for reviewing the manuscript, Dr L Tucker for developing the Id 16/6 ELISA protocol, Dr A Puccetti for helpful advice while carrying out this investigation, Ambrin Dar for technical assistance, and Professor I M Roitt and Mrs A Alavi for constructive criticism of the manuscript. JSA was a Fulbright Scholar and thanks the Fulbright Commission and the Peel Medical Research Trust for financial assistance. RAW was supported by the Wellcome Trust and the Emily Le Rossignol Fund. AAL was the recipient of a fellowship in basic sciences from the Canadian Arthritis Society and is a Sandoz scholar in medicine. ACS was supported by public health service grants AR - 20358 and AR-40576. The anti-PR4 reagent was kindly donated by Dr W Williams.

These data have previously been published in Ann Rheum Dis (1993) **52: 199-205.**

REFERENCES

(1)   Steere AC. Lyme disease. N Engl J Med (1989); 321 : 586-96.

(2)   Grodzicki RL, Steere AC. Comparison of immunoblotting and indirect enzyme-linked immunosorbent assay using different antigen preparation for diagnosing early Lyme disease. J Infect Dis (1988); 157 : 790-9.

(3)   Mackworth-Young CG, Harris EN, Steere AC et al. Anti-cardiolipin antibodies in Lyme disease. Arthritis Rheum (1988); 31: 1052-6.

(4)   Shoenfeld Y, Isenberg DA, Rauch J, Madaio MP, Stollar BD, Schwartz RS. Idiotypic cross-reactions of monoclonal human lupus autoantibodies. J Exp Med (1983); 158 : 718-30.

(5)   Williams W, Zumla A, Behrens R et al. Studies of a common idiotype PR4 in autoimmune rheumatic disease. Arthritis Rheum (1988); 31 : 1097-104

(6)   Watts TA, Ravirajan CT, Staines NA, Isenberg DA. A human fetal monoclonal DNA-binding antibody shares idiotypes with fetal and adult murine monoclonal DNA-binding antibodies. Immunology (1990); 69 : 348-54.

(7)   Chen PP, Liu MR, Sinha S, Carson DA. A 16/6 idiotype-positive anti-DNA antibody is encoded by a conserved V_H gene with no somatic mutation. Arthritis Rheum (1988); 31 : 1429-31.

(8)   Isenberg DA, Collins C. Detection of cross-reactive anti-DNA antibody idiotypes on renal tissue bound immunoglobulins from lupus patients. J Clin Invest (1985); 67 : 287-94.

(9)   Beradi VP, Weeks KE, Steere AC. Serodiagnosis of early Lyme disease : analysis of IgM and IgG antibody responses by using an antibody-capture enzyme immunoassay. J Infect Dis (1988); 158 : 754-60.

(10)   Long AA, Ginsberg JS, Brill-Edwards P et al. The relationship of antiphospholipid antibodies to thromboembolic disease in system lupus erythematosus : a cross sectional study. Thromb Haemost (1991); 66 : 520-4.

(11)   Kalish R, Leong J, Steere AC. Delay in the immune response to outer-surface
         proteins (Osp) A and B of Borrelia burgdorferi : correlation with arthritis and
         treatment failure in susceptible idividuals. Arthritis Rheum (1991); 34 : S43.

(12)   Watts RA, Isenberg DA. Idiotypes and anti-idiotypes : what are they trying to tell us?
         Ann Rheum Dise (1988); 47 : 705-7.

(13)   Naparstek Y, Mackworth-Young C, Breitman L, Schwartz R. Sequential anti-
         idiotypes define reciprocal idiotypes on the same anti-DNA antibody. Clin
         Immuno Immunopathol (1989); 50 : S106-16.

(14)   Puccetti A, Koizumi T, Migliorini P, Andrew-Schwartz J, Barrett KJ, Schwartz RS.
         An immunoglobulin light chain from a lupus-prone mouse induces
         autoantibodies in normal mice. J Exp Med (1990); 171 : 1919-30.

(15)   Van-Eden W, Holoshitz J, Nevo Z, Frankel A, Klajman A, Cohen IR. Arthritis
         induced by a T-lymphocyte clone that responds to mycobacterium
         tuberculosis and to cartilage proteoglycan. Proc Natl Acad Sci USA (1985);
         82 : 5117-20.

CELLULAR IMMUNE REACTIONS TO *BORRELIA BURGDORFERI* -

THE T-CELL-MACROPHAGE AXIS

Gerd R. Burmester, Thomas Häupl and Michael Rittig

Address correspondence to:
Prof. Gerd R. Burmester, M.D., Institute of Clinical Immunology and
Rheumatology, Department of Medicine III, University of Erlangen-
Nürnberg, Krankenhausstr. 12, DW-8520 Erlangen, Germany
Tel. x-49-9131-859131
Fax. x-49-9131-854770

Even though Lyme borreliosis (LB) has been clearly identified as an infectious disease, there are still many riddles to be solved, especially with regard to the immune response to this organism. These include the question why only a small proportion of individuals originally exposed to *B. burgdorferi* progress to later disease stages, the genetic background and the apparent dissociation between the cellular and humoral immune reactivity with an extensive, but nevertheless restricted and apparently non-protective antibody repertoire.

Most patients with LB develop an early and strong cellular immune reactivity to *B. burgdorferi* that in early stages often preceeds an only slowly evolving humoral response[1.] Thus, while in early cases serology may still be negative, in later stages, especially arthritis, nearly all patients are seropositive with only rare, but clinically important exceptions. In studying the immune response, two major groups of proteins of *B. burgdorferi*, the outer surface proteins A (OspA) and B (OspB) as well as flagellin, appear to be of particular interest. The spirochete specific OspA is a 31 kD protein which is expressed in large amounts on the outer membrane of *B. burgdorferi*. It is antigenetically variable between different isolates of *B. burgdorferi* [2]. Despite its abundance on the bacterial surface, in humans the humoral response to OspA as detected by immuno-blotting is quite rare, and, if present, develops only at later stages[3]. In contrast, antibodies to flagellin, the internally localized flagellar protein of *B. burgdorferi,* are detectable in nearly all patients with LB even in early stages of the disease[4].

The comparison of the humoral immune responses with the cellular immune reactivity to these two major proteins of *B. burgdorferi*, OspA and flagellin, in patients with LB led to several interesting observations (figure 1)[5]. Thus, patients with LB showed a significantly elevated T cell response to whole *B. burgdorferi* bacteria as compared to patients with other inflammatory joint diseases and to normal controls. This response was already detectable in very early disease stages and did not correlate with the presence of antibodies against *B. burgdorferi*. The *B. burgdorferi* specific T cell responses observed included those directed against OspA and flagellin. However, there was a considerable variation of proliferation values ranging from a lack of T cell proliferation to very high responses. Irrespective of the disease stage, *in vitro* T cell responses to either whole *B. burgdorferi* or to the recombinant proteins did not correlate with the presence of specific antibodies in the patient sera. This dissociation of the humoral and the cellular immune

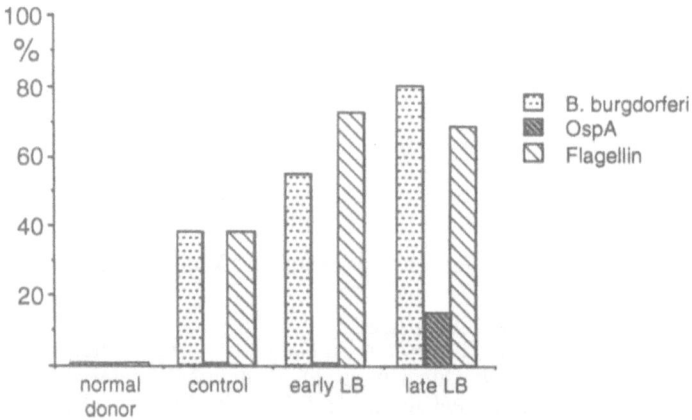

Figure 1A: Humoral immune reactivity to *Borrelia burgdorferi* in normal donors (ND) non LB-patients (control) and patients with early or late manifestations of Lyme Borreliosis. Reactivity to *Borrelia burgdorferi* was determined by the immunofluorescence assay and by immunoblotting against OspA and flagellin. Percentages of postive individuals are listed. The relatively high frequency of positive donors in the control population is apparently influenced by the referral pattern to our outpatient clinic.

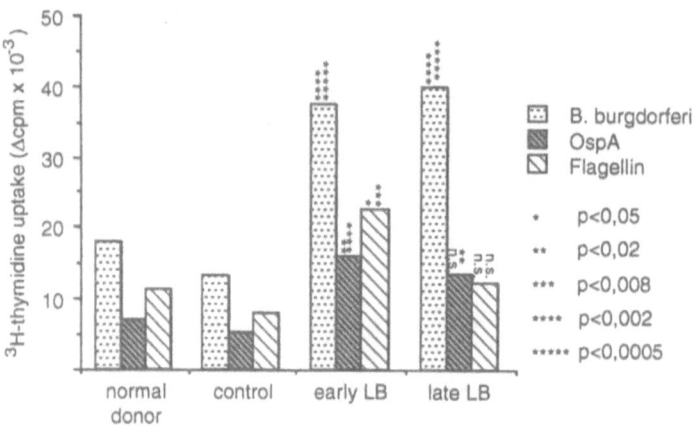

1B: Parallel cellular immune response against *Borrelia burgdorferi*, OspA and flagellin in the same groups as described above.

response may be explained by several mechanisms. Initial experiments using T cell clones in chronic Lyme patients suggest a selective activation of a certain T cell subset that produces a restricted pattern of cytokines which are incompetent to activate B cells[6]. This selective expansion of responding cells may depend on the HLA profile as suggested by the high association of chronic Lyme arthritis refractory to antibiotic treatment with certain HLA-DR2 and DR4 alleles[7]. In their studies on inbred mice, Schaible et al. also demonstrated a clear association of resistance or susceptibility to arthritical symptoms with certain MHC II genes[8].

In addition to an unusual T and B cell response there is a peculiar uptake mechanism of B. burgdorferi by phagocytes which may explain some of the subsequent specific immune reactions. All types of phagocytes internalize the spirochetes preferentially via a special uptake mechanism designated "coiling phagocytosis" (figure 2A). In this phagocytic mechanism, sheath-like cell protrusions wrap around segments of spirochetes in multiple turns, thereby creating cylindrical coil-shaped complexes which are engulfed subsequently. Spirochetes internalizied by this mechanism are released into the free cytoplasm and disintegrate without visible participation of lysosomes. This unususal degradation of phagocytosed material does not fit in the established model of antigen processing and presentation of extracellular antigen via MHC class II. Instead, the cytosolic degradation suggests a MHC class I restricted antigen presentation that might explain an insufficient activation of the specific immune response and the chronification of this infectious disease. In the event of conventional phagocytosis, comparable to the coiling phagocytosis, only sections of spirochetes are trapped in large-sized vacuoles wereas the remaining segments of the spirochetes project out of the phaocyte (figure 2B). Some spirochetes already start to disintegrate when the phagosomes are still connected with the cell surface, possibly implying the leakage of lysosomal material into the surrounding environment and causing an exaggerated inflammatory rection. Studies using this model will provide new insights in membrane processing during phagocytosis which may be important for the presentation of bacterial antigens in the context of MHC class I antigens[9].

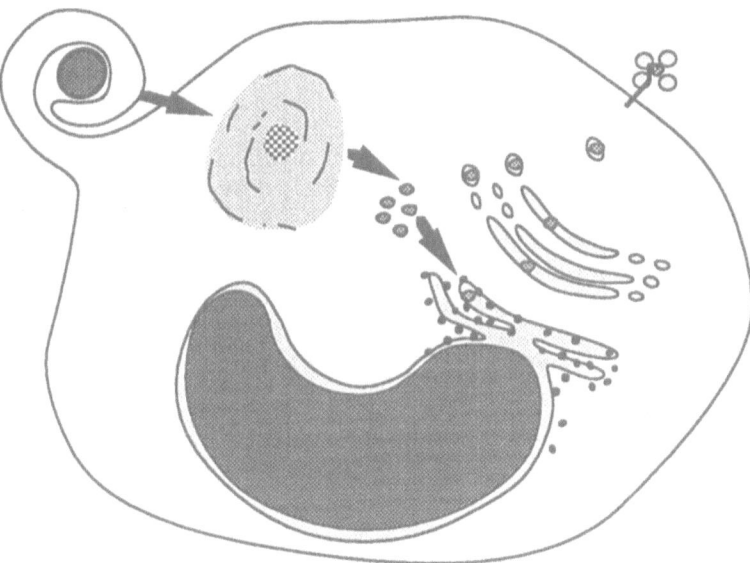

Figure 2A: Coiling phagocytosis. This mode of antigen uptake offers the possibility of antigen presentation by HLA class I antigen due to cytosolic processing.

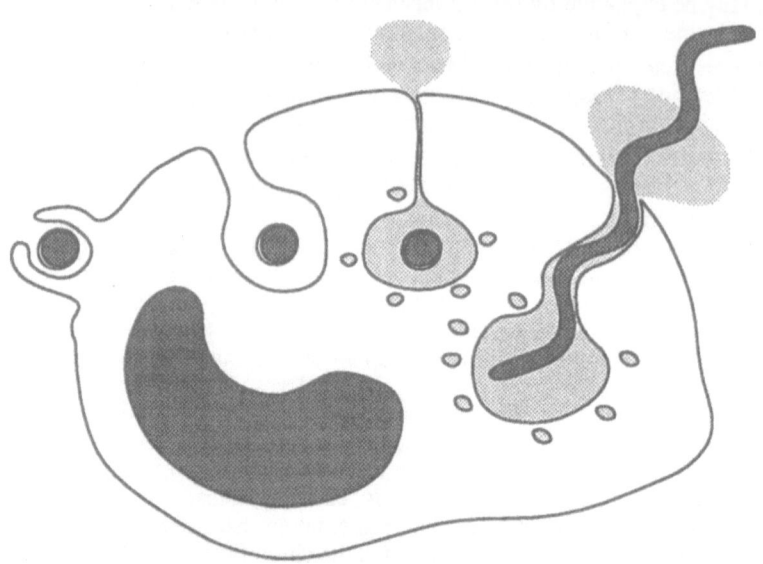

Figure 2B: Conventional phagocytosis may result in the leakage of lysosomal enzymes into the surrounding tissue with inflammatory reactions.

It is an interesting finding that in the late stages of spirochetal diseases, such as syphilis and notably Lyme disease, only very few infectious organisms are detectable in the lesional sites. However, there is frequently a strong local immune response with a sometimes vigorous synovial inflammatory hypertrophy in Lyme arthritis which is histologically indistinguishable from RA synovitis[10]. Apparently, only a few bacteria are sufficient to attract vast numbers of cells to the areas affected. Early and significant antibody production against common antigens like flagellin[11] or heat shock proteins[12, 13] after Borrelia infection suggest a reactivation of a previously established immune response against other pathogens with similar antigens. Although the amount of antigen for a recall immune response is sufficient, typical antigens of Borrelia burgdorferi like OspA or OspB are recognized only by few patients. It may be either a general immunosuppressive state that does not allow for a effective immune activation and complete elimination of the microbe - or, the immune system is stimulated into the wrong direction and uses false or inadequate means for elimination. Alternatively, *Borrelia burgdorferi* may evade into sites hardly accessible to immune cells, antibodies or antibiotics. Possible sites of evasion are the central nerve system[14] and as shown recently, dense connective tissue like tendons and ligaments[15] . Thus, the study of Lyme disease, notably its arthritic manifestations, have taught important lessons about an inadequate immune response to eliminate microbial organisms in a chronic infection. It is very likely that these observations will provide important insights also into other arthritides which are triggered by the contact with infectious agents.

REFERENCES

1. A. Krause, V. Brade, C. Schoerner, W. Solbach, J.R. Kalden, and G.R. Burmester. 1992. T cell proliferation induced by Borrelia burgdorferi in patients with Lyme borreliosis: Autologous serum required for optimum stimulation. *Arthritis Rheum* . 34: 393-402
2. A.G. Barbour, R.A. Heiland, and T.R. Howe. 1985. Heterogeneity of major proteins in Lyme disease borreliae: a molecular analysis of North American and European isolates. *J Infect Dis*. 152:478-484
3. B. Wilske, V. Preac-Mursic, G. Schierz, W. Gueye, P. Herzer, and K. Weber. 1988. Immunochemical analysis of the immune response in late manifestation of Lyme borreliosis. *Zentralbl Bakteriol Hyg A* . 267:549-558
4. L. Zöller, S. Burkhard, and H. Schäfer. 1991. Validity of Western immunoblot band patterns in the serodiagnosis of Lyme borreliosis. *J Clin Microbiol*. 29:174-182

5. A. Krause, G.R. Burmester, A. Rensing, C. Schoerner, U.E. Schaible, M.M. Simon, P. Herzer, M.D. Kramer, and R. Wallich. 1992. Cellular immune reactivity to recombinant OspA and flagellin from Borrelia burgdorferi in patients with Lyme borreliosis. Complexity of humoral and cellular immune responses. *J Clin Invest.* 90:1077-1084

6. H. Yssel, M.-C. Shanafelt, C. Soderberg, P.V. Schneider, J. Anzola, and G. Peltz. 1991. Borrelia burgdorferi activates a T helper type 1-like T cell subset in Lyme Arthritis. *J Exp Med.* 174:593-601

7. A.C. Steere, E. Dwyer, and R. Winchester. 1990. Association of chronic Lyme disease with HLA-DR4 and HLA-DR2 alleles. *N Engl J Med.* 123:219-223

8. U.E. Schaible, M.D. Kramer, R. Wallich, T. Tran, and M.M. Simon. 1991. Experimental Borrelia burgdorferi infection in inbred mouse strains: antibody response and association of H-2 genes with resistance and susceptibility to development of arthritis. *Eur.J.Immunol.* 21:2397-2405

9. M. Rittig, A. Krause, T. Häupl, U.E. Schaible, M. Modolell, M.D. Kramer, E. Lütjen-Drecoll, M.M. Simon, and G.R. Burmester. 1992. Coiling phagocytosis is the preferential phagocytic mechanism for *Borrelia burgdorferi. Infect Immun.* 60:4205-4212

10. A.C. Steere, P.H. Duray, and E.C: Butcher. 1988. Spirochetal antigens and lymphoid cell surface markers in Lyme synovitis. Comparison with rheumatoid synovium and tonsillar lymphoid tissue. *Arthritis Rheum.* 31:487-495

11. J. E., Craft, D.K. Fischer, G. T. Shimamoto, and A. C. Steere. 1986. Antigens of Borrelia burgdorferi recognized during Lyme disease. Appearance of a new immunoglobulin M response and expansion of the immunoglobulin G response late in the illness. J. Clin. Invest. 78:934-939.

12. B.J. Luft, P.D. Gorevic, W. Jiang, P. Munoz, and R.J. Dattwyler. 1991. Immunologic and structural characterization of the dominant 66- to 73-kDa antigens of Borrelia burgdorferi. J.Immunol. 146:2776-2782

13. J.L. Coleman, and J.L. Benach. 1992. Characterization of antigenic determinants of Borrelia burgdorferi shared by other bacteria. *J.Infect.Dis.* 165:658-666

14. V. Preac-Mursic, K. Weber, H.W. Pfister, B. Wilske, B. Gross, A. Baumann, and J. Prokop. 1989. Survival of *Borrelia burgdorferi* in antibiotically treated patients with Lyme borreliosis. *Infection* 17: 355-359

15. T. Häupl, G. Hahn, A. Krause, M. Rittig, Ch. Schoerner, U. Schönherr, J. R. Kalden, G. R. Burmester, Persistence of *Borrelia burgdorferi* in ligamentous tissue from a patient with chronic Lyme borreliosis. *Arthritis Rheum.* in press

OSPB SEQUENCE VARIATION OF *BORRELIA BURGDORFERI* ALONG THE COAST OF MAINE

Diane A. Caporale,[1] Thomas D. Kocher,[1] Robert P. Smith,[2] Peter W. Rand,[2] and Eleanor H. Lacombe[2]

[1]Department of Zoology
University of New Hampshire
Durham, NH 03824

[2]Research Department
Maine Medical Center
Portland, ME 04102

ABSTRACT

We sequenced the OspB genes of 51 *Borrelia burgdorferi* isolates from five sites along the coast of Maine. Seven unique nucleotide sequences were observed, with an average pairwise difference of 0.47%. Five unique amino acid sequences were represented, differing at up to five residues. Multiple strains were observed at three sites, and most strains were found at more than one site. Multiple strains were recovered from a single rat on Monhegan Island. High strain diversity of *B. burgdorferi* may play a role in determining the abundance of this spirochete.

INTRODUCTION

Lyme disease has become a significant health problem in the United States over the last decade. First recognized as a distinct syndrome in the early 1970's, it is now the most common vector-borne disease in the U.S. It was not until the 1980's that the causative agent was shown to be a spirochete bacterium, *Borrelia burgdorferi* (Burgdorfer et al., 1982). The pathogen is transmitted among warm-blooded hosts by *Ixodes* ticks. Over 6,000 cases of human infection are being reported in the U.S. each year, concentrated in the Northeast, upper Midwest and the Pacific Northwest (CDC, 1993). The outer surface

Lyme Borreliosis, Edited by J.S. Axford and
D.H.E. Rees, Plenum Press, New York, 1994

of pathogens plays an important role in host-parasite interactions and in the ability to establish an infection in a variety of hosts (Schwan et al., 1988). Lyme disease vaccines being developed today are directed at the major outer surface proteins (denoted as Osp) on *B. burgdorferi* (Fikrig et al., 1992, Fikrig et al., 1990). The outer surface proteins are encoded by the OspA and OspB genes, which are tandemly arrayed on linear plasmids and are cotranscribed (Howe et al., 1986; Barbour and Garon, 1988; Bergstrom et al., 1989;). Although there appears to be higher strain heterogeneity in Europe, some genetic variation has been observed in the U.S. (Barbour et al., 1985; Marconi and Garon, 1992). In a previous study, we sequenced the OspA and OspB genes of several U.S. strains and found significant variation in the OspB gene (Caporale and Kocher, in prep).

In the U.S. human symptoms range from arthritis to neurological and cardiac abnormalities. It is not known whether there are geographically localized strains in the U.S. which may cause different patterns of disease. Although many investigators are looking at Osp variation at broad geographic scales across the U.S. and in Europe, local genic variation within a spirochetal population has not been addressed. The detailed study of variation among strains from one geographic area may be helpful in identifying virulence factors, and in following the spread of the spirochete among animal populations.

Previous work by MMC researchers has demonstrated the presence of Lyme disease on islands in the Gulf of Maine (Rand et al., 1993; Smith et al., 1993). Lyme disease foci on the islands involve different mammalian hosts, and are presumably more isolated from gene flow than mainland populations. On the mainland, the white-footed mouse, *P. leucopus*, is the main reservoir for the bacterium. In contrast, the major host on Monhegan Island is the Norway rat, *R. norvegicus*, and on Isle au Haut it is the field mouse, *P. maniculatus*. It was our expectation that these islands would contain single strains, possibly adapted to the unique set of conditions on the islands.

Here we report the OspB DNA sequence of 51 *B. burgdorferi* isolates from five sites along the coast of Maine. The sequences allow us to define strains and determine their frequency at each site. We also isolated borrelia from tick larvae from Monhegan Island rats, to examine whether multiple variants simultaneously infect a single mammalian host.

MATERIALS AND METHODS

Larval, nymphal and adult-staged *Ixodes dammini* ticks were collected from Monhegan Island, Isle au Haut, and Wells Reserve. The ticks were collected either by flagging or by direct removal from rodents captured in Sherman live-traps (Model 3310A, H.B. Sherman Traps, Inc., Tallahassee, Fl). Deer-derived adult ticks from Ogunquit (mainland) and bird-derived ticks from Appledore Island were also collected.

Ear biopsies were obtained from the captured rodents for the isolation of *B. burgdorferi*, following the method of Sinsky and Piesman (1989). Spirochetes were cultured from tissues in BSKII medium (Barbour, 1984). All ticks were screened for the spirochete by a direct fluorescent antibody procedure using a fluorescein isothiocyanate-conjugated polyclonal rabbit antibody to *B. burgdorferi* as described (Donahue et al., 1987). Borrelia taken from tick midguts were also cultured in BSKII medium.

DNA was isolated from 51 cultures using a standard phenol extraction protocol (Sambrook et al., 1989). The OspB gene was amplified using the polymerase chain reaction (PCR) (Saiki et al., 1985). Three primer sets directed at the OspB gene (Caporale and Kocher, in prep) were used for amplification. Reactions were performed in a Perkin-Elmer (Model 480) thermal cycler. Components were denatured at 93^0C for 30 seconds, annealed at 50^0C for 1 min, and extended at 72^0C for 2 min, for a total of 35 cycles. The double-stranded PCR amplification products were separated on 1% SeaPlaque agarose gels. The bands were treated with agarase and cycle sequenced using Taq polymerase-mediated incorporation of dye-labeled dideoxy terminators (Applied Biosystems). Samples were run through Centri-Sep columns (Princeton Separations) to remove unincorporated nucleotides. Purified samples were analyzed in an Applied Biosystems 373A DNA sequencer, and aligned with New York strains B31 and 19535 using the Seq Ed™ DNA sequence editor program (ABI) on a MacIntosh IIci computer.

RESULTS

The OspB nucleotide sequences of all 51 isolates were aligned. A total of 7 different OspB gene sequences were observed from coastal areas in Maine (Figure I). Five strains, each defined as having a unique OspB amino acid sequence, were observed. The range of variation was found to be between 1 and 7 nucleotide differences, which translated to a maximum of 5 amino acid differences. The average pairwise difference in the OspB gene was 0.47%. The OspB nucleotide sequence of strain I was identical to strain B31 (Bergstrom et al., 1989), with the exception of one nucleotide substitution (guanine) at position 593 .

There appears to be a clustering of nucleotide substitutions in four regions of the OspB gene (Figure 1). The 5' terminus, which translates to the amino end of the protein, appears to be highly conserved. This hydrophobic region is membrane-embedded and

Strain	258 [1]	279	288	376	382	385	510	526	729	758	801
I [2]	T	T	T	G	G	G	C	G	T	A	A
IIa										C	
IIb	C		C							C	
IIIa	C			A	A					C	
IIIb	C		C	A	A					C	G
IV	C	C		A	A		T				
V	C			A	A	A		A	C	C	
subst [3]	S	S	S	R	R	R	S	R	S	R	S

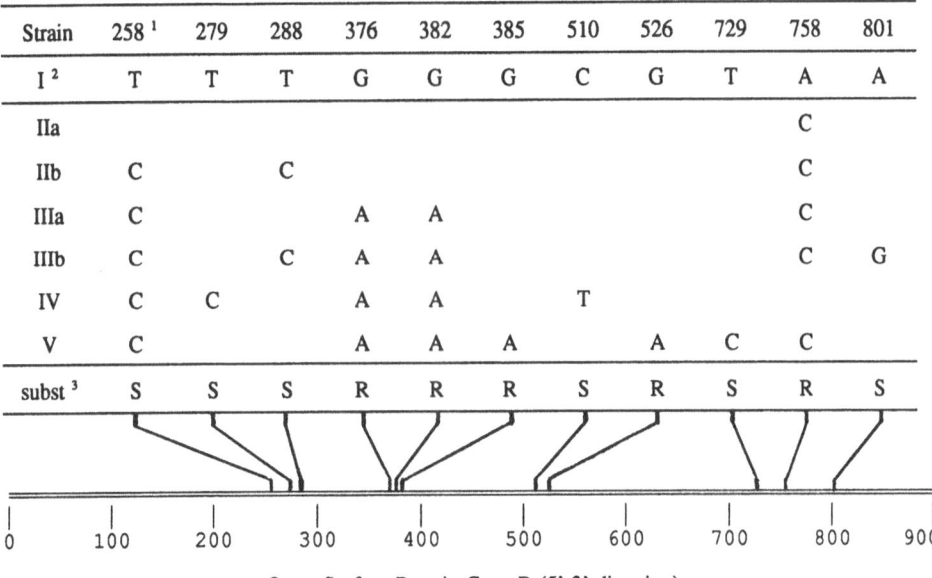

Outer Surface Protein Gene B (5'-3' direction)

Figure 1. OspB sequence differences between five strains located near the Maine coastline. [1] Each number represents a site containing a nucleotide substitution. [2] Strain I is used as the reference sequence, and only changes from this sequence are recorded in Figure 1. [3] Synonymous (S) and nonsynonymous (R) base substitutions were observed. Synonymous (silent) substitutions are those which do not alter the amino acid type at that position. Nonsynonymous (replacement) substitutions are those with an alternate amino acid at that position.

under functional constraint. The first position contains only silent substitutions. The amino acid replacements were located in the second half of the protein, which is exposed to the external environment.

As seen in Table I, 3 different strains were observed from Isle au Haut, 4 strains from Monhegan Island, and 5 strains from Wells Reserve. Strains I and IIa (which contain only one amino acid replacement) appear to be the most abundant strains and are in equal frequency within each of these three locations. Isle au Haut, Wells and Ogunquit contained one strain unique to each area. Strain I was identified in culture from a bird-derived larva from Appledore Island. Cultures from two bird-derived larvae from Wells contained strain IV, which was also observed on Monhegan Island.

Table I. Frequency and distribution of seven strains located near the coast of Maine.

SITE	I	IIa	IIb	IIIa	IIIb	IV	V
ISLE AU HAUT	3	4	0	2	0	0	0
MONHEGAN ISLAND	11	11	0	0	0	2	1
LAUDHOLM FARMS, WELLS	4	6	0	0	2	2	1
OGUNQUIT	0	0	1	0	0	0	0
APPLEDORE ISLAND	1	0	0	0	0	0	0

To examine the possibility that several strains might coinfect a single mammalian host, we isolated borrelia from multiple ticks from each of 5 rats on Monhegan Island (Table II). In order to be sure that the borrelia were derived from the same host the ticks must be at the larval stage (which get their first blood meal from that host). The nymphs taken off of rat #1 could have attained its strain from a previous blood meal, and not from this rat. But the larvae taken off of rats #2-5 must have acquired their particular strains from those rats. Therefore, rats #3, 4 and 5 must be infected with both strains I and II.

Table II. *B. burgdorferi* strain diversity within rats from Monhegan Island.

Rat #	Vector Stage	Strain #
1	2 nymphs	I
	1 nymph	IIa
2	3 larvae	IIa
3	3 larvae	I
	1 larva	IIa
4	2 larvae	I
	1 larva	IIa
5	1 larva	I
	2 larvae, 2 nymphs	IIa

DISCUSSION

Lyme disease foci on the islands in the Gulf of Maine involve different mammalian hosts whose populations are presumably more isolated from gene flow than mainland populations. Our hypothesis was that isolates from these islands would contain single strains, possibly adapted to the unique set of conditions on the islands. After sequencing the OspB gene of 51 isolates, we found an unexpectedly high level of diversity. Five strains were found on the mainland in Wells, four strains from Monhegan Island and two strains from Isle au Haut. Strains I and IIa were found to be the most abundant strains and in equal frequency at these three sites. *R. norvegicus, P. maniculatus,* and *P. leucopus* are competent to carry these two strains. In contrast, strains IIb, IIIa and IIIb were found to be unique to their respective locations. The possibility remains that these strains are associated with the unique host present at these three sites.

The distribution of each strain was investigated to assess the spread of Lyme disease along the coast of Maine. Four strains were identified at more than one location, suggesting high levels of gene flow. Mammalian hosts are not thought capable of frequent dispersal among these sites. However, since migratory birds have been found to be competent in carrying *B. burgdorferi* (Anderson et al., 1986), their dispersal may play a significant role in the spread of Lyme disease along the coast of Maine.

To investigate the possibility that multiple strains may coexist within a given host, isolates derived from larvae taken off of the same rat were compared. Strains I and IIa have been found to coexist in the Norway rat. Serial, transient infection may be occurring within the rats as they encounter new strains. In time, one of two strains may eventually over-populate the other strain and "take over" the host. Alternately, these strains may each persist for long periods in the host. In either case, PCR procedures may only be detecting the most abundant strain present. The number of strains detected by PCR may be an underestimation of the number of strains actually coexisting in a host.

Although seven unique OspB sequences have been observed in Maine, the biological significance of this diversity is presently unknown. Our next goal is to test for cross reactivity with monoclonal antibodies between these 5 strains by immuno-double diffusion. This will provide insight into possible immunological differences among the strains. We will also be looking for pathological differences between strains in a mouse and rat model. The heart, spleen and joint regions will be the targets of investigation. The results of these studies could lead to a better understanding of the pathogenesis of Lyme disease.

REFERENCES

Anderson, J.F., Johnson, R.C., Magnarelli, L.A., and Hyde, F.W., 1986, Involvement of birds in the epidemiology of the Lyme disease agent *Borrelia burgdorferi, Infect. Immun.* 51:394-396.

Barbour, A.G., 1984, Isolation and cultivation of Lyme disease spirochetes, *Yale J. Biol. Med.* 57:521-525.

Barbour, A.G., and Garon, C.F., 1988, The genes encoding major surface proteins of *Borrelia burgdorferi* are located on a plasmid, *Ann. N.Y. Acad. Sci.* 539:144-153.

Barbour, A.G., Heiland, R.A., and Howe, T.R., 1985, Heterogeneity of major proteins in Lyme disease borreliae: a molecular analysis of North American and European isolates, *J. Infect. Dis.* 152:478-484.

Bergstrom, S., Bundoc, V.G., and Barbour, A.G., 1989, Molecular analysis of linear plasmid-encoded major surface proteins, OspA and OspB, of the Lyme disease spirochete *Borrelia burgdorferi, Mol. Micr.* 3:479-486.

Burgdorfer W., Barbour, A.G., Hayes, S.G., Benach J.L., Grunwaldt, E., and Davis, J.P., 1982, Lyme disease - a tick-borne spirochetosis?, *Science* 216:1317-1319.

Caporale, D.A., and Kocher, T.D., Sequence variation in the outer surface protein genes of *Borrelia burgdorferi*, in prep.

Craven, R., and Dennis, D. (eds), 1993, "Lyme Disease Surveillance Summary", Centers for Disease Control (CDC) Fort Collins. 4:1-5.

Donahue, J.G., Piesman, J., and Spielman, A., 1987, Reservoir competence of white-footed mice for Lyme disease spirochetes, *Am. J. Trop. Med. Hyg.* 36:92-96.

Fikrig, E., Barthold, S.W., Kantor, F.S., and Flavell, R.A., 1990, Protection of mice against the Lyme disease agent by immunizing with recombinant OspA, *Science* 250:553-555.

Fikrig, E., Barthold, S.W., Persing, D.H., Sun, X., Kantor, F.S., and Flavell, R.A., 1992, *Borrelia burgdorferi* strain 25015: characterization of outer surface protein A and vaccination against infection, *J. Immun.* 148:2256-2260.

Howe, T.R., LaQuier, F.W., and Barbour, A.G., 1986, Organization of genes encoding two outer membrane proteins of the Lyme disease agent *Borrelia burgdorferi* within a single transcriptional unit, *Infect. Immun.* 54:207-212.

Marconi, R.T., and Garon, C.F., 1992, Phylogenetic analysis of the genus *Borrelia*: A comparison of North American and European isolates of *Borrelia burgdorferi*, *J. Bacteriol.* 174:241-244.

Rand, P.W., Lacombe, E.H., Smith, Jr., R.P., Rich, S.M., Kilpatrick, C.W., Dragoni, C.A., and Caporale, D.A. 1993, Competence of *Peromyscus maniculatus* (Rodentia:Cricetidae) as a reservoir host for *Borrelia burgdorferi* (Spirochaetares: Spirochaetaceae) in the wild, *J. Med. Entomol.* in press.

Saiki, R.K., Gelfand, D.H., Stoffel, S., Scharf, S.J., Higuchi, R., Horn, G.T., Mullis, K.B., and Erlich, H.A., 1988, Primer-directed enzymatic amplification of DNA with a thermostable DNA polymerase, *Science* 239:487-491.

Sambrook, J., Fritsch, E.F., and Maniatis, T., 1989, "Molecular Cloning: A Laboratory Manual", 2nd ed.Cold Spring Harbor Laboratory Press, New York E.3-4.

Schwan, T.G., Burgdorfer, W., and Garon, C.F., 1988, Changes in infectivity and plasmid profile of the Lyme disease spirochete, *Borrelia burgdorferi*, as a result of in vitro cultivation, *Infect. Immun.* 56:1831-1836.

Sinsky, R.J., and Piesman, J., 1989, Ear punch biopsy method for detection and isolation of *Borrelia borgdorferi* from rodents, *J. Clin. Microbiol.* 27:1723-27.

Smith, Jr., R.P., Rand, P.W., Lacombe, E.H., Telford III, S.R., Rich, S.M., Piesman, J, and Spielman, A., 1993, Norway rats as reservoir hosts for Lyme disease spirochetes on Monhegan Island, Maine, *J. Infect. Dis.* in press.

MULTIPLE AMINO ACID SEQUENCE ALIGNMENT OF THE MAJOR OUTER SURFACE PROTEINS OSPA AND OSPB OF VARIOUS *BORRELIA BURGDORFERI* STRAINS

Waltraud Fellinger, Markus Reindl, Georg Stöffler and Bernhard Redl

Institut für Mikrobiologie (Med. Fakultät)
Fritz Pregl Str. 3
A-6020 Innsbruck

INTRODUCTION

OspA and OspB are major outer membrane proteins of *Borrelia burgdorferi,* encoded by two linear plasmid located genes (*ospA* and *ospB*), which are cotranscribed by a single mRNA species (Barbour and Garon, 1987; Howe et al., 1986). Since these proteins are surface-exposed, they may play a role in the host-parasite interaction and have therefore been studied in detail (Barbour et al., 1983; Barbour, 1984; Barbour et al., 1984). Several sera from patients in the late or chronic phase of Lyme borreliosis contain antibodies against OspA and OspB proteins. It could be demonstrated, that monoclonal antibodies directed against OspA protect scid (severe combined immundeficiency) mice acutely infected with *B. burgdorferi* from clinical symptoms (Schaible et al., 1990). Furthermore, it has been shown that an active immunization might be possible, because C3H/HeJ mice immunized with recombinant OspA were protected from infection with *B. burgdorferi* (Fikrig et al., 1990).

Biochemical and immunological studies have revealed a considerable variability of the Osp proteins in apparent molecular mass and in reactivity with monoclonal antibodies between isolates from North America and Europe (Barbour et al., 1985 and 1986; Wilske et al., 1988). This heterogeneity is more prominent among European isolates than among North American isolates. Furthermore, OspB proteins show more strain heterogeneity (Bergström et al., 1989).

In an attempt to learn more about the variation of OspA and B and for future evaluation of the gene products, we have cloned and sequenced the complete *osp*-operon of the European strain B29 (Fellinger et al., 1992) and have compared the deduced amino acid sequences with different *B. burgdorferi* strains.

RESULTS and DISCUSSION

Cloning and sequencing of the *osp*-operon of B29

Total DNA of *B. burgdorferi* was isolated, partially digested with Sau3A and ligated with the BamHI-digested, dephosphorylated cosmid pHC79 and packaged into λ-phages. Based on the nucleotide sequence of the *ospA* gene from *B. burgdorferi* B31 (Bergström et al., 1989), a labeled oligodeoxynucleotide-probe was used for screening the resulting colonies. The DNA of hybridizing clones was restricted with several restriction endonucleases. A 2.2 kb-HaeIII-fragment, which contained the entire *osp*-operon, was cloned into pBluescript and sequenced by the dideoxy chain termination method.

Amino acid sequence analysis

The multiple alignment of amino acid sequences was performed using the Clustal V software (Higgins et al., 1992). The amino acid sequences were deduced from the *ospA* genes of strains B31, N40, 19857, ZS7, Goe2 and ZQ1 with the assigned accession numbers X14407, M57248, X68059, X16467, X60300 and X66065 in the EMBL/GenBank. The OspA amino acid sequences of *B. burgdorferi* 25015, PKo and PBi were derived from previously published sequences (Fikrig et al., 1992; Zumstein et al., 1992), as well as the whole Osp amino acid sequences of the strains ACA-1 and IP90 (Jonsson et al., 1992).

When comparing the amino acid sequence of OspA of *B. burgdorferi* B29 with OspA proteins of other strains, the degree of identity varied from 67.8% (with strain 19857) to 99.6% (with strain Goe2) (Table 1).

Table 1. Amino acid sequence identity (in %) of different OspA proteins (a) and OspB proteins (b).

	Borrelia burgdorferi isolate	European strains							North American strains			
		Goe2	ACA-1	IP90	PBi	PKo	ZQ1	ZS7	B31	N40	25015	19857
a)	B29	98.5	80.7	92.7	86.9	80.7	96.7	80.3	80.3	80.3	79.6	67.8
	Goe2		81.0	93.4	87.6	81.0	98.2	80.7	80.7	80.7	80.3	68.1
	ACA-1			81.0	83.2	99.6	81.8	77.7	77.7	77.7	80.6	70.5
	IP90				91.2	80.7	94.5	79.2	79.2	79.2	79.9	68.1
	PBi					83.2	89.1	80.2	80.2	80.2	82.4	68.4
	PKo						81.8	77.3	77.3	77.3	80.2	70.5
	ZQ1							80.3	80.7	80.3	80.7	68.8
	ZS7								80.3	98.9	99.6	85.0	72.0
	B31									99.3	85.7	72.7
	N40										85.0	72.4
	25015											72.4
b)	B29	-	62.8	66.3	-	-	-	-	62.7	-	-	-
	ACA-1	-		66.2	-	-	-	-	65.4	-	-	-
	IP90	-			-	-	-	-	66.9	-	-	-

Based on restriction fragment length polymorphism of the *ospA* gene (Wallich et al., 1992) or a serotyping system using the immunological reactivity of OspA with monoclonal antibodies (Wilske et al., 1993), a classification of *B. burgdorferi* strains has been established.

In our studies, the genotyping, respectively serotyping system was confirmed by the degree of homology. The aligned amino acid sequences represent five of six, respectively seven (according to Wilske et al.) subtypes: the North American strains B31 and N40 and the European strain ZS7 belong to OspA subtype I, B29, Goe2 and ZQ1 represent genotype II (according to Wallich) or serotype 6 (according to Wilske). The German strain PBi is a member of serotype 4, the North American strain 19857 is a single representative of genotyp III. ACA-1 and PKo belong to genotype IV (according to Wallich) or serotype 2 (according to Wilske). The North American strain 25025 and the Russian strain IP90 were not to classify, although the latter shows a remarkable high degree of homology with strains of genotype II (from 92.7 to 94.5%) and might even be closer related to these strains. Representatives of one particular OspA subtype exhibits sequence homologies from 96.7 to 99.6%, whereas those from different types show less homology (67.8 to 81.8%). According to Wilske et al. (1993), serotype 1 corresponds to *B. burgdorferi* sensu strictu, serotype 2 corresponds to group VS461 and serotypes 3 to 7 correspond to *Borrelia garinii* sp. nov..

When inspecting the amino acid alignment of different OspA proteins in more detail, it becomes evident, that the N-terminal region shows the highest degree of homology (Fig. 1).

```
        1
B29     MKKYLLGIGLILALIACKQNVSSLDEKNSVSVDLPGGMTVLVSKEKDKDGKYSLDATVDKLELKGTSDKN
GOE2    MKKYLLGIGLILALIACKQNVSSLDEKNSVSVDLPGGMTVLVSKEKDKDGKYSLEATVDKLELKGTSDKN
ACA-1   MKKYLLGIGLILALIACKQNVSSLDEKNSASVDLPGEMKVLVSKEKDKDGKYSLKATVDKIELKGTSDKD
IP90    MKKYLLGIGLILALIACKQNVSSLDEKNSVSVDLPGGMQVLVSKEKDKDGKYSLMATVDKLELKGTSDKN
PBI     MKKYLLGIGLILALIACKQNVSSLDEKNSVSVDLPGEMKVLVSKEKDKDGKYSLMATVDKLELKGTSDKS
PKO     MKKYLLGIGLILALIACKQNVSSLDEKNSASVDLPGEMKVLVSKEKDKDGKYSLKATVDKIELKGTSDKD
ZQ1     MKKYLLGIGLILALIACKQNVSSLDEKNSVSVDLPGGMKVLVSKEKDKDGKYSLEATVDKLELKGTSDKN
ZS7     MKKYLLGIGLILALIACKQNVSSLDEKNSVSVDLPGEMNVLVSKEKNKDGKYDLIATVDKLELKGTSDKN
B31     MKKYLLGIGLILALIACKQNVSSLDEKNSVSVDLPGEMKVLVSKEKNKDGKYDLIATVDKLELKGTSDKN
25015   MKKYLLGIGLILALIACKQNVSSLDEKNSVSVDLPGEMKVLVSKEKDKDGKYSLMATVDKLELKGTSDKN
N40     MKKYLLGIGLILALIACKQNVSSLDEKNSVSVDLPGEMNVLVSKEKNKDGKYDLIATVDKLELKGTSDKN
19857   MKKYLLGIGLILALIACKQNVSSLDEKNSVSVDVPGGMKVLVSKEKNKDGKYDLMATVDNVDLKGTSDKN
        **************************** *** .** * ******* .***** * ****...*******
```

```
        71
B29     NGSGTLEGEKTDKSKVKLTIADDLSQTKFEIFKEDGKTLVSKKVTSKDKSSTEEKFNEKGETSEKTIVRA
GOE2    NGSGTLEGEKTDKSKVKLTIADDLSQTKFEIFKEDGKTLVSKKVTLKDKSSTEEKFNEKGETSEKTIVRA
ACA-1   NGSGVLEGTKDDKSKAKLTIADDLSKTTFELFKEDGKTLVSRKVSSKDKTSTDEMFNEKGELSAKTMTRE
IP90    NGSGTLEGEKTDKSKAKLTIAEDLSKTTFEIFKEDGKTLVSKKVTLKDKSSTEEKFNAKGEASEKTIVRA
PBI     NGSGTLEGEKSDKSKAKLTISEDLSKTTFEIFKEDGKTLVSKKVNSKDKSSIEEKFNAKGELSEKTILRA
PKO     NGSGVLEGTKDDKSKAKLTIADDLSKTTFELFKEDGKTLVSRKVSSKDKTSTDEMFNEKGELSAKTMTRE
ZQ1     NGSGTLEGEKTDKSKVKLTIAEDLSKTTFEIFKEDGKTLVSKKVTLKDKSSTEEKFNEKGEISEKTIVRA
ZS7     NGSGVLEGVKADKSKVKLTISDDLGQTTLEVFKEDGKTLVSKKVTSKDKSSTEEKFNEKGEVSEKIITRA
B31     NGSGVLEGVKADKSKVKLTISDDLGQTTLEVFKEDGKTLVSKKVTSKDKSSTEEKFNEKGEVSEKIITRA
25015   NGSGVLEGVKADKSKVKLTVSDDLSTTTLEVLKEDGKTLVSKKRTSKDKSSTEEKFNEKGELVEKIMARA
N40     NGSGVLEGVKADKSKVKLTISDDLGQTTLEVFKEDGKTLVSKKVTSKDKSSTEEKFNEKGEVSEKIITRA
19857   NGSGILEGVKADKSKVKLTVADDLSKTTLEVLKEDG-TVVSRKVTSKDKSTTEAKFNEKGELSEKTMTRA
        **** *** * **** ***...**. * .*.**** *.**.*   ***...  ** *** * . *
```

```
        141
B29     NGTRLEYTDIKS-DGSGKAKEVLKD-FTLEGTLAADGKTTLKVTEGTVVLSKNILKSGEITVALDDSDTT
GOE2    NGTRLEYTDIKS-DGSGKAKEVLKD-FTLEGTLAADGKTTLKVTEGTVVLSKNILKSGEITVALDDSDTT
ACA-1   NGTKLEYTEMKS-DGTGKAKEVLKN-FTLEGKVAND-KVTLEVKEGTVTLSKEISKSGEVTVALNDTNTT
IP90    NGTRLEYTDIKS-DKTGKAKEVLKD-FALEGTLAADGKTTLKVTEGTVVLSKHISNSGEITVELNDSDTT
PBI     NGTRLEYTEIKS-DGTGKAKEVLKD-FALEGTLAAD-KTTLKVTEGTVVLSKHIPNSGEITVELNDSNST
PKO     NGTKLEYTEMKS-DGTGKAKEVLKN-FTLEGKVAND-KVTLEVKEGTVTLSKEIAKSGEVTVALNDTNTT
ZQ1     NGTRLEYTDIKS-DGSGKAKEVLKD-FTLEGTLAADGKTTLKVTEGTVVLSKNILKSGEITVALDDSDTT
ZS7     DGTRLEYTEIKS-DGSGKAKEVLKS-YVLEGTLTAE-KTTLVVKEGTVTLSKNISKSGEVSVELNDTDSS
B31     DGTRLEYTGIKS-DGSGKAKEVLKG-YVLEGTLTAE-KTTLVVKEGTVTLSKNISKSGEVSVELNDTDSS
25015   NGTILEYTGIKS-DGSGKAKETLKE-YVLEGTLTAE-KATLVVKEGTVTLSKHISKSGEVTAELNDTDST
N40     DGTRLEYTEIKS-DGSGKAKEVLKG-YVLEGTLTAE-KTTLVVKEGTVTLSKNISKSGEVSVELNDTDSS
19857   NGTTLEYSQMTNEDNAAKAVETLKNGIKFEGNLAS-GKTAVEIKEGTVTLKREIDKNGKVTVSLNDTAS-
        .** ***.. .. * ..** ** **    .** ..   * ... **** * ..* ..* .. *.*. .
```

```
        211
B29     QATKKTGNWDSKSSTLTISVNSQKTKNLVFTKEDTITVQKYDSAGTNLEGKAVEITTLKELKAALK
GOE2    QATKKTGKWDSKTSTLTISVNSQKTKNLVFTKEDTITVQKYDSAGTNLEGKAVEITTLKELKDALK
ACA-1   QATKKTGAWDSKTSTLTISVNSKKTTQLVFTKQDTITVQKYDSAGTNLEGTAVEIKTLDELKNALK
IP90    QATKKTGTWDSKTSTLTISVNSRKTKNLVFTKEDTITVQKYDSAGTNLEGKAVEITTLKELKDALK
PBI     QATKKTGKWDSNTSTLTISVNSKKTKNIVFTKEDTITVQKYDSAGTNLEGNAVEIKTLDELKNALK
PKO     QATKKTGAWDSKTSTLTISVNSKKTTQLVFTKQDTITVQKYDSAGTNLEGTAVEIKTLDELKNALK
ZQ1     QATKKTGKWDSKTSTLTISVNSQKTKNLVFTKEDTITVQKYDSAGTNLEGKAVEITTLKELKDALK
ZS7     AATKKTAAWNSGTSTLTITVNSKKTKDLVFTKENTITVQQYDSNGTKLEGSAVEITKLDEIKNALK
B31     AATKKTAAWNSGTSTLTITVNSKKTKDLVFTKENTITVQQYDSNGTKLEGSAVEITKLDEIKNALK
25015   QATKKTGKWDAGTSTLTITVNNKKTKALVFTKQDTITSQKYDSAGTNLECTAVEIKTLDELKNALR
N40     AATKKTAAWNSGTSTLTITVNSKKTKDLVFTKENTITVQQYDSAGTNLEGTAVEIKTLDELKNALR
19857   -GSKKTASWQESTSTLTISANSKKTKDLVFLTNGTITVQNYDSAGTKLEGSAAEIKKLDELKNALR
        ..***.*. .*****.*...** .**  . *** *.*** **.*** * **  * *.* **.
```

Figure 1. Alignment of OspA sequences. Stars represent identical amino acids, points indicate conservative amino acid substitutions.

The North American strain 19857 exhibits considerable differences, especially in the central part of the protein (20 amino acid substitutions concern only strain 19857, although 12 of them are conservative substitutions). This may indicate, that strain 19857 separated earlier in evolution. Alltogether, there are 153 identical amino acids and 53 conservative substitutions in the aligned OspA proteins.

A lower degree of sequence identity was observed by comparing OspB proteins. The homology of the amino acid sequence of B29 with other strains varies from 62.7% (with strain B31) to 66.3% (with strain IP90). OspB of the European strain Goe2 shows 97.5% amino acid sequence identity with strain B29 in 122 amino acids overlap (data not shown). Goe2 is not sequenced completely and as there is a frameshift referring to the *ospA* gene, it cannot be expressed together with OspA (Eiffert et al., 1992). Although OspB from strain B29 is expressed, the high degree of homology may indicate, that Goe2 and B29 are very closely related and separated recently.

Apart from the hydrophobic signal peptide, which is removed by the signal peptidase II during processing, no identity between the different OspB proteins is found in the N-terminal part (Fig. 2). The central region of the protein is highly conserved, especially the C-terminal part, extending from position 248 to 272. There are 78 conservative amino acid substitutions and 140 identical amino acids in the OspB proteins.

Figure 2. Alignment of OspB sequences. Stars represent identical amino acids, points indicate conservative amino acid substitutions.

In general, OspB proteins show a higher degree of heterogeneity than OspA proteins. Due to this fact, OspB may be difficult to handle in serological tests, in which specific antigens are used.

The significance of the *B. burgdorferi* subtypes is not known so far, but it is to recognize, that Lyme disease in Europe more frequently progresses to neuroborreliosis and the severe acrodermatitis chronica atrophicans (ACA), whereas in the United States, arthritis is a more common manifestation of late Lyme disease. Wilske et al. (1988) established, that immune sera from ACA patients reacted with many serotype 2 proteins but only with a few proteins from other serotypes. In their studies, OspA serotype 6 organisms were frequently found in ticks but were rarely isolated from patients. The reduced pathogenicity of OspA serotype 6 organisms should be explained by strain Goe2, which does not express OspB. As strain B29 is also a representative of this serotype and OspB is expressed in this isolate, we cannot generalize, that serotype 6 is less pathogenic.

The homology between the respective OspA and OspB proteins of different strains is significantly higher than that between OspA and OspB within each strain (Table 2). Thus, it is evident from the results, that the ancestral *osp*-operon diverged into the two genes before the strains seperated. As the European isolates revealed a higer degree of heterogeneity, we can assume, that *B. burgdorferi* may have spread in Europe before a few types reached North America.

Table 2. Amino acid sequence comparison between the respective OspA and OspB proteins of the different strains.

Borrelia burgdorferi isolate	% amino acid identity
B29	44.9
ACA-1	49.6
IP90	45.5
B31	55.6

In view of the variability of OspA and OspB in *B. burgdorferi*, these data are of great importance for all those, who work on these proteins as a possible vaccine candidate.

REFERENCES

Barbour, A.G., Burgdorfer, W., Grunwaldt, E., and Steere, A.C., 1983, Antibodies of patients with Lyme disease to components of the *Ixodes dammini* spirochete, *J. Clin. Invest.* 72: 504-515.

Barbour, A.G., 1984, Immunochemical analysis of Lyme disease spirochetes, *Yale J. Biol. Med.* 57: 581-586.

Barbour, A.G., Tessier, S.L., and Hayes, S.F., 1984, Variation in a major surface protein of Lyme disease spirochetes, *Infect. Immun.* 45: 94-100.

Barbour, A.G., Heiland, R.A., and Howe, T.R., 1985, Heterogeneity of major proteins in Lyme disease borreliae: a molecular analysis of North American and European isolates, *J. Infect. Dis.* 152:478-484.

Barbour, A.G., and Schrumpf, M.E., 1986, Polymorphisms of major surface proteins of *Borrelia burgdorferi*, *Zbl. Bakteriol. Hyg. A* 263: 83-91.

Barbour, A.G., and Garon, C.F., 1987, Linear plasmids of the bacterium *Borrelia burgdorferi* have covalently closed ends, *Science* 237: 409-411.

Bergström, S., Bundoc, V.G., and Barbour, A.G., 1989, Molecular analysis of linear plasmid-encoded major surface proteins, OspA and OspB, of the Lyme disease spirochaete *Borrelia burgdorferi*, *Mol. Microbiol.* 364: 479-486.

Eiffert, H., Ohlenbusch, A., Fehling, W., Lotter, H., and Thommssen, R., 1992, Nucleotide sequence of the *ospAB* operon of a *Borrelia burgdorferi* strain expressing OspA but not OspB, *Infect. Immun.* 60: 1864-1868.

Fellinger, W., Redl, B., and Stöffler, G., 1992, Sequence of the complete *osp* operon encoding two major outer membrane proteins of a European *Borrelia burgdorferi* isolate (B29), *Gene* 120: 127-128.

Fikrig, E., Barthold, S.W., Kantor, F.S., and Flavell, R.A., 1990, Protection of mice against the Lyme disease agent by immunizing with recombinant OspA, *Science* 250: 553-556:

Fikrig, E., Barthold, S.W., Persing, D.H., Sun, X., Kantor, F.S., and Flavell, R.A., 1992, *Borrelia burgdorferi* strain 25015: characterization of outer surface protein A and vaccination against infection, *J. Immunol.* 148: 2256-2260.

Higgins, D.G., Bleasby, A.J., and Fuchs, R., 1992, Clustal V: improved software for multiple sequence alignment, *Cabios* 8: 189-191.

Howe, T.R., LaQuier, F.W., and Barbour, A.G., 1986, Organization of genes encoding two outer membrane proteins of the Lyme disease agent *Borrelia burgdorferi* within a single transcriptional unit, *Infect. Immun.* 54: 207-212.

Jonsson, M., Noppa, L., Barbour, A.G., and Bergström, S., 1992, Heterogeneity of outer membrane proteins in *Borrelia burgdorferi:* comparison of the *osp* operons of three isolates of different geographic origins, *Infect. Immun.* 60: 1845-1853.

Schaible, U.E., Kramer, M.D., Eichman, K., Modolell, M., Museteanu, C., and Simon, M.M., 1990, Monoclonal antibodies specific for the outer surface protein A (OspA) of *Borrelia burgdorferi* prevent Lyme borreliosis in severe combined immunodeficiency (scid) mice, *Proc. Natl. Acad. Sci. USA* 87: 3768-3772.

Wallich, R., Helmes, C., Schaible, U.E., Lobet, Y., Moter, S.E., Kramer, M.D., and Simon, M.M., 1992, Evaluation of genetic divergence among *Borrelia burgdorferi* isolates by use of OspA, *fla*, Hsp60 and Hsp70 gene probes, *Infect. Immun.* 60: 4856-4866.

Wilske, B., Praec-Mursic, V., Schierz, G., Kuhbeck, R., Barbour, A.G., and Kramer, M.1988, Antigenic variability of *Borrelia burgdorferi*, *Ann. N.Y. Acad. Sci.* 539: 126-143.

Wilske, B., Preac-Mursic, V., Göbel, U.B., Graf, B., Jauris, S., Soutschek, E., Schwab, E., and Zumstein, G., 1993, An OspA serotyping system for *Borrelia burgdorferi* based on reactivity with monoclonal antibodies and OspA sequence analyses, *J. Clin. Microbiol.* 31: 340-350.

Zumstein, G., Fuchs, R., Hofmann, A., Praec-Mursic, V., Soutschek, E., and Wilske, B., 1992, Genetic polymorphism of the gene encoding the outer surface protein A (OspA) of *Borrelia burgdorferi*, *Med. Microbiol. Immunol.* 181: 57-70.

LYME DISEASE IN AN EXPERIMENTAL CAT MODEL

Michael D. Gibson[1], Colin R. Young[2], M. Tawfik Omran[2], Kathy Palma[3],
John F. Edwards[2], and Julie A. Rawlings[4]

[1]Department of Veterinary Anatomy & Public Health
[2]Department of Veterinary Pathobiology
[3]Department of Entomology
 Texas A&M University
 College Station, TX 77843
[4]Texas Department of Public Health
 Austin, TX 78756

ABSTRACT

In spite of broad species susceptibility from dogs to horses, attempts to experimentally reproduce Lyme Disease in laboratory cats have been disappointing. We have examined the susceptibility and/or development of Lyme borreliosis (LB) in experimental cats under Category II Containment. Three groups of cats were injected with 10^6 live *Borrelia burgdorferi* (Bb) spirochaetes in a single intradermal site. The three strains of Bb used were: Bb31 reference strain, Bb1579 strain isolated from an *Amblyomma americanum* tick and Bb532 strain isolated from a pool of five cat fleas. The cats were examined daily, bled bi-weekly, and one cat per group was sacrificed monthly for histopathological study. Initial results indicate that all three strains of Bb are immunogenic in cats and caused some lesions which are characteristic of LB. However spirochaetes could not be recovered from ticks or cat fleas that fed upon these experimentally infected cats. Differential white blood cell counts indicted that the cats underwent infection, histological data indicated that there was multiple organ infection in test cats including regional lymph nodes, lungs and liver. The presence of whole intact or fragmented spirochaetes at the site of lesion(s) was determined using Steiner's silver stain for spirochaetes in tissue sections. The significance of ticks, cat fleas and cats in the transmission of Lyme Disease will be discussed.

INTRODUCTION

Lyme disease is a multisystem disease resulting in dermatologic, rheumatologic, neurologic and cardiac abnormalities which develop in stages with various clinical pictures (1) following infection by a newly discovered spirochaete, *Borrelia burgdorferi* (2).

Since its first diagnosis in Lyme, Connecticut, USA, Lyme disease has been identified in all continents throughout the world (3). It is found over most of Europe (4) Asia (5,6) and some parts of Africa (7,8). The clinical picture of Lyme disease is similar in some respects to other spirochetal diseases such as syphilis, relapsing fever and leptospirosis (9). The early stage is characterized by distinctive Erythema Chronicum Migrans, annular skin lesions, regional adenopathy these may be accompanied by flu like symptoms and signs. A few weeks later the second or disseminated stage, characterized by multiple secondary erythema migrans, lymphocytoma, migratory musculoskeletal pain, acute arthritis, early neurologic manifestations (lymphocytic meningitis, cranial neuritis, radiculoneuritis, encephalitis and myelitis), generalized adenopathy, splenomegaly and cardiac abnormalities (carditis, conduction defects and arrythmia). During the third or late stage severe malaise and fatigue, Acrodermatitis chronica atrophicans, chronic arthritis, late neurologic manifestations (chronic progressive encephalomyelitis, focal encephalitis, mild encephalopathy, distal axonopathy and neuropathy in patients with acrodermatitis), chronic arthritis and chronic fatigue are manifested.

Borrelia burgdorferi belongs to the order *Spirochaetales* that forms two families of gram negative helical bacteria with common morphologic features. *Borrelia* species are the only pathogenic spirochaetes transmitted to vertebrates by blood sucking arthropods (9,10). *Borrelia burgdorferi* like other *Borrelia* stains well with Giemsa (11) and can be demonstrated in the tissue using silver impregnation methods (12,13,14).

Borrelia spirochaetes are believed to be primarily transmitted by ticks. Transmission of spirochaete from the vector to the host occurs either by way of saliva or regurgitation (15). The tick feeds as an obligate blood and tissue fluid feeding ectoparasite, on terrestrial vertebrates including birds (16), during each of its three successive postembryonic stages or instars i.e. larva, nymph and adult (17).

Different species of ticks are involved in the transmission of Lyme disease. *Ixodes scapularis, Dermacentor variabilis* and *Amblyomma americanum* are the principal vectors of transmission in the USA (18,19). Outside the USA *Borrelia* spirochaetes have been demonstrated in other ticks and blood sucking insects such as *I. pacificus* (20), *I. ricinus* in Europe and North Africa (21) and *I. persulcatus* in Asia (5,22). Furthermore, some blood sucking arthropods such as mosquitoes, horseflies (18,23) and cat fleas *Ctenocephalides felis* (24) have been reported to be infected with *Borrelia burgdorferi*, although their involvement in the transmission of Lyme disease is still understudy. Oral infection cannot be ruled out, since mice and ducks have been experimentally infected by the oral route (23).

The list of wild animals which can act as a reservoir for *Borrelia burgdorferi* differs according to geographical location. The agent has been recovered from rodents, rabbits and medium size mammals, wolves, coyotes, foxes, deer as well as large animals such as buffalos and bears (25). Migratory birds may play an important role both as a local reservoir of infection and long-distance dispersal agents for *Borrelia burgdorferi* infected ticks (26). This potential reservoir of infection is underscored by the fact that over 95 species of migratory birds have been shown to be infected with *Borrelia burgdorferi*.

Disease signs have also been demonstrated in domestic animals. Arthritis sometimes occurs in naturally infected horses and cows. Infection in horses led to swelling of legs, weight loss, dermatitis, conjuctivitis, uveitis, nasal discharge and cough, while mastitis, spontaneous abortion, myocarditis, interstitial nephritis, glomerulonephritis and pneumonitis have been demonstrated in infected cows (27,28,29,30,31).

Many different animal models of Lyme borreliosis have been investigated, however an animal model demonstrating the complete clinical picture of the human Lyme borreliosis has not yet been established. For example naturally infected white-footed mice (32) and experimentally infected rabbits (33,34) develop skin lesions and spirochaetaemia, while experimentally infected laboratory rats (35,36) and irradiated hamsters (37,38) develop arthritis.

As Lyme disease is zoonotic it is prudent to study closely the animals, particularly pets, in the proximity of man which may act as a reservoir for the disease. Dogs can naturally be infected with *Borrelia burgdorferi* and show signs of Lyme disease (39), likewise with cats (40,41). Since other blood sucking insects such as fleas (24) may act as vectors for transmitting Lyme borreliosis this study was initiated to determine whether cats could be i) infected with different isolates of *Borrelia burgdorferi*, ii) will different isolates cause any pathological changes in cats and iii) does the clinical picture in cats resembles human Lyme borreliosis?

MATERIALS & METHODS

Borrelia burgdorferi

Three strains of *Borrelia burgdorferi* were used in this study; Bb 31 - a reference strain isolated from *Ixodes scapularis* from Shelter Island, New York (42); Bb1579 - isolated from *A. americanum* or Lone Star tick from Uvalde County in Texas (24); Bb532 - isolated from a pool of 5 cat fleas *Ctenocephalides felis* from Fort Bend County in Texas (24).

Cats

Twenty uninfected normal healthy cats (obtained from the College of Veterinary Medicine, Auburn University, Auburn, Alabama) were divided into four groups each containing 5 cats. Group 1 cats were used as non-infected control cats. Groups 2,3 and 4 were used as test cats. Groups 2,3, and 4 each was injected with 10^6 live *Borrelia burgdorferi* of a different isolate intradermally in a single site in the sacral region. Ten days post injection each group of cats was split into two subgroups. Uninfected cat fleas or Lone Star ticks were placed onto either of these subgroups for a period of 7 days. At this time the ticks and cat fleas were removed and examined by IFA for the presence of *Borrelia burgdorferi*. There after the cats were bled every other week, and monthly one cat per group was sacrificed, post mortem and histopathological studies were carried out on all tissues and organs.

Ticks and Fleas

The *Amblyomma americanum* or Lone Star tick and the cat fleas *Ctenocephalides felis* were obtained from the Department of Entomology, Texas A&M University, College Station, Texas.

Blood Smears Staining

Hemacolor® stain (EM Diagnostic Systems, Inc., New Jersey, USA) was used. It is a stain set for rapid manual staining of blood smears with colour pattern similar to that of Wright Stain and Wright-Giemsa stains. The blood cells in the blood smear stain in the following manner: The erythrocytes cytoplasm stains pink; polymorphonuclear neutrophils stain with a purple nucleus, purplish-grey cytoplasm with lilac-violet pink granules; eosinophils stain with a purple nucleus, purplish-grey cytoplasm with red/orange granules; basophils stain with a dark-blue nucleus and dark-purple cytoplasm; monocytes stain with violet nucleus and blue-grey cytoplasm; lymphocytes stains with dark-blue nucleus and blue-grey cytoplasm.

Immunofluorescent Antibody (IFA)

All arthropods were evaluated for spirochaetal infection as follows: arthropods were washed by immersion in 70% ethanol for 1 min, 30% ethanol for 1 min, and 3% hydrogen peroxide for 30 seconds, then transferred to and rinsed twice in sterile PBS. The midgut and all other internal organs from each tick were compressed on a glass slide and identification of spirochaetes was assessed by IFA for reactivity to monoclonal antibodies H9724, H5332, H68, and H4825 (kindly donated by Dr. Alan Barbour, University of Texas Health Science Center, San Antonio, TX) as described else where (43). Specimen slides were prepared for IFA examination by incubation for 30 min at 34°C, washed with PBS, then incubated with fluorescein isothiocyanate conjugated goat anti-mouse Ig (Cappel, Organon Teknika Corp., West Chester PA). For positive controls, slides prepared with the B31 strain of *Borrelia burgdorferi* were included in each run of specimens.

RESULTS

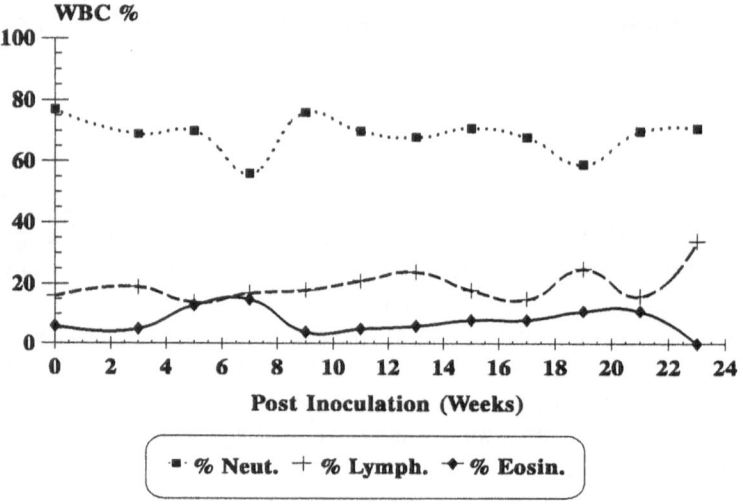

Figure 1. The differential WBC in control cat. There are no marked changes in the neutrophils, lymphocytes, and eosinophils counts.

In 13 out of 15 infected cats the white blood corpuscles differential count showed cycles of reduction in the neutrophil count accompanied by an increase in the lymphocyte and eosinophil counts compared to the control (Figures 1-4). During the course of infection of these cats we noted the appearance of an 'atypical cell' in the blood films. Following staining with Hemacolor® this 'atypical cell' had the following characteristics; 6-7 μ in diameter, a single compact round dark blue nucleus 1.5-3 μ in diameter, blue-grey cytoplasm (Figure 5).

In 8 cases of 15 test cats the cerebrum, cerebellum and thalamus- showed mild ellipsoid formation, mild to moderate spongiosis (Figure 6), minimal perivascular reaction (Figure 7) and foci of perivascular meningeal reaction and meningeal periarteriolar cuffing.

The liver in all test cats showd moderate to severe periportal fatty degeneration, lymphocyte infiltration (Figure 8), mild reticuloendothelial hyperplasia, minimal bile duct proliferation, nonsuppurative pericholangitis, and vacuolar degeneration with bile duct proliferation.

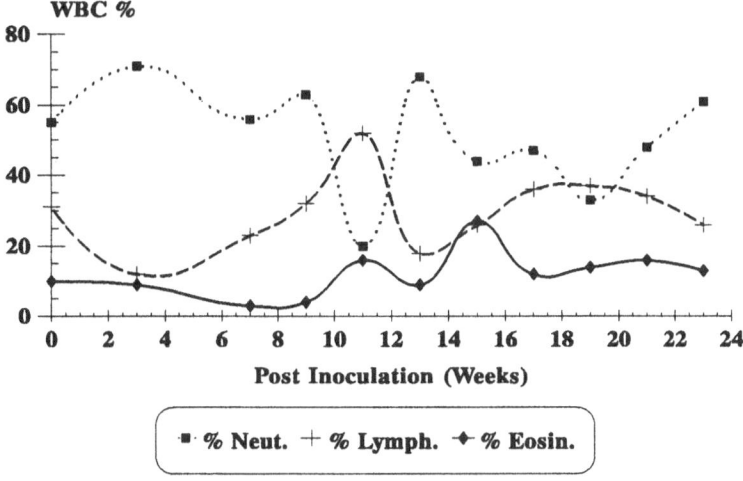

The cat exhibited lameness

Figure 2. The differential WBC in the cat infected with Bb31. There is drop in the neutrophils accompanied by rise in the lymphocytes and eosinophils counts at weeks 11,15, and 19.

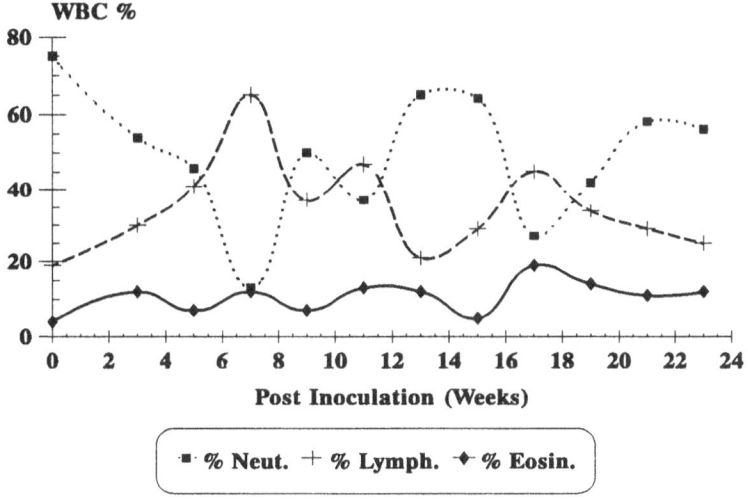

Figure 3. The differential WBC in the cat infected with Bb532. There is drop in the neutrophils accompanied by rise in the lymphocytes and eosinophils counts at weeks 7, 11, 13, and 17.

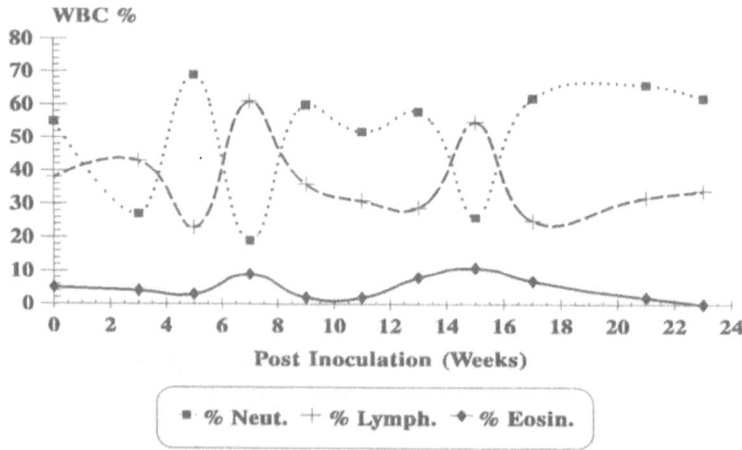

The cat exhibited lameness

Figure 4. The differential WBC in the cat infected with Bb1579. There is drop in the neutrophils accompanied by rise in the lymphocytes and eosinophils counts at weeks 7 and19.

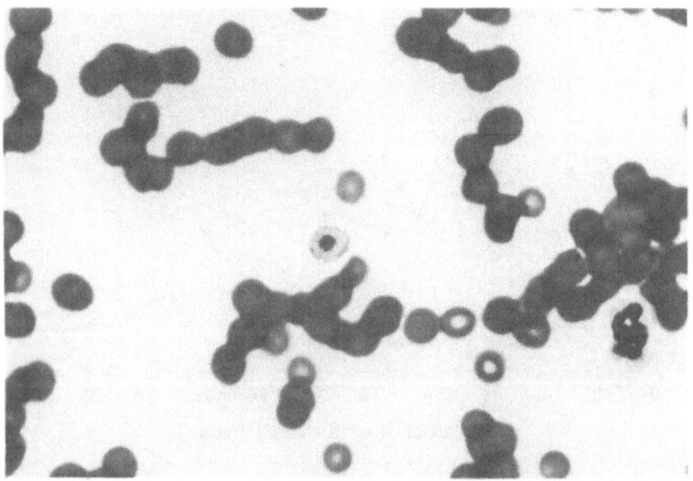

Figure 5. "Atypical cell": compare the cell size to that of the RBS and notice the compact nucleus (Hemacolor® stain x1000).

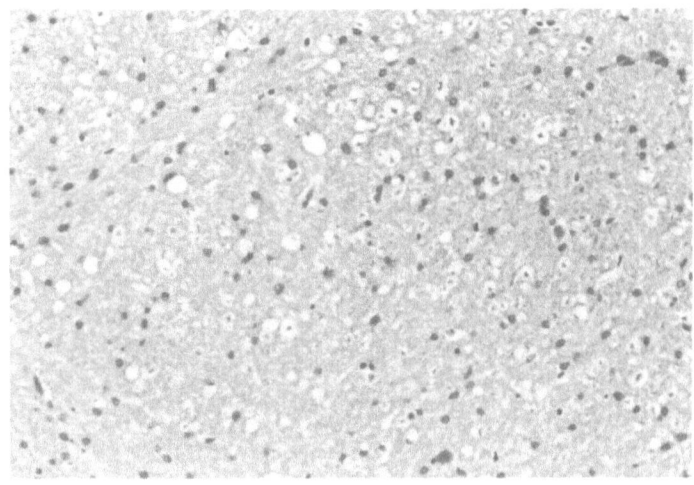

Figure 6. Spongiosis of the cerebellum in an infected cat with Bb1579, notice the ellipsoid formation in the brain tissue.

The Stifle joint capsule showed occasional mild synovial cell hyperplasia with multiple aggregates of lymphocytes.

In the kidneys the following findings were observed: foci of interstitial lymphocytic nephritis; peri-glomerulitis; severe segmental renal degeneration; interstitial nephritis with chronic ischaemic nephropathy; mild multifocal acute pyelitis. The adrenal glands showed occasional zona fasciculate microvesicular swelling.

In all the test cats the spleen showed mild lymphoid hyperplasia, mild plasmacytosis lymphoid hyperplasia, and occasional megakaryocytes.

Regional lymph nodes such as axillary-, retropharyngeal-, pharyngeal and submandibular lymph nodes showed mild lymphoid hyperplasia and plasmacytosis. The mesenteric lymph nodes showed occasional moderate to severe lymphoid hyperplasia.

Test cats lungs exhibited the following: mild alveolar emphysema, mild multifocal interstitial pneumonitis (Figure 9), plasmacytosis, and lymphoid hyperplasia.

DISCUSSION

Earlier studies on cats have shown that cats seroconvert and develop antibodies against *Borrelia burgdorferi* (40,41,44). In some instances infected cats show a history of fever, fatigue, anorexia, and lameness (40) whilst in another study no clinical signs nor gross or histological abnormalities were found (41). In one study the author was successful in isolating *Borrelia burgdorferi* from the blood of a cat following ocular inoculation (41).

The 'atypical cells', found in our study differ from the enlarged whilst cells (occasionally similar to Reed-Sternberg cells of Hodgkin's disease) or the atypical enlarged plasmacytoid whilst cells (sometimes binucleate) reported for certain cases of human Lyme Borreliosis (45). The size and colour pattern of these cells suggest that they may be lymphocytes. To the best of our knowledge these 'atypical cells' have not been reported before for any animal species infected with Bb. The presence of this 'atypical cell' in other animal species infected with Bb is currently under study.

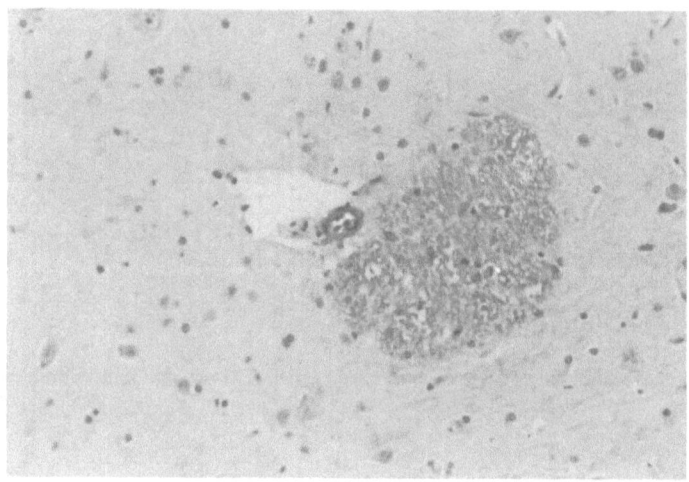

Figure 7. Perivascular round cell infiltration and spongiosis in the cerebellum of an infected cat with Bb532.

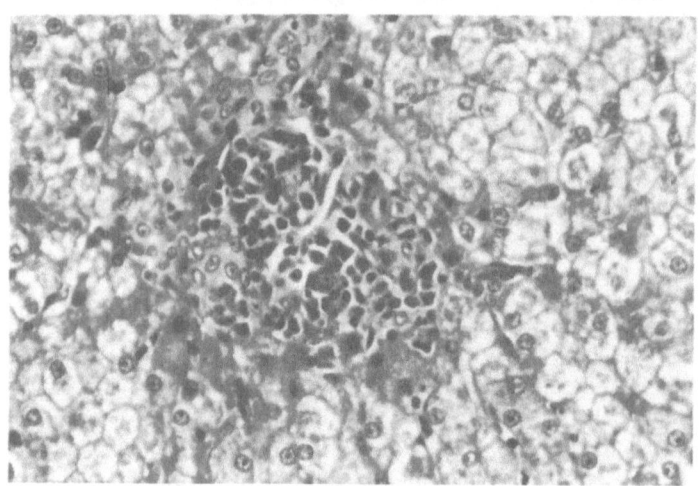

Figure 8. Round cell infiltration and fatty degeneration in the liver of an infected cat with Bb31.

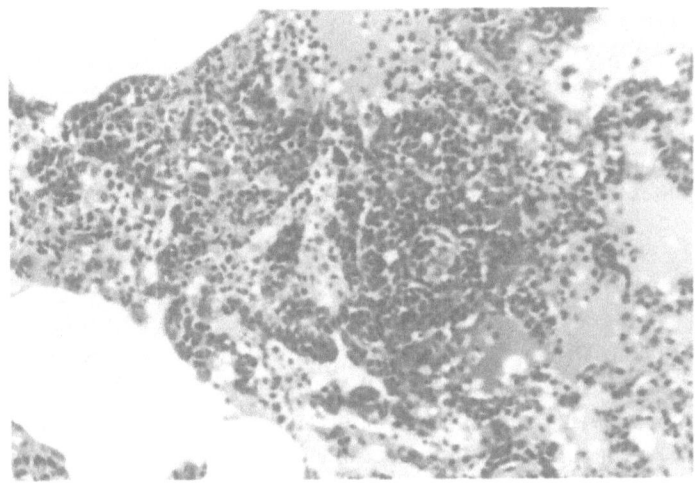

Figure 9. Round cell infiltration and pneumonitis in the lung of an infected cat with Bb532.

Our findings that CNS changes in the cat in the form of perivascular lymphocytic infiltration and mild spongiosis is comparable to findings in humans (46,47,48,49).

The lymphoid hyperplasia, plasmacytosis, and immunoblasts found in most of the regional lymph nodes including and not restricted to cervical-, axillary-, inguinal-, submandibular-, paraaortic-, pharyngeal-, and retropharyngeal lymph nodes parallel findings observed in human Lyme borreliosis (46,50).

The moderate to severe cytoplasmic swelling and fatty degeneration of the hepatocytes and pericholangitis accompanied with lymphocytic and plasma cell infiltration was similar to findings in human Lyme borreliosis (51).

The mild splenic hyperplasia, occasional megakaryocytes, lymphoid hyperplasia, and plasmacytosis found in the test cats reflect changes found in human Lyme borreliosis (52).

The synovial membrane hypertrophy with multiple aggregates of lymphocytes and plasma cells coupled with other signs are suggestive of Lyme arthritis.

To our knowledge the interstitial lymphocytic infiltration, interstitial nephritis, periglomerulitis, severe renal degeneration and interstitial nephritis, nephropathy, and acute pyelitis found in the kidneys of some infected cats has not yet been reported as a distinct histopathological profile in human Lyme borreliosis.

The mild alveolar emphysema, mild multifocal interstitial pneumonitis, plasmacytosis, and lymphoid hyperplasia observed in the lungs of our infected cats is supportive of spirochaetaemia as previously reported for case of human and feline Lyme borreliosis (41).

This plethora of histopathological changes in the cats which may seem not to conform with other previous studies (40,41) may be explained by the fact that in our work we used a high dose of *Borrelia burgdorferi* for the inoculation of the cats and more than one isolate of *Borrelia* (not restricted to any particular isolate which may or may not be pathogenic to the experimental cats) was used in our study. Furthermore the time study on our infected cats extended over a considerable period covering some six months.

In conclusion our study demonstrated that cats were susceptible to infection with different isolates of *Borrelia burgdorferi* with minimal clinical signs but the histopathological changes conformed with that reported in human Lyme borreliosis and cats may act as reservoirs for *Borrelia burgdorferi*. These findings warrant further studies concerning the role of cats and their ectoparasites in the epidemiology of Lyme disease.

ACKNOWLEDGEMENTS

The authors gratefully acknowledge Ms. Sarah Christo, Department of Veterinary Anatomy & Public Health, Texas A&M University, for expert histopathological slide preparation. Special thanks to Dr. Pete Teel, Department of Entomology, Texas A&M University, for supplying the ticks used in this study. Also we wish to thank Ciba-Geigy Co., Paravax and the Lyme Disease Foundation for their enthusiasm and financial support in the pursuit of this project.

REFERENCES

1. Steere AC. Lyme Disease. *N Engl J Med* 1989;**321**:586-596.

2. Steere AC, Grodzicki RL, Kornblatt AN. The spirochetal etiology of Lyme disease. *N Engl J Med* 1983;**308**:733

3. Schmid GP. The global distribution of Lyme disease. *Rev Infect Dis* 1985;**7**:41-50.

4. Dekonenko EJ, Steere AC, Berardi VP. Lyme borreliosis in the Soviet Union: a cooperative US-USSR report. *J Infect Dis* 1988;**158**:748-753

5. Ai CX, Wen YX, Zhang YG. Clinical manifestations and epidemiological characteristics of Lyme disease in Hailin county, Heilongjiang Province, China. *Ann NY Acad Sci* 1988;**539**:302-313.

6. Isogai E, Isogai H, Sato N. Antibodies to *Borrelia burgdorferi* in dogs in Hokkaido. *Microbiol Immunol* 1990;**34**:1005-1012

7. Fivaz BH and Petney TN. Lyme disease--a new disease in southern Africa? *J S Afr Vet Assoc* 1989;**60**:155-158.

8. Haberberger RL Jr., Constantine NT, Shwan TG and Woody JN. Lyme disease agent in Egypt. *Trans R Soc Trop Med* 1989;**83**:556.

9. Schmid GP. Epidemiology and clinical similarities of human spirochetal diseases. *Rev Infect Dis* 1989;**Supp 6**:S1460-S1469.

10. Barbour AG and Hayes SF. Biology of *Borrelia* species. *Microbiol Rev* 1986;**50**:381-400.

11. Burgdorfer W. Discovery of the Lyme disease spirochete and its relation to tick vectors. *Yale J Biol Med* 1984;**57**:515-520.

12. Duray PH, Kusnitz A and Rayan J. Demonstration of the Lyme disease spirochete by a modified Dieterle stain methods. *Lab Med* 1985;16:685-687.

13. de Koning J, Bosma RB and Hoogkamp-Korstanje AA. Demonstration of spirochetes in patients with Lyme disease with a modified silver stain. *J Med Microbiol* 1987;23:261-267.

14. Swisher BL. Modified Steiner procedure for microwave staining of spirochaetes and nonfilamentous bacteria. *J Histotech* 1987;**10**:241-243.

15. Burgdorfer W. Vector/host relationships of the Lyme disease spirochete *Borrelia burgdorferi*. *Rheum Dis Clin North Am* 1989;**15**:748-753.

16. Anderson JF. Mammalian and avian reservoirs for *Borrelia burgdorferi*. *Ann NY Acad Sci* 1988;**539**:180-191.

17. Bosler EM. Tick Vectors and Hosts. In: Coyle PK, ed. *Lyme Disease*. St. Louis: Mosby Year Book, 1992:18-26.

18. Magnarelli LA and Anderson JF: Ticks and Biting Insects Infected with The Etiologic agent of Lyme disease, *Borrelia burgdorferi*. *J Clin Microb* 1988;**26**:1482-1486.

19. Schulze TL, Bowen GS, Bosler EM. *Amblyomma americanum*: A potential vector of Lyme disease in New Jersey. *Science* 1984;**224**:601-603.

20. Piesman J and Stinsky RJ. Ability of *Ixodes pacificus*, *Dermacentor variabilis* and *Amblyomma americanum* (Acri: Ixodidae) to acquire, maintain and transmit Lyme disease spirochetes, *Borrelia burgdorferi*. *J Med Entomol* 1988;**25**:336.

21. Rousselle C, Floret D, Cochat P, Reignier F and Wright C. Encéphalite aiguë à *Borrelia burgdorferi* (maladie de Lyme) chez un enfant algérien. *Pediatrie* 1989;**44**:265-269.

22. Kawabata M. Lyme disease in Japan and its possible incriminated tick vector, *Ixodes persulcatus*. *J Infect Dis* 1987;**156**:854.

23. Post JE. Lyme Disease in Large Animals. *NJ Med* 1990;**87**:575-577.

24. Teltow GJ, Fournier PV and Rawlings JA. Isolation of *Borrelia burgdorferi* from arthropods collected In Texas. *Am J Trop Med Hyg* 1991;**44**:469-474.

25. Cohen D, Bosler EM, Bernard W, Meir D 2d, Eisner R and Schulze TL. Epidemiologic studies of Lyme disease in horses and their public health significance. *Ann NY Acad Sci* 1988;**539**:244-257.

26. Weisbrod AR and Johnson RC. Lyme disease and migrating birds in Saint Croix River Valley. *Appl Environ Microbiol* 1989;**55**:1921-1924.

27. Burgess EC, Gendron-Fitzpatrick A and Wright WO. Arthritis and systemic disease caused by *Borrelia burgdorferi* infection in a cow. *JAVMA* 1987;**191**:1468-1470.

28. Burgess EC and Mattison M. Encephalitis associated with *Borrelia burgdorferi* infection in a horse. *JAVMA* 1987;**191**:1457-1458.

29. Burgess EC. *Borrelia burgdorferi* infection in Wisconsin horses and cows. *Ann NY Acad Sci* 1988;**539**:235-243.

30. Bosler EM, Cohen DP, Schulze TL. Host responses to *Borrelia burgdorferi* in dogs and horses. *Ann NY Acad Sci* 1988;**539**:244-257.

31. Post JE, Shaw EE and Wright SD. Suspected borreliosis in cattle. *Ann NY Acad Sci* 1988;**539**:488.

32. Magnarelli LA, Anderson JE and Chappell WA. Geographic distribution of humans raccoons, and white-footed mice with antibodies to Lyme disease spirochetes in Connecticut. *Yale J Biol Med* 1984;**57**:619-626.

33. Burgdorfer W. The New Zealand White rabbit: An experimental host for infecting ticks with Lyme disease spirochetes. *Yale J Biol Med* 1984;**57**:609-612.

34. Kornblatt AN, Steere AC and Brownstein DG. Experimental Lyme disease in rabbits: Spirochetes found in Erythema Migrans and blood. *Infect Immun* 1984;**46**:220-223.

35. Barthold SW, Moody KD, Terwilliger GA. An animal model for Lyme arthritis. *Ann NY Acad Sci* 1988;**539**:264-273.

36. Barthold SW, Moody KD, Terwilliger GA. Experimental Lyme arthritis in rats infected with *Borrelia burgdorferi*. *J Infect Dis* 1988;**157**:842-846.

37. Schmitz JL, Schell RF and Hejka A. Induction of Lyme arthritis in LSH hamsters. *Infect Immun* 1988;**56**:2336-2342.

38. Hejka A, Schmitz JL, England DM. Histopathology of Lyme arthritis in LSH hamsters. *Am J Path* 1989;**134**:1113-1123.

39. Lissman BA. Lyme disease in small animals. *NJ Med* 1990;**87**:573-574.

40. Magnarelli LA, Anderson JF, Levine HR and Levy SA. Tick parasitism and antibodies to *Borrelia burgdorferi* in cats. *JAVMA* 1990;**197**:63-66.

41. Burgess EC. Experimentally induced infection of cats with *Borrelia burgdorferi*. *Am J Vet Res* 1992;**53**:1507-1511.

42. Burgdorfer W., Barbour AG, Hayes SF, Benach JL, Grunwald E and Davis JP. Lyme disease - a tick-borne spirochetosis? *Science* 1982;**216**:1317-1319.

43. Barbour AG, Tessier SL and Todd WJ. Lyme disease spirochetes and ixodid tick spirochetes share a common surface antigenic determinant defined by monoclonal antibody. *Infect Immun* 1983;**41**:795-804.

44. Rawlings JA. Lyme disease in Texas. *Zentralbl Bakteriol Mikrobiol Hyg [A]* 1986;**263**:483-487.

45. Duray PH. Target organs of *Borrelia burgdorferi* infections: Functional responses and histology. In: Scutzer SE, ed. *Lyme disease molecular and immunologic approaches*. New York, Cold Spring Harbor Laboratory Press, 1992:11-30.

46. Duray PH and Steere AC. Clinical pathologic correlation of Lyme disease by stage. *Ann NY Acad Sci* 1988;**539**:65-79.

47. Duray PH. Histopathology of clinical phases of human Lyme disease. *Rheum Dis Clin North Am* 1989;**15**:691-710.

48. Meurers B, Kohlhepp W, Gold R, Rohrbach E, Mertens HG. Histopathological findings in the central and peripheral nervous systems in neuroborreliosis: A report of three cases. *J Neurol* 1990;**237**:113-116.

49. Pachner AR, Duray PH, and Steere AC. Central nervous system manifestations of Lyme disease. *Arch Neurol* 1989;**46**:790-795.

50. Kirsch M, Ruben RL, Steere AC, Duray PH, Norden CW, and Winkelstein A. Fatal adult respiratory distress syndrome in a patient with Lyme disease. *JAMA* 1988;**259**:2737-2739.

51. Goellner MH, Agger WA, and Duray PH. Hepatitis due to recurrent Lyme disease. *Ann Intern Med* 1988;**108**:707.

52. Cimmino MA, Azzolini A, Tobia F, and Pesce CM: Spirochetes in the spleen of a patient with chronic Lyme disease. *Am J Clin Pathol* 1988;**91**:95.

BORRELIA BURGDORFERI INFECTION IN MICE: ASPECTS OF INFLAMMATION AND IMMUNE RESPONSES

Naveed Honarvar[1], Elena Böggemeyer[1], Chris Galanos,[1] Manuel Modolell[1], Dietmar Vestweber[2], Reinhard Wallich[3], Michael D. Kramer[4], Ulrich E. Schaible[1], and Markus M. Simon[1]

Max-Planck-Institut für Immunbiologie Freiburg, [2]Hans-Spemann-Labor der Max-Planck-Gesellschaft, Freiburg, [3] Abteilung Angewandte Immunologie, Deutsches Krebsforschungszentrum, Heidelberg, and [4]Institut für Immunologie und Serologie der Universität Heidelberg, Heidelberg, Germany

INTRODUCTION

Lyme disease is caused by the spirochete Borrelia burgdorferi (B.burgdorferi) and to date the most common vector borne infectious disease of the temperate climate[1]. The multifacetal nature of the illness which affects skin, joints, nervous system and heart as well as the inability of many patients to control infection, in spite of their specific immune responses, has been a considerable challenge for clinicians and scientists[1]. Two aspects of B.burgdorferi infections, i.e., the pathogenesis of Lyme disease with its chronic course of tissue destruction and the role of the immune response in protection, have attracted attention in the past.

Recent studies in the mouse model[2-5] have shown that experimental as well as natural inoculation of immunodeficient recipients lead to the development of chronic arthritis, myositis and carditis with concomitant persistence of spirochetes within the affected tissues and infiltration predominantly of macrophages, into the inflamed tissue[6]. The fact that similar inoculations of immunocompetent mice did not cause disease in many cases indicated that specific cellular and humoral immune responses to various antigens observed in these recipients are able to control the infection[3]. Adoptive transfer experiments revealed that immune sera and in particular antibodies to the outer surface lipoproteins (Osp), OspA and OspB, were able to confer protection against spirochetemia and the development of disease whereas B.burgdorferi-specific T cells alone had no effect[3,7,8].

In spite of the fact that recipients of all immunocompetent inbred strains of mice tested so far are able to develop protective antibody responses, irrespective of their genotype, some strains of mice, in particular those of the H-2^k-haplotype develop disease (susceptible), whereas those of the H-2^d-haplotype do not (resistant)[3,9]. The association of H-2^k gene products with disease suggests a critical role of T cells in the pathogenesis of arthritis and carditis in immunocompetent mice. Most notably, histopathological studies in experimentally inoculated AKR-N mice (H-2^k) revealed that the cellular infiltrations in the joints and heart analyzed up to 150 days post inoculation mainly consisted of Mac-1 positive cells (macrophages and/or granulocytes) and if at all only few T- and B cells[3]. These findings indicate that arthritis and carditis is initiated independently of T and/or B cells and that the first steps leading to inflammation of organs are similar both in the absence or presence of the immune system. However T cells and/or B cells may contribute to the further development of infection. Recent studies have provided evidence for B-cell mitogen activity of Borrelia burgdorferi[10-12] which may have an impact on both inflammatory and immune responses.

In the following we will discuss recent results which may have some relevance for our understanding of the B.burgdorferi-mediated process(es) leading to inflammation as well as of the regulation of specific immune responses by nonspecific spirochetal signals.

RESULTS AND DISCUSSION

It is hypothesized that during infection the spirochetes disseminate from the site of the tick bite, most probably via the blood, and subsequently colonize various tissues by binding to and penetrating vessel walls[13]. It is further reasoned that B.burgdorferi organisms specifically bind to and activate endothelial cells (EC) during this process to express adhesion molecules and cytokines which promote chemotaxis and attachment of inflammatory cells and their subsequent extravasation, together with spirochetes, into the underlying tissue. In fact, previous studies have shown that a) B.burgdorferi can adhere to cultured EC in vitro and penetrate cell monolayers by passing through intercellular tight junctions[14] as well as through the host cell cytoplasm[15,16] and b) that these processes can be inhibited by pretreatment of spirochetes with either proteases or monoclonal antibodies to OspA and OspB[16].

In an attempt to further elucidate the role of EC in B.burgdorferi infection, we have now applied a polyoma virus tranformed murine brain endothelioma cell line (bEnd3)[17] to study the capacity of spirochetes to induce adhesion molecules in vitro. We have tested the expression of intracellular as well as surface membrane bound structures - E-selectin, P-selectin, VCAM-1, ICAM-1 - on the intact monolayer of bEnd3 cells by immunocytochemistry as well as by ELISA. As summarized in Table 1, preincubation of bEnd3 cells with either viable or killed spirochetes or sonicated preparations thereof resulted in the induction of E-selectin, ICAM-1 and VCAM-1 as well as an increased expression of P-selectin when compared to untreated bEnd3 cells[19]. In contrast to LPS which induced a strong and homogeneous expression of all four molecules on all bEnd3 cells (Table 1) the intensity of expression of the respective structures induced by B.burgdorferi preparations was much less pronounced and only seen on a fraction of bEnd3 cells. Quantitative and kinetic analyses of the expression of E-selectin, P-selectin, ICAM-1 and VCAM-1 in response to spirochetal preparations using a cell surface ELISA yielded data corroborating those obtained with immunocytochemistry staining. It is important to note that maximal

Table 1. B.burgdorferi induced expression of adhesion molecules on endothelia cells in vitro and in vivo

Structure	in vitro (bEnd3)			in vivo (joint/heart)		
	Medium	B.burgdorferi	LPS	n.i.	B.burgdorferi	LPS[a]
E-selectin	-	++	+++	-	++	+++
P-selectin	++	++	+++	+	++	+++
ICAM-1	-	+	+	+	++[b]	+++
VCAM-1	-	+++	+++	+/-	+	++
vWF[c]	+++	+++	+++	++	++	++

For in vitro studies confluent cell layers of the endothelioma cell line bEnd3 were incubated with either live or inactivated B.burgdorferi (6×10^7 cells/ml), a sonicate thereof (25 µg/ml), LPS (1 µg/ml) or medium for 4 h (E-selectin, P-selectin) or 16 h (ICAM-1, VCAM-1), respectively. Prior to staining, bEnd3 cell layer was fixed with acetone, which allows the detection of both intracellular- and cell surface-associated structures. For in vivo studies immunodeficient C.B-17-SCID mice and immunocompetent AKR/N mice were either not infected, subcutaneously inoculated with 10^8 viable B.burgdorferi ZS7 spirochetes or intravenously with 40 µg LPS. Mice were killed at 4 h (LPS) or 17, 27 and 96 days (B.burgdorferi) after inoculation. Organs were frozen in liquid nitrogen, cryosections were prepared and further processed for immunohistology as described[6]. For staining the following antibody preparations were used: rat anti E-selectin (21KC10)[18], rabbit anti P-selectin immune serum[18], rat anti ICAM-1 monoclonal antibody (25ZC7; Hahne and Vestweber, unpublished), rat anti VCAM-1 monoclonal antibody (MK-2)[20] and rabbit anti-von Willebrandt-Factor (vWF) immune serum. As secondary antibodies biotin-labelled goat anti-rat IgG and IgM or donkey anti-rabbit antibodies (Jackson, Dianova, Hamburg, Germany) together with the streptavidin peroxidase complex were used. For comparison of the various staining intensities, the following scores were used: [(+++) very strong expression; (++) strong expression; (+) moderate expression; (+/-) weak expression; (-) no expression][19].
[a], shown for expression in heart tissue only
[b], in addition to endothelia also infiltrating inflammatory cells stained positive for ICAM-1
[c], vWF served as control for a constitutively expressed structure associated with endothelial cells

expression of the respective molecules upon stimulation with spirochetes (~50 h post incubation) was significantly delayed (25-40 h) compared to LPS (maximal expression 4-24 h post incubation; Table 1 and Figure 1, shown for E-selectin)[19]. The differential capacity of LPS and B.burgdorferi preparations to induce adhesion molecules on EC is most probably due to the fact that spirochetes are free of conventional LPS but contain chemically distinct molecules with common (mitogenic, 10-12) and different biological activities[10]. It is therefore to be expected that LPS⁺ bacteria and B.burgdorferi (LPS⁻) have differential capacities to activate EC and to induce inflammatory responses.

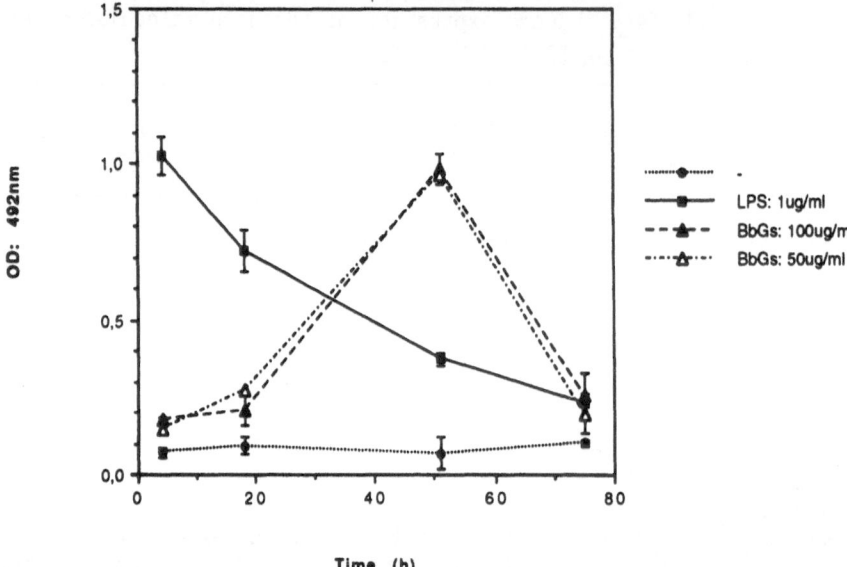

Figure 1. Induction of expression of E-selectin on the cell surface of bEnd3 cells (detected with monoclonal antibody [mAb] 21KC10[18]) 4 h, 18 h, 51 h and 70 h after incubation with B.burgdorferi sonicate (BbGS 50 µg/ml; 100 µg/ml, LPS (1 µg/ml) or with medium alone (negative) as analyzed by ELISA[18].

Preliminary experiments showing that the attachment of a B lymphoma cell line (L1-2)[21] which expresses VLA-4 the receptor for VCAM-1, to the bEnd3 cell layer was significantly increased after B.burgdorferi sensitization compared to the unsensitized control suggest that the induction of a variety of adhesion molecules on endothelia by spirochetes may contribute to the attraction of inflammatory cells such as neutrophils and monocytes/macrophages to vessel walls of infected organs. That similar processes may also occur in vivo is indicated by the finding that B.burgdorferi infection of SCID and AKR/N mice leads to inflammatory reactions in the joints and the heart which are characterized by pronounced neo-vascularization of the synovium and the expression of increased levels of E-selectin, P-selectin, ICAM-1 and VCAM-1 on the vessel walls as well as the underlying tissues in both organs (Table 1 and Figure 2, shown for ICAM-1). Whether these molecules are induced in vivo directly by spirochetes or indirectly by inflammatory molecules has yet to be determined.

Studies on cellular immune responses in Lyme disease patients[22] as well as in experimentally inoculated mice[3,23] have lead to variable and conflicting results. Previous experiments indicate the concomitant development of antigen specific- and non-specific responses of peripheral blood lymphocytes from Lyme disease patients[24] and lymphocyte populations from B.burgdorferi infected mice[10-12]. In the course of analyzing T cell responses to spirochetes we[12] and others[10,11] have found that lymphocytes from both infected and naïve mice exhibit similar proliferative responses to B.burgdorferi preparations in vitro. Further studies demonstrated that the

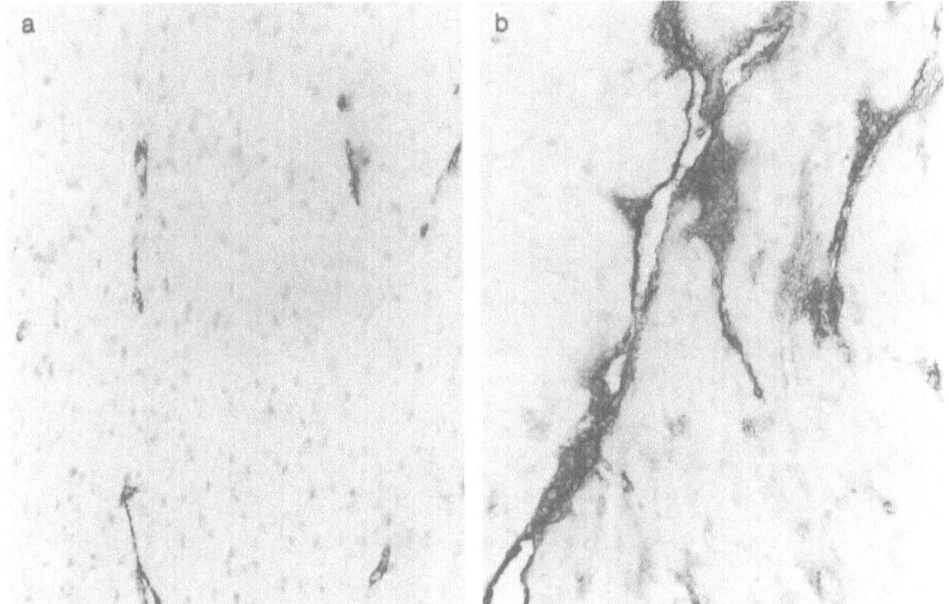

Figure 2. Expression of ICAM-1 in the tibiotarsal joint of a SCID mouse either non infected (a) or inoculated with 10^8 B.burgdoreri ZS7 (day 96 post inoculation; b). Organs were frozen in liquid nitrogen, cryosections were prepared and further processed for immunohistological staining using a rat monoclonal antibody to ICAM-1 (mAb 25ZC7, Hahne and Vestweber, unpublished) followed by a species-specific secondary antibody and the immune peroxidase complex. (Magnification 200x).

non-specific response is due to a spirochetal(s) structure with the potential to stimulate naïve B cells to proliferate (Figure 3 and 10-12) and to respond with IgM and IgG secretion at frequencies similar to those obtained with LPS[12]. The finding that B cells from both LPS susceptible (C57BL/10ScSn) and resistant strains (C57BL/10ScCr) of mice respond to the B.burgdorferi associated mitogens(s) suggests that the respective structure(s) is distinct from LPS[12]. It is now clear from our own and independent studies[11,12,25,26] that spirochetes express more than one structure with potent mitogenic activity for B cells; these include OspA, OspB as well as a 14 kD lipoprotein and a 3 kD glycolipid(s) contained within a phenol-chloroform-petroleum-ether fraction of B.burgdorferi.

Although the biological relevance of the B.burgdorferi associated mitogens is not known to date, it is possible that they contribute to the pathogenesis of the disease as well as the quality and quantity of T- and B cell responses induced during infection. The first assumption is supported by recent studies, demonstrating that the mitogen induces adhesion molecules and/or inflammatory cytokines in

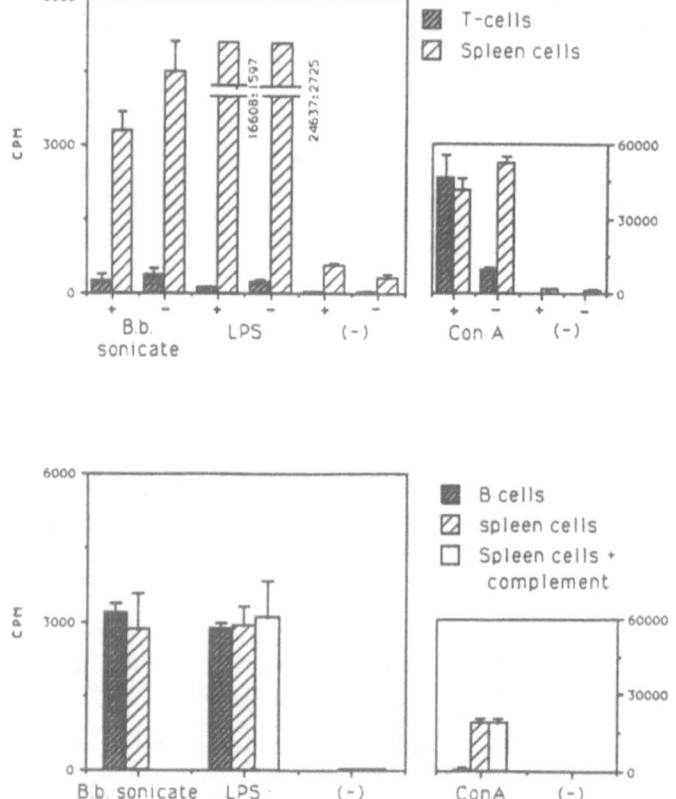

Figure 3. Induction of proliferation of selected and unselected BALB/c lymphocytes by B.burgdorferi. B cells were enriched via complement-mediated lysis of T cells pretreated with monoclonal antibodies to Thy1.2 and L3T4[12]. For T cell enrichment spleen cells were passed twice over nylon wool columns. The individual cell populations were cultured (2×10^5 cells/well) together with either a sonicate of B.burgdorferi (B.b. sonicate) (2.5 µg/ml), LPS (1 µg/ml), ConA (5 µg/ml) or alone (-). The T cells and spleen cells were cultured with or without irradiated feeder cells (splenic adherent cells or -). As a control spleen cells were treated with complement alone and cultured with LPS or ConA. [³H]-thymidine was added for the final 18 h of culture. B-cell proliferation was measured on day 3 and T-cell proliferation on day 4 of incubation.

EC[19,26] and macrophages in vitro[27]. The second assumption is indicated by the fact that antibodies to various B.burgdorferi antigens and with various isotypes, i.e. IgG2b, IgG3, can be induced in infected immunodeficient mice previously reconstituted with naïve or pre-sensitized B cells in a T cell independent manner[8].

ACKNOWLEDGEMENT

We would like to thank Ms G. Nerz, M. Prester, T. Tran and T. Stehle for expert technical help, L. Lay for photographs and R. Brugger and G. Prosch for typing the manuscript. This work was supported in part by a grant of BMFT (O1KI9104; M.M.S.).

REFERENCES

1. Steere, A.C., 1989, Lyme disease, N. Engl. 321:586.
2. Simon, M.M., Schaible, U.E., Wallich, R., and Kramer, M.D., 1991a, A mouse model for Borrelia burgdorferi infection: approach to a vaccine against Lyme disease, Immunol. Today 12:11.
3. Schaible, U.E., Wallich, R., Kramer, M.D., Museteanu, C., Rittig, M., Moter, S., and Simon, M.M. Role of the immune response in Lyme disease: lessons from the mouse model, in: Lyme Disease: Molecular and Immunologic Approaches, Cold Spring Harbor Laboratory Press (1992).
4. Fikrig, E., Barthold, S.W., Sears, J.E., Telford III, S.R., Spielman, A., Kantor, F.S., and Flavell, R.A. A recombinant vaccine for Lyme disease, in: Lyme Disease: Molecular and Immunologic Approaches, Cold Spring Harbor Laboratory Press (1992).
5. Barthold, S.W., de Souza, M., Fikrig, E., and Persing, D.H. Lyme Borreliosis in the laboratory mouse, Cold Spring Harbor Laboratory Press (1992).
6. Schaible, U.E., Gay, S., Museteanu, C., Kramer, M.D., Zimmer, G., Eichmann, K., Museteanu, U., and Simon, M.M., 1990b, Pathogenesis of Lyme Borreliosis in the severe combined immunodeficiency (SCID) mice, Am. J. Pathol. 137:811.
7. Schaible, U.E., Kramer, M.D., Eichmann, K., Modolell, M., Museteanu, C., and Simon, M.M., 1990a, Monoclonal antibodies specific for the outer surface protein A (OspA) of Borrelia burgdorferi prevent Lyme Borreliosis in severe combined immunodeficiency (SCID) mice, Proc. Natl. Acad. Sci. 87:3768.
8. Schaible, U.E., Wallich, R., Kramer, M.D., Nerz, G., Museteanu, C., and Simon, M.M., 1993, Protection against Borrelia burgdorferi infection in SCID mice is confered by presensitized spleen- and partially B cells but not by T cells alone, submitted for publication.
9. Schaible, U.E., Kramer, M.D., Wallich, R., Tran, T., and Simon, M.M., 1991a, Experimental Borrelia burgdorferi infection in inbred mouse strains: Antibody response and association of H-2 genes with resistance and susceptibility to development of arthritis, Eur. J. Immunol. 21:2397.
10. Schoenfeld, R., Araneo, B., Ma, Y., Yang, L., and Weiss, J.J., 1992, Demonstration of a B-lymphocyte mitogen produced by the Lyme disease pathogen, Borrelia burgdorferi, Infect. Immun. 60:455.
11. de Souza, M., Fikrig, E., Smith, A.L., Flavell, R.A., and Barthold, S.W., 1992, Nonspecific proliferative responses of murine lymphocytes to Borrelia burgdorferi antigens, J. Infect. Dis. 165:471.

12. Honarvar, N., Schaible, U.E., Galanos, C., Wallich, R., and Simon, M.M., 1993, Quantitative analysis of naïve murine B cells responding to a Borrelia burgdorferi-associated mitogen(s), submitted for publication.
13. Simon, M.M., Kramer, M.D., Wallich, R., and Schaible, U.E. Lyme arthritis: Pathogenic principles emerging from studies in man and mouse, in: The Immunology of the Connective Tissue Diseases, G. Panayi, ed., Kluwer Acad. Publ., Lancaster, UK (1993).
14. Szezepanski, A., Furie, M.B., Benach, J.L., Lane, B.P., and Fleit, H.B., 1990, Interaction between Borrelia burgdorferi and endothelium in vitro, J. Clin. Invest. 85:1637.
15. Ying, M., Sturrock, A., and Weis, J.J., 1991, Intracellular localization of Borrelia burgdorferi within human endothelial cells, Infect. Immun. 59:671.
16. Comstock, L.E., Fikrig, E., Shoberg, R.J., Flavell, R.A., and Thomas, D.D., 1993, A monoclonal antibody to OspA inhibits association of Borrelia burgdorferi with human endothelial cells. Infec. and Immun. 61:423.
17. Montesano, R., Pepper, M.S., Möhle-Steinlein, U., Risau, W., Wagner, E.F., and Orci, L., 1990, Increased proteolytic activity is responsible for the aberrant morphogenetic behavior of endothelial cells expressing the middle T oncogene. Cell 62:435.
18. Hahne, M., Jäger, U., Isenmann, S., Hallmann, R., and Vestweber, D., 1993, Five tumor necrosis factor-inducible cell adhesion mechanisms on the surface of mouse endothelioma cells mediate the binding of leukocytes, J. Cell Biol. 121:655.
19. Böggemeyer, E. In vitro Expression von Adhäsionsmolekülen auf Endothelzellen der Maus nach Inkubation mit dem Erreger der Lyme Borreliose: Borrelia burgdorferi, Diploma thesis, Freiburg i.Br. (1993).
20. Miyake, K., Medina, K., Ishihara, K., Kimoto, M., Auerbach, R., and Kincade, P.W., 1991, A VCAM-like adhesion molecule on murine bone marrow stromal cells mediates binding of lymphocyte precursors in culture. J. Cell. Biol. 114:557.
21. Sher, B.T., Bargatze, R., Holzmann, B., Gallatin, W.M., Matthews, D., Wu, B., Picker, L., Butcher, E.C., and Weissman, I.L., 1988, Homing receptors and metastasis, Adv. Can. Res. 51:361.
22. Dattwyler, R.J.. Thomas, J.A., Benach, J.L., and Golightly, M.G., 1986, Cellular immune response in Lyme disease: the response to mitogens, live Borrelia burgdorferi, NK cell function and lymphocytte subsets, Zentralbl. Bakteriol. Mikrobiol. Hyg. Ser. A 263:151.
23. de Souza, M.S., Smith, A.L., Beck, D.S., Terwilliger, G.A., Fikrig, E., and Barthold, S.W., 1993, Long-term study of cell-mediated responses to Borrelia burgdorferi in the laboratory mouse, Infect. Immun. 61:1814.
24. Zoschke, D.D., Skemp, A.A., and Defosse, D.L., 1991, Lymphoproliferative responses to Borrelia burgdorferi in Lyme disease, Ann. Intern. Med. 114:285.
25. Katona, L.I., Beck, G., and Habicht, G.S., 1992, Purification and immunological characterization of a major low-molecular-weight lipoprotein from Borrelia burgdorferi. Infect. Immun. 60:4995.

26. Ma, Y., and Weis, J.J., 1993, *Borrelia burgdorferi* outer surface lipoproteins OspA and OspB possess B-cell mitogenic and cytokine-stimulatory properties, *Infect. Immun.* 61:3843.

27. Modolell, M., Schaible, U., Corraliza, I., Rittig, M., and Simon, M.M., 1992, The role of macrophages in experimental Lyme Borreliosis, *J. Immunopathol.* 60:23.

CHEMOTAXONOMY OF *BORRELIA*

M. Anne Livesley and Patricia A. Nuttall

NERC Institute of Virology and Environmental Microbiology
Mansfield Road
Oxford
OX1 3SR
U.K.

ABSTRACT

Using FAME profiles we have demonstrated the inter-species differentiation of certain members of the spirochaete family- *Leptospira*, *Serpulina* and *Borrelia* species. Now we propose that intra-species differentiation of *Borrelia* species is possible using this phenotypic method. The delineation of *B. burgdorferi* into *B. burgdorferi* Sensu Stricto, *B. garinii* and Group VS461 as proposed by 16SrRNA sequencing is maintained. Isolates of *B. hermsii* form a separate group and *B. turicatae* and *B. parkeri* cluster together but within one group which includes all the *Borrelia* in our study. Previous studies of *B. burgdorferi* have not included examples of the relapsing fever borreliae, *B. hermsii B. turicatae* and *B. parkeri*. This does not allow the comparison of the results of FAME analysis with such genetic studies at present.

The *B. garinii* group contains the reference isolate, 20047 and a number of Swiss isolates. Group VS461 contains mostly European isolates whilst *B. burgdorferi* Sensu Stricto contains the reference isolate, B31 and several other American isolates. We propose that this clustering has arisen through convergent evolution of the *Borrelia* leading to the production of similar fatty acids by genotypically dissimilar groups.

INTRODUCTION

Many methods have been used to compare different isolates of *Borrelia burgdorferi*. unique, apparently isolate specific cleavage patterns of restriction digested whole *B. burgdorferi* DNA have been obtained (Le Febvre *et al.*,1989). Plasmid profiles have been found to differ between isolates but also to change upon prolonged cultivation (Schwan *et al.*,1988). The use of polymerase chain reaction (PCR) has shown differences in the reactivity of different isolates of *B. burgdorferi* with chromosome -specific primers. Similarly variation has been demonstrated by DNA hybridization analysis of restriction patterns of the genes coding for rRNA (Postic *et al.*, 1990) and reactivity of *B. burgdorferi* with monoclonal antibodies (Wilske *et al.*, 1988). More recently phylogenetic analysis based on 16S rRNA sequences of North American and European isolates of *B. burgdorferi* has been performed (Marconi and Garon, 1992) . Further attempts to classify *B. burgdorferi* isolates have involved the use of multilocus enzyme electrophoresis (Boerlin *et al.*,1992), polymorphism of outer surface proteins (Peter and Bretz, 1992) and arbitrarily primed PCR (Welsh *et al.*, 1992). Most of the methods used to classify *B. burgdorferi* currently rely upon one or a few genes or their products. Three DNA groups (described as genospecies) have been delineated based on rRNA gene restriction and protein electrophoresis patterns and patterns of reactivity with murine monoclonal antibodies: *B. burgdorferi* sensu stricto, *B. garinii* sp. nov. and Group VS461 (Baranton *et al.*, 1992).

A comparatively recent method of bacterial classification is based on analysis of fatty acid methyl esters (FAME). Fatty acid synthesis involves the action of multiple genes and the derived products constitute a stable phenotype when the bacteria are grown in controlled culture conditions (Welch,1991). The use of fatty acid profiles as a key for classification is known as chemotaxonomy and has been used successfully with *Xanthomonas maltophilia, Klebsiella terrigena, Erwinia herbicola* and *Microbacterium lacticum* (Thompson *et al.*,1992), numerous gram negative bacteria (Stoakes *et al.*, 1991) and *Leptospira* species (Cacciopuoti *et al.*, 1991). Although borreliae were thought to scavenge fatty acids, FAME analysis has been employed recently to distinguish members of the spirochaete family including *Borrelia* species (Livesley *et al.*,1993).

METHODS

Culture of *Borrelia* strains

Borrelia species were grown in modified Barbour-Stoenner-Kelly medium (Barbour,1984). The medium contained bovine serum albumin fraction V (Boehringer Corporation Limited) and CMRL 1066 without sodium bicarbonate and glutamine (Gibco BRL). The media were sterilized using 0.22μm filters. Immediately prior to use, 10%(v/v) rabbit serum and 20ml of 7%(w/v) gelatin per 100ml were added. The media were dispensed into 10x100mm tubes where 6ml occupied approximately 75% of the volume available. The *Borrelia* were incubated at 34°C and harvested by centrifugation (1000g,15 min) when grown to a highly motile condition as determined by dark field microscopy. Fifty milligrams (wet weight) of spirochaetes were required for each fatty acid profile. The spirochaetes were washed three times in phosphate buffered saline (pH 7.4) and finally resuspended in 0.2ml phosphate buffered saline.

Fatty acid methyl ester extraction

Long chain fatty acids from bacteria require derivatization before gas liquid chromatography. FAME derivatives were prepared using the following procedure (Kloepper *et al.*, 1992): spirochaetes (50mg, collected as described above) were saponified with 1ml of 3.75M NaOH in 50% aqueous methanol by heating at 100°C for 30 min. Free fatty acids in the preparation were methylated with 2.0ml of 6.00M HCl in methanol, heated at 80°C for 10 min followed by rapid cooling to room temperature. FAMEs were extracted from the aqueous phase with a 1:1 mixture of hexane and methyl tert butyl ether and the acidified lower phase discarded. The extract was washed and neutralized with 3.0ml 0.3N NaOH, separation being aided by the addition of a few drops of saturated sodium chloride. The final FAME extract was the upper organic phase.

Gas liquid chromatography

FAMEs were analysed with a Hewlett Packard series II gas chromatography model 5890A and the information processed using a Hewlett Packard integrator. The information was analysed using the Microbial Identification Software (MIS)

Table 1 Sources and suppliers of the *Borrelia* isolates used in this study. The group number corresponds with that shown in Figure 1.

Borrelia isolate	Source	Country	Supplier	Group
B. burgdorferi ACA-1	ACA	Sweden	S. Cutler	1
B. burgdorferi 328	I. dammini	U.S.A	S. Cutler	1
B. burgdorferi HB19	Human blood	U.S.A	S. Cutler	1
B. burgdorferi HB4	Human blood	U.S.A	S. Cutler	1
B. burgdorferi B31 type strain	I. dammini	U.S.A	ATCC 35210	1
B. burgdorferi JD1	I. dammini	U.S.A	B. Johnson	1
B. turicatae	O. turicata	U.S.A	T. Schwan	2
B. parkeri	O. parkeri	U.S.A	T. Schwan	2
B. burgdorferi Ir210	I. ricinus	C.I.S.	E. Korenberg	3
B. burgdorferi 21305	I. dammini	U.S.A	S. Cutler	3
B. burgdorferi M32	I. ricinus	Netherlands	S. Carter	3
VS461	I. ricinus	Switzerland	D. Postic	3
B. burgdorferi M27	I. ricinus	Netherlands	S. Carter	3
B. hermsii CON-1	Human	U.S.A	T. Schwan	4
B. hermsii HS1	O. hermsii	U.S.A.	T. Schwan	4
B. burgdorferi NE550	I. ricinus	Switzerland	L. Gern	5
B. burgdorferi NE2	I. ricinus	Switzerland	L. Gern	5
B. burgdorferi NE4R4	I. ricinus	Switzerland	L. Gern	5
B. burgdorferi NE58P8	I. ricinus	Switzerland	L. Gern	5
B. burgdorferi NE58	I. ricinus	Switzerland	L. Gern	5
B. garinii 20047	I. ricinus	France	D. Postic	5

Table 2 .Summary table of the fatty acids present in the *Borrelia* used in this study.

Shorthand Designation	B. burgdorferi Sensu Stricto(1)	B. turicatae B. parkeri(2)	VS461 (3)	B. hermsii (4)	B. garinii (5)
12:0			X		X
14:0	X	X	X	X	X
15:0 iso					X
15:0 anteiso					X
15:0	X	X	X	X	X
16:1			X	X	X
16:0 iso				X	
16:1 cis 9	X	X	X	X	X
16:0	X	X	X	X	X
17:0 anteiso	X	X	X	X	
17:1		X		X	X
17:0	X	X	X	X	X
18:1 cis 9	X	X	X	X	X
18:0	X	X	X	X	X
20:4 cis 5,8,11,14		X	X	X	X

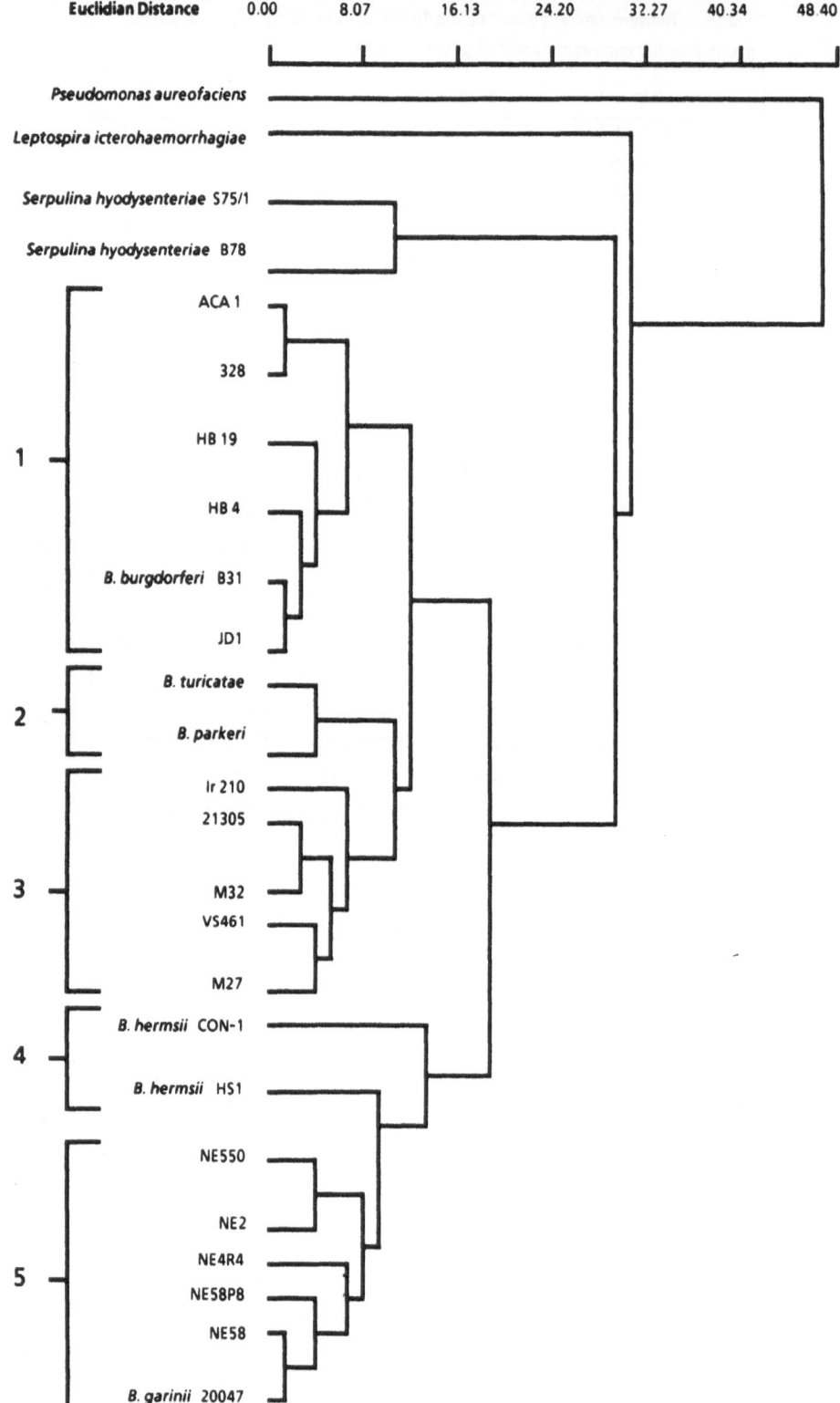

Figure 1

Dendrogram of the FAME profile relatedness of reference isolates of *Borrelia* used in this study. Isolates of *Leptospira interrogans* (serogroup *icterohaemorrhagiae*) and *Serpulina hyodysenteriae* were included as representative examples of other spirochaetes. *Pseudomonas aureofaciens* was included in the dendrogram as an "outgroup".

which calibrates the gas chromatograph with a commercial mixture of FAMEs at the beginning of each analysis run and after every tenth sample. A library was constructed based on three replicates per isolate. The MIS software contains a statistical facility for the generation of dendrograms which groups samples according to qualitative and quantitative relatedness of FAMEs. Similarity and relatedness are expressed as Euclidian distance values (Austin and Priest, 1986).

RESULTS

We have shown that the fatty acid methyl ester (FAME) profile of *Borrelia*, *Leptospira* and *Serpulina* species are stable and reproducible Using the MIS software we generated dendrograms to show the FAME profile relatedness of the spirochaetes, thus demonstrating inter-species differentiation of the members of the spirochaetes used in this study (Livesley *et al.*, 1993).

This dendrogram suggested that FAME analysis could have applications to the intra-species differentiation of *Borrelia* using the isolates shown in Table 1.

A large group of isolates from a number of geographical locations was then analysed. Five groups were visualized (Figure 1) The delineation of *B. burgdorferi* into *B. burgdorferi* Sensu Stricto, *B. garinii* and Group VS461 was maintained, but additional groups corresponding to *B. hermsii* isolates and *B. turicatae* and *B. parkeri* were apparent. The fatty acids present in each of the reference isolates of Groups 1-5 is shown in Table 2.

DISCUSSION

These results suggest that fatty acid methyl ester profiles provide a method for intra- species differentiation of *Borrelia* species. *B. burgdorferi* is delineated into three groups corresponding to *B. burgdorferi* Sensu Stricto, *B. garinii* and Group VS461.

The groupings of *B. burgdorferi* were approximately geographical, with the exceptions of ACA-1 and 21305. The group corresponding to *B. burgdorferi* Sensu Stricto contained the reference isolate B31 and several other American isolates. *B.garinii* contained the reference isolate 20047 (a French isolate) and a number of Swiss isolates. Furthermore studies of the outer membrane proteins of the Swiss isolates shown to be in Group 5 have indicated that they are most likely to be members of the *B. garinii* group (Peter and Bretz, 1992). Group VS461 contained Dutch , Russian and Swiss isolates.

Two other FAME clusters result from this analysis corresponding to isolates of *B. hermsii* and *B.turicatae* and *B. parkeri*. These clusterings may arise as a consequence of convergent evolution leading to the production of fatty acids similar to those of *B. burgdorferi*. The positions of *B. hermsii* isolates in Group 4 and *B. turicatae* and *B. parkeri* in Group 2 suggest that the distinction of *Borrelia* into separate species is a tenuous one as previously proposed on the basis of DNA-DNA hybridization studies (Hyde and Johnson, 1984).

ACKNOWLEDGMENTS

We thank the Health and Safety Executive for funding M.A.L.

REFERENCES

Austin, B. and Priest, F.G., 1986, "Modern Bacterial Taxonomy" Van Nostrand Reinhold, U.K.

Barbour, A.G. , 1984, Isolation and cultivation of lyme disease spirochaetes. *Yale J. Biol. Med.* 57:521-525.

Baranton, G., Postic, D., Saint Girons, I., Boerlin, P., Piffaretti, J.C., Assous, M. and Grimont, P.A.D., 1992, Delineation of *Borrelia burgdorferi* sensu stricto, *Borrelia garinii* sp. nov. and group VS461 associated with Lyme borreliosis. *Int. J. Syst. Bact.* 42:378-383.

Boerlin, P., Peter, O., Bretz, A.G., Postic, D., Baranton, G. and Piffaretti, J.C., 1992,Population genetic analysis of *Borrelia burgdorferi* by multilocus enzyme electrophoresis. *Infect. Immun.* 60:1677-1683.

Cacciopuoti, B., Ciceroni, L. and Attard Barbini, D., 1991, Fatty acid profiles- a chemotaxonomic key for classification of strains of the *Leptospiraceae*. *Int. J. Syst. Bact.* 41:295-300.

Hyde, F.W. and Johnson R.C., 1984, Genetic relationship of Lyme disease spirochaetes to *Borrelia, Treponema* and *Leptospira* spp. *J. Clin. Micro* 20:151-154.

Kloepper, J.W., Rodrigo, R., McInroy, J.A., and Young, R.W., 1992,Rhizosphere bacteria antagonistic to soybean cyst (*Heterodera glycines*) and root knot (*Meloidogyne incognita*) nematodes: Identification by fatty acid analysis and frequency of biological control activity. *Plant and Soil* 139:75-84.

LeFebvre, R.B., Perng, G.C. and Johnson, R.C., 1989, Characterization of *Borrelia burgdorferi* isolates by restriction endonuclease analysis and DNA hybridization. *J. Clin. Micro.* 27:636-639.

Livesley, M.A., Thompson, I.P., Bailey, M.J. and Nuttall,P.A., 1993,Comparison of the fatty acid profiles of *Borrelia, Serpulina* and *Leptospira* species. *J. Gen. Micro.* 139:889-895.

Marconi, R.T. and Garon, C.F., 1992, Phylogenetic analysis of the genus *Borrelia* a comparison of North American and European isolates of *Borrelia burgdorferi. J.Bact.*174:241-244.

Peter, O. and Bretz, A.G.,1992, Polymorphism of outer surface proteins of *Borrelia burgdorferi* as a tool for classification. *Zent. Bakt.* 277:28-33.

Postic, D., Edlinger, C., Richaud, C., Grimont, F., Dufresne, Y., Perolat, P., Baranton, G. and Grimont, P.A.D., 1990, Two genomic species in *Borrelia burgdorferi. Res. Micro.* 141:465-475.

Stoakes, L., Kelly, T., Schieven, B., harley, D., Ramos, M., Lannigan, R., Groves, D. and Hussain, Z., Gas liquid chromatographic analysis of cellular fatty acids for identification of gram negative anaerobic bacilli. *J. Clin. Micro.* 29:2636-2638.

Thompson, I.P., Bailey, M.J., Ellis, R.J. and Purdy, K.J., 1992, Subgrouping of bacterial populations by cellular fatty acid composition. *FEMS Microbiol. Ecol.* 102:75-84.

Welch, D., 1991, Applications of cellular fatty acid analysis. *Clin. Microbiol. Rev.* 4:422-438.

Welsh, J., Pretzman, C., Postic, D., Saint Girons, I., Baranton, G. and McLelland M.,1992, Genetic fingerprinting by arbitrarily primed polymerase chain reaction resolves *Borrelia burgdorferi* into three distinct phyletic groups. *Int. J. Syst. Bact.* 42:370-377.

Wilske, B., Preac-Mursic, V., Schierz, G., Kuhbeck, R., Barbour, A.G. and Kramer, A.G., 1988, Antigenic variability of *Borrelia burgdorferi. Ann. N.Y. Acad. Sci.* 539:126-143.

PHENOTYPIC AND GENOTYPIC ANALYSIS OF CHINESE

BORRELIA BURGDORFERI FROM VARIOUS SOURCES

Li Muqing, Wang Jianhui and Zhang Zhefu

Department of Spirochetosis
Institute of Epidemiology and Microbiology
PO BOX 5, Changping, Beijing 102206

INTRODUCTION

Lyme borreliosis, a multisystem disorder caused by the spirochete B. burgdorfer, is a worldwide tick-borne disease[1]. Many clinical cases of B. burgdorferi infections have been described in humans and animals. Borrelia burgdorferi is transmitted to humans and animals by the ticks Ixodes dammini and Ixodes pacificus in North American, by Ixodes ricinus in Europe, and by Ixodes persulcatus in Asian parts of Russia[2]. Although there are many similarities between the clinical manifestations of North American Lyme disease and the European disorders. Some of the signs, such as acrodermatitis chronic atophicans, were primarily in Europe. Others, such as arthritis, were diagnosed more frequently in American patients[3]. It has been speculated that genetic and phenotypic differences between European and American isolates may account for the multifaceted clinical appearance of B.burgdorferi infections in different geographic regions[3]. Different isolates have been distinguished on the basis of their reactivities with monoclonal antibodies and the apparent weights of their major surface proteins as well as their plasmid profiles, North American strains have shown less heterogeneity in these traits than have European isolates[4]. In general, it is apparent that many characters of B.burgdorferi appear to be relative to their geographic distributions.

In China, an epidemiological investigation had been conducted from 1987 to 1992. Lyme disease were also found to be a widespread infectious disease in China. Ixodes persulcatus were proved to be the primary vector in transmission of B.burgdorferi to human in North China, where its distribution correlates directly with that of cases of Lyme disease in human[5,6]. Which species of ticks is the principle vector in South China is unclear, H.concinna may play an important role in transmission cycle of Lyme disease[7]. Complex clinical symptoms were observed in China and it appeared some relativity with their different geographic distributions. More signs on skin were seen in North China and more arthritic, neurologic manifestations has been reported in South China[5,8]. In this study, we used several typing methods to analyzed 22 B.burgdorferi strains isolated from ticks and animals. Our goals were the following: (i) to understand the features of Chinese isolates; (ii) to reveal correlations between plasmid contents and protein components; (iii) to asses the diversity among B.burgdorferi isolates from various sources and to try to explain the multifaceted clinical appearance of B.burgdorferi infections in different geographic regions.

MATERIALS AND METHODS

Bacterial Strains and Cultivation

All strains were kept for more than five passages before inclusion in this study. Strain designations and sources are listed in Table 1. Bacteria were cultivated to late logarithmic growth in BSKII medium at 33°C[9] and harvested by centrifugation at 5,000 x g for 30 min. at room temperature.

DNA Preparation

The plasmid-enriched DNA were prepared by the method of Barbour[4]. Whole-cell DNA were isolated by the procedure of Thomas Adam et al.[3], the chromosome-enriched method: A harvest of 10^9 borreliae (10 ml of a stationary-phase culture) was suspended in 240ul of TES(50mM Tris pH8.0, 50mM EDTA, 15%sucrose) in a 1.5ml microcentrifuge tube. To the suspension was added 60ul of lysozyme solution (2mg/ml) and incubated on ice for 15 min, then added 300 ul of 1% sodium deoxycholate in TES and 7ul of dimethyl pyrocarbonate (DEP) and shake the tube gently at room temperature for 10 min. To the lysate was added 250ul of 7.5M ammonium acetate, and the tube was centrifuged at 11,000 x g for 10 min. at room temperature. The supernatant was transferred to another tube and continued to extract plasmid-enriched DNA. The pellet was resolved in 290ul TES buffer(0.1M Nacl, 10mM Tris pH8.0, 1mM EDTA) and 300ul of 1% sodium deoxycholate in TES buffer, 10ul of proteinase K (10mg/ml) were added and incubated for 1-2 hours at 37°C. The mixture was then extracted twice with phenol-chloroform and then with two cycles of chloroform. The nucleic acids were precipitated by addition of 600ul of isopropanol and placement of the tube on ice for 20min. The precipitate was recovered and rinsed with 75% ethanol by centrifugation at 11,000xg for 10min.at 4°C, dried and suspended it in 20ul of TE (10mM Tris, 1mM EDTA, pH8.0).

Low-Percentage Agarose Gel Electrophoresis

DNA samples were applied to 15-cm-long, 0.2% agarose gels (Agarose MP, Boehringer Mannheim Biochemica). The electrophoresis was run with o.5xTBE buffer at 5.5V/cm for 20 min. and 1.4V/cm for another 20 hours.

Restriction Endonuclease Digestion and Southern Blot

DNA was digested with HindIII as recommended by the manufacturer. Resulting DNA fragments were separated on 1.0% agarose gels, stained with EB, and transferred to nitrocellulose filters by the method of Southern. Hybridization and subsequent washing were done under stringent conditions[10].

Table 1. Borrelia burgdorferi strains and reactivities of Borrelia Proteins with monoclonal antibodies.

Strain	Source	Location	MAB		
			H5332	H6831	H9724
B31(1)	I.dammini	New York	+	+	+
IP21(2)	I.persulcatus	St.Petersburg	+	+	+
CS3	C.sinensis	Hunan	+	+	+
CS4	C.sinensis	hunan	+	+	+
SZ21	H.longiconis	Beijing	+	+	+
SZ22	H.longiconis	Beijing	+	+	+
PD89	Patient	Mudanjiang	+	-	+
SH2	H.bispinosa	Sichuan	+	-	+
SH3	h.bispinosa	Sichuan	+	-	+
SH6	H.bispinosa	Sichuan	+	-	+
IM25	I.persulcatus	Yakeshi	+	-	+
IM26	I.persulcatus	Yakeshi	-	-	+
CH4	I.persulcatus	Yakeshi	+	-	+
CH7	I.persulcatus	Yakeshi	+	-	+
L8	I.persulcatus	Liaoning	+	-	+
M7	I.persulcatus	Mudanjiang	+	-	+
SR1	R.coxingi	Sichuan	+	-	+
FP1	Patient	Sichuan	+	-	+
H20	I.persulcatus	Mudanjiang	+	-	+
XJ23	I.persulcatus	Xinjiang	+	-	+
R9	Patient	Mudanjiang	+	-	+
XJ3	I.persulcatus	Xinjiang	-	-	+
PD91	patient	Inner Mongolia	+	-	+

(1)Strain B31 was kindly provided by Dr.Quan, CDC, USA.
(2)Strain IP21 was kindly provided by Prof. Korenberg, Russian Academy of Medicine, Russia.

Protein Analysis

Freshly harvested spirochetes were centrifuged and washed three times with PBS-MgCl$_2$ (0.01M PBS, 5mM MgCl$_2$, pH7.2). The final pellet was resuspended in 1XSDS gel loading buffer and then denatured the proteins by heating them to 100°C for 3 min. Aliquots (20ul, containing about 30ug of protein) were separated by SDS-PAGE[10].

RESULTS

Protein Analysis

All isolates were subjected to phenotype analysis by comparing the protein profiles of B.burgdorferi whole-cell lysates. They are characterized by several major peptide band, the strongest appearing in the range of about 12 to 36KD. Despite some minor overall variations, patterns of low-molecular-weight proteins allow a preliminary grouping. GroupI strains B31,CS3,CS4,SZ21, SZ22 and IP21(Fig.1a, lane 1-5,10; Fig. 1b,lane 17), characterized by two bands of 31 and 34kD, representing OspA and OspB proteins, respectively. GroupII strains PD89, SH2, SH6, IM25, CH4, CH7 (Fig.1b,lane2-5,15,16), show several major peptide bands of 34,31, 24,19,12KD. the band of 34KD in groupII strains appear to be weaker than that in groupI strains and no reactivity with monoclonal antibody for OspB, H6831(data not shown). GroupIII strains M7,L8,SR1,FP1,H20, XJ23(Fig. 1b, lane 8,9, 11-14) are distinguished by four peptides, about 36,32,21, 17KD in size. The other four strains IM26,R9,PD91,XJ3 (Fig.1b,lane 1,6,7, 10) show their individual protein profiles respectively.

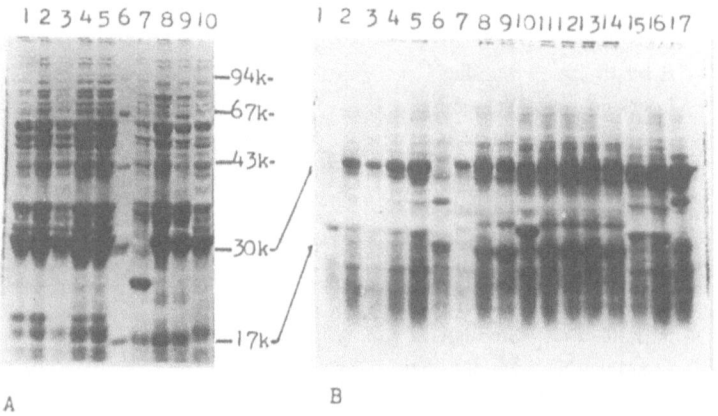

Fig.1. SDS-PAGE of whole-cell proteins, Coomassie blue stain. (A) 12% gel, lanes 1-10: CS3, CS4, B31, SZ21, SZ22, MW marker, R9, SR1, FP1, IP21. (B) 15% gel, lanes1-17: IM26,PD89,SH2,SH6,IM25,R9,PD91,M7,L8, XJ3,SR1,FP1,H20,XJ23,CH4, CH7,IP21.

Plasmid Profiles Analysis of B.burgdorferi

All isolates were subjected to low-percentage agarose gel electrophoresis and compared the plasmid profiles. Fig.2 show the plasmid profiles of 21 strains. Each strain contained 3-7 discernible plasmids ranging in size from 15 to 53kb. All strains except one(CH4,lane 9) had a 49kb large linear plasmid. Strain CH4, isolated from Ixodes persulcatus in Inner Mongolia, had an approximately 53kb large plasmid and a 48kb large plasmid. No small supercoiled plasmid were observed and plasmid loss did not occurred in the course of continuous laboratory cultivation. Plasmid profile analysis also agreed grouping these strains as done in protein analysis. GroupI strains were seen three discernible plasmids, Strains CS3(Fig.2,lane 11), a South China isolate, SZ21 and SZ22(lane 12, 13) isolated from North China (Beijing), shared the same plasmid profile as that of strain B31-86^4(lane 8). Strain IP21 (lane 4,GroupI)plasmid profile was observed to be different to that of other strains in groupI. Two plasmids, 38kb and 24kb, existed in strain B31-82,4 were not seen in the four groupIII strains CS3,CS4,SZ21,SZ22,(5 passages). It may be speculated that loss of plasmid likely occurred in vivo or the two plasmids had ever not existed in the four isolates. Strains in groupII and groupIII had more plasmids than that in groupI, and several different plasmid contents were observed among these isolates, e.g., PD89(groupII, lane 2),CH4(groupII,lane9), M7(groupIII,lane 14). We have also found that strains in groupIII FP1,SR1,M7,L8, H20,XJ23(Fig.2,lane 5,6,14,16,17, 18), had same plasmid profiles. The another intersting result is that, strain R9(Fig.2,lane 1), isolate from a patient with Chronic meningitis in Mudanjiang was found to have the same plasmid profile as the third groupIII strains,but obviously different protein profiles (Fig.1a,lane 7; Fig.1b,lane 6). It may be suggested that there are some inner relativity between the antigenic variation and virulence of B.burgdorferi.

Restriction Endonuclease Analysis and DNA hybridization

The DNAs from B.burgdorferi strains were characterized by REA and hybridization studies. Restriction patterns of HindIII-digested (Fig,3) or EcoRI-digested(data not shown) B.burgdorferi DNA were difficult to interpret.The many individual bands did not allow identification of particular clusters. However,from whole ranges of restriction fragments we can observe the differences among these isolates.A close correlationship had been observed between the plasmid

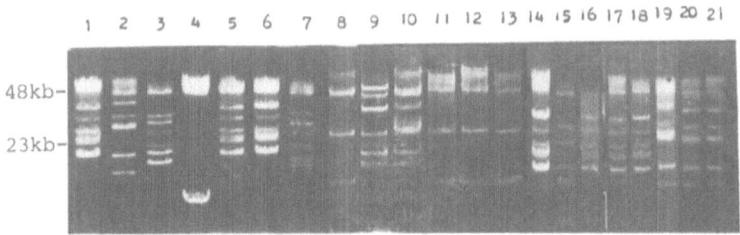

Fig.2. Low-percentage agarose gel electrophoresis of plasmid-enriched DNA fractions of B.burgdorferi isolates. Lanes 1-21: R9, PD89, PD91, IP21, FP1, SR1, IM26, B31, CH4, IM25, CS3, SZ21, SZ22, M7, XJ3, L8, H20, XJ23, SH2, SH3, SH6.

profile and REA of whole-cell DNA. Fig.3 shows that those strains with same plasmid profiles had the same restriction bands of whole-cell DNAs, such as SZ21, SZ22, CS3, CS4 and B31(groupI); R9, FP1, SR1, M7, L8, XJ23, H20(groupIII); PD89, SH2, SH3, SH6 (groupII); CH4, CH7, IM25(groupII); Meanwhile, we have found that some strains with different plasmid compositions, had very similar restriction patterns of whole-cell DNA, e.g., IP21 and B31(groupI), CH4 and PD89(groupII), XJ3 and PD91.

Six strains were subjected to REA and southern hybridization analysis in order to compare the restriction and homology patterns of chromosome-enriched DNA and plasmid-enriched DNA (Fig. 4). Several different restriction bands of HindIII-digested plasmid-enriched DNA allow to interpret clearly. Three strains R9 (Fig.A, lane1,7), FP1(groupIII, Fig.4a,lane 4,10),SR1(groupIII,Fig.4a,lane 5,11), showed identical restriction patterns of chromosome-enriched DNA and plasmid- enriched DNA. GroupI stains IP21(Fig.4a lane 3,9) and B31 (Fig.4a lane 6,12) had very similar restriction patterns of chromosome-enriched DNA but obviously different patterns of plasmid-enriched DNA. A southern hybridization of these isolates is represented in Fig.4b. The homology shared between these organisms wse visualized by southern blot hybridization using labeled whole-cell DNA from the reference strain B31. The

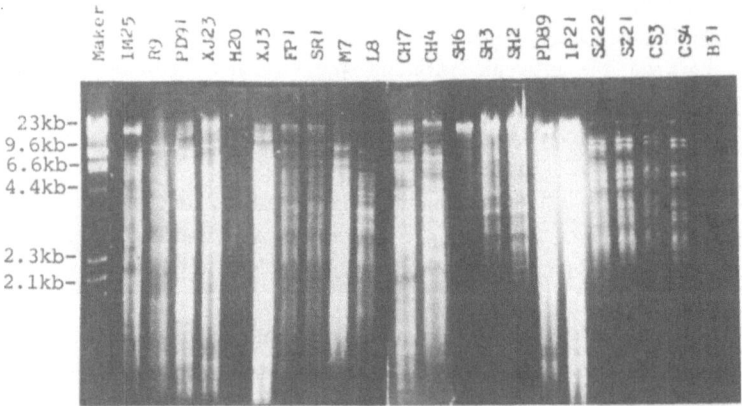

Fig.3. Restriction enzyme analysis of HindIII-digested DNA from B. burgdorferi isolates. Ethidium bromide-stained 1.0% agarose gel. The sizes of DNA markers are given in kilobase pairs.

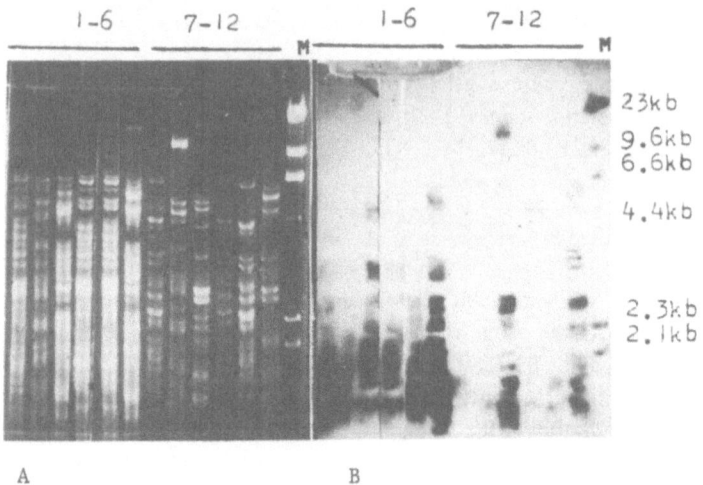

Fig.4. Restriction endonuclease (HindIII) digest and autoradiography of chromosome-enriched DNA and plasmid-enriched DNA of six B. burgdorferi isolates. (A) HindIII digest of these strains, fractionated on a 0.8% agarose gel. (B) Autoradiography of a southern blot of the autoradiography in panel A probed with radiolabeled B.burgdorferi strain B31. Lanes 1-6:Chromosome-enriched DNA. Lanes 7-12: Plasmid-enriched DNA. Strains:R9 (lane 1,7),PD89(lane 2,8),IP21(lane 3,9),SR1(lane 4,10) FP1(lane 5,11), B31(lane 6,12).

stronger homology were observed among the chromosome-enriched DNA of IP21, B31 (groupI strains, Fig.4b,lane9,12), and showed some similarity between these isolates. The four strains R9 (Fig.4b, lane 7),PD89(groupII strain, Fig.4b,lane 8), FP1 and SR1(groupIII strains,Fig.4b,lane 10,11), showed less homology of plasmid-enriched DNA with reference strains B31.It may suggest that some strains of this genus with different plasmid contents have the same or very similar chromosome DNA sequences, or may imply that some changes of plasmids e.g., DNA rearrangement or DNA exchange, have occurred in spirochete B.burgdorferi. Restriction patterns of Whole-cell DNA were difficult to discriminate the difference among B.burgdorferi isolates. It might cover the real DNA features of B.burgdorferi, especially the plasmid contents. Fig.5 showed a result of restriction digest and autoradiography of plasmid-enriched DNA from eight B.burgdorferi strains. It was clear to reflect the difference of plasmid contents described in plasmid profile analysis and showed apparently homology among these isolates plasmids.

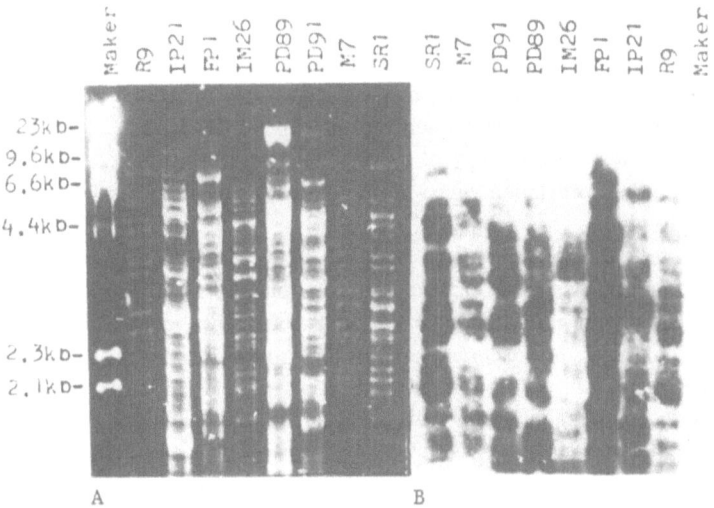

Fig.5. HindIII digest and autoradiography of plasmid-enriched DNA from eight B.burgdorferi isolates. (A) HindIII digest of these strains , fractionated on a 1.0% agarose gel. (B) Autoradiography of a southern blot of the autoradiography in panel A probed with radiolabeled B.burgdorferi strain R9.

DISCUSSION

A series of antigenic variations were described to occur in B.burgdorferi, the etiological agent of Lyme borreliosis. Several studies have demonstrated changes in surface proteins[11,12],infectivity[13]and plasmid profile[4] during in vitro cultivation. However, the phenotypic stability was found and loss of plasmid were not observed in our study. The lack of small circular plasmid, e.g., the 7.6kb plasmid, carrying genes or regulatory elements possibly involved in phenotypic variation[13] may be an important reason. The second reason may be that we have not cloned these strains with continuous in vitro passage. As we know, there are phenotypic heterogeneity among European and American isolates[14] B.burgdorferi strains isolated from China also showed several types, but difference did not appeared obviously between North and South China where the clinical symptoms of Lyme disease were observed to be different[5]. Vectors carrying the Lyme disease spirochete are different between North and South China[7]. It is probably related with clinical manifestations of Lyme disease in the two regions. Protein and plasmid analysis defined Chinese isolates used in this study as three major types. GroupI strains had the same protein profiles and similar plasmid profiles as the American isolate B31-86[4], Most of the 22 Chinese isolates showed the features of European isolates. The results of monoclonal antibody testing also supported the groupings. Table 1 shows the IFA reactions of monoclonal antibodies H5332, H6831 and H9724 against these isolates. Except the first group strains, all Chinese isolates showed no reaction with MAB H6831 for OspB.

Of particular interest are the differences of protein profiles between the third group strains and strain R9. R9, an isolate from a patient with Chronic meningitis, was characterized to own identical DNA feature of the third group strains by plasmid profile analysis, REA and southern blot analysis. This result demonstrated antigenic heterogeneity of the pathogenic agent, B.burgdorferi, and it was inferred that antigenic variation may occur in human infection, and the antigenic change may play an important role to cause Lyme disease. The mechanism behind antigenic variation is unclear. In the related Borrelia species, B.hermsii, which switches surface proteins to avoid the immune response of the host during relapsing fever, changes in serotype-specifying proteins are the consequence of the translocation of an antigen-encoding gene from a silent to an expression locus[15]. We can speculate that DNA rearrangements may also have been the basis for antigenic changes in B. burgdorferi, the variation may occur more easily under selective agents, e.g., antibodies, in vivo.

Several methods were used to analyzed B.burgdorferi Chinese isolates in this study. In contrast, restriction enzyme patterns of whole-cell DNA were difficult to interpret. The many individual bands did not allow identification of particular clusters. In addition to that, we have found some strains with different plasmid profiles had the same or very similar restriction pattern in this study. The whole-cell DNA restriction fragments may cover that of plasmid. The protein analysis and plasmid profile analysis were demonstrated to be useful tools for distinguishing strains. In our study, some strains with the same protein profiles were found to contain different plasmid profiles, but in these strains with the same plasmid profiles, only one isolate was found to show a different protein profile. The two methods, especially plasmid profile analysis, as described by Barbour[4], may be the easiest as well as the most discriminating method to differentiate one isolate from another.

REFERENCES

1. Burgdorfer, W. and Barbour, A.G., et al., Lyme disease, a tick-borne spirochetosis? Science, 216:1317-1319 (1982).
2. Boerlin et al., Population genetic analysis of Borrelia burgdorferi isolates by multilocus enzyme electrophoresis, Immun. 60:1677-1683 (1992).
3. Adam, T., et al., Phenotypic and genotypic analysis of Borrelia burgdorferi isolates from various sources, Infect. and Immun. 58-2579-2585 (1991).
4. Barbour, A.G., Plasmid analysis of Borrelia burgdorferi, the Lyme disease agent, J.Chin., Microbiol. 26:475-478 (1988).

5. Zhang, Z.F., et al., Investigation of Lyme disease in nineteen provinces, autonomous regions and cities in China, Chinese Journal of Vector Biology and Control. 3 (Suppl.2):1-4 (1992).

6. Dou, G.L., A study on the dissemination of Lyme disease by Ixodes perolcatus, Chinese Journal of Vector Biology and Control, 1:117-119 (1990).

7. Zhang, Z.F., et al., Survey on tick vectors of Lyme disease spirochetes in China, Chinese Journal of Vector Biology and Control. 3 (Suppl. 2):140-143 (1992).

8. Feng, W.X., et al., Clinical study on Lyme Borreliosis, Chinese Journal of Vector Biology and Control. 3 (Suppl. 2):119-128 (1992).

9. Barbour, A.G., Isolation and cultivation of Lyme disease spirochetes, Yale J. Biol. Med. 57:71-75 (1984).

10. Maniatis, T., Fritsch, E.F., and Sambrook, J., "Molecular Cloning: A Laboratory Manual," Cold Spring Harbor Laboratory, New York (1989).

11. Bisset, M.L. and Hill, W., Characterization of Borrelia burgdorferi strains isolated from Ixodes pacificus ticks in California, J. Clin. Microbiol. 25:2296-2301 (1987).

12. Schwan, T.C. and Burgdorfer, W., Antigenic changes of Borrelia burgdorferi, as a result of in vitro cultivation, J. Infect. Dis. 156:852-853 (1988).

13. Schwan, T.C., Burgdorfer, W., and Caron, C.F., Changes in infectivity and plasmid profile of the Lyme disease spirochete, Borrelia burgdorferi, as a result of in vitro cultivation, Infect. Immun. 56:1831-1836 (1988).

14. Barbour, A.G., Heiland, R.A., and Howe, T.R., Heterogeneity of major proteins in Lyme disease borrelia: A molecular analysis of North American and European isolates, J. Infect. Dis. 152:478-485 (1985).

15. Barbour, A.G. and Bundoc, V.G., Clonal polymorphisms of outer membrane protein OspB of Borellia burgdorferi, Infect. Immun. 57:2733-2741 (1989).

LYME DISEASE IN AN EXPERIMENTAL MOUSE MODEL

Sunitha Reddy,[1] Mike D. Gibson,[2] Julie Rawlings,[3] George Stoica,[4] Colin R. Young[4]

[1]Department of Biology, Texas A&M University
[2]Department of Vet. Anatomy and Public Health, Texas A&M University
[3]Texas Department of Health, Austin, Texas
[4]Department of Vet. Pathobiology, Texas A&M University

ABSTRACT

This study sought to define the route of inoculation necessary to establish infection or disease in susceptible C3H/HeJ mice and also, to determine the virulence of four *Borrelia burgdorferi* isolates. Our results indicate that experimental Lyme disease in mice can be caused by different isolates of *B. burgdorferi* isolated from different animal species and through different routes of infection. C3H/HeJ mice were used in our experiments since they are known to develop severe polyarthritis and carditis.[31] We report borrelia induced arthritis, and, lesions in the heart, spleen, liver, skin and joints, and also the presence of whole or intact spirochetes in some of these organs. In our efforts to find a suitable animal model of Lyme disease we have been successful in demonstrating dermatological, cardiac and arthritic lesions of Lyme disease in C3H/HeJ mice. Based on these results we find that mice inoculated with low-passage tick isolates of *B. burgdorferi* develop spirochetaemia, multisystemic infection, and a high prevalence of polyarticular arthritis resembling Lyme disease in man. This paper provides a chronology of our progress in development of this potentially useful model.

INTRODUCTION

Lyme disease also known as Lyme borreliosis is a progressive infectious disease caused by the tick borne spirochete *Borrelia burgdorferi* .[1,2] It was first isolated from ticks in 1981[1] and from patients in 1982.[3] Although now known as Lyme disease, the first case of Lyme disease was described in 1883 by Buchwald in Germany as an idiotypic skin atrophy[4] in a patient. Similar cases were then described by Herxheimer and Hartman in 1902 who called the disease Acrodermatitis chronica atrophicans (ACA).[5] In Sweden, Afzelius described in 1909,[6] a migrating annular skin lesion in a patient that developed at the site of a tick bite. He called it Erythema migrans (EM) which was later denominated as erythema chronicum migrans (ECM) on the basis of its long, apparent duration by Lipschutz in

1913.[7,8] In France, Garin and Bujadoux associated erythema with neurological manifestations in 1922[9] in a patient who developed limb palsy with meningitis following a tick bite and an expanding erythematous skin lesions, then not recognized as ECM. This suggested that a spirochete could be responsible for this 'tick paralysis". Bannwarth[10,11] in 1941 describes several patients with lympphocytic meningitis and neuritis that frequently involved the facial nerve. This neurological entity came to be known as lymphocytic meningoradiculoneuritis, or Garin-Bujadoux, Bannwarth syndrome. These disorders are now recognized as components within the spectrum of lyme disease. The relationship of ECM with ACA was established by Ludwig in 1956.[12]

It was not until 1975 that lyme disease became recognized in the United States.[13] Until then, ECM was only described as a rare disease contracted during traveling in European countries. In 1975, the rheumatology section of Yale university and the centers for disease control headed by Dr. Allan C. Steere conducted a survey in Lyme, Connecticut, after being confronted with a large number of cases diagnosed as having juvenile rheumatoid arthritis (JRA).[13] They found that before the onset of arthritis 25% of the patients had been affected by a dermatitis of the ECM type as described in Europe. On further evaluation, they found that patients suffering from ECM lesions which was followed by arthritis, also developed neurological and/or cardiac manifestations. Due to the involvement of different organs and organ systems, this multisystemic disease was renamed Lyme disease[14] in 1977. Steere then described the different stages extensively.[14] Further, *Ixodes dammini* was identified as the vector, after some patients had saved the ticks by which they had been bitten. Epidemiologic studies show that the distribution of the disease is similar to the distribution of the tick.[8,15] In 1982, Willy Burgdorfer identified the bacterium responsible for lyme disease. The bacterium, a spirochete, belonging to the genus *Borrelia* was a new species and therefore was called *Borrelia burgdorferi* after him. The wild type strain isolated from *I. dammini* was called B31.[2,16,17] Now it is known that several other species of the genus *Ixodes* are vectors to *B. burgdorferi*,[18,19,20,21] including *Amblyomma americanum* , another genus of the Ixodidae family.[22] Epidemiological studies show that Lyme borreliosis can be transmitted in rare cases by horse flies, deer flies, and most recently the spirochete has been isolated from mosquitoes and cat fleas.[22,23]

Clinically, Lyme disease can be divided into three stages.[14] The first and most obvious stage is characterized by the ECM rash which develops two to thirty days after the bite of an infected tick. The second stage (not always expressed) usually occurring weeks to months after the bite, is marked by neurological complications and migratory muscoloskeletal pain. The third stage, beginning months or years after the bite, typically involves the onset of arthritis. Attacks of arthritis usually last from a few days to a few weeks at a time and primarily effect the knees and other large joints.

Lyme disease is a veterinary as well as a human problem. Descriptive stages of the disease have not been found in free ranging animals. However, spontaneous cases of lyme borreliosis have been documented in cattle and horses,[24] dogs,[25] rats,[26] and hamsters[27] with varying degrees of dermatological, musculoskeletal, neurologic, renal, and/or cardiac lesions.

Although a broad range of species susceptibility exists, attempts to experimentally reproduce Lyme disease in animal models have often been disappointing. Many animal models such as rabbits,[5] dogs,[7] albino and syrian hamsters,[8] outbred laboratory mice, the severe combined immunodeficiency (scid) mouse,[28] white footed (*Peromyscus*) mice,[29] guinea pigs,[30] and outbred rats reported seroconversion and spirochete recovery, but minimal lesion involvement. Mice have been extensively investigated as an experimental animal model because the mouse immune system is by far the best understood. Genetically determined murine models are useful for providing a deeper insight into lyme disease pathogenesis.

MATERIALS AND METHODS

Experimental Animals

Three week old C3H/HeJ mice were commercially bought from Jackson laboratories, Bar Harbour, Maine. All mice were pathogen free and shipped in different crates. They were housed separately for the infectivity experiments, five per cage, in isolation cubicles with an air filtered environment maintained at 20-22 ^{0}C. All mice were pre-bled before inoculation and were found to be sero-negative and culture negative for *B. burgdorferi* antigen. Mice were divided into two experimental sections:

(i) groups of five mice were split into seven groups based on seven different routes of infection namely; Sub-cutaneous S/C (Group 1), Intra-venous I/V (Group 2), Intra-peritoneal I/P (Group 3), Intra-cranial I/C (Group 4), Hind-footed intra-dermal, F/P (Group 5), Base of tail intra-dermal (Group 6), and finally, Intra-muscular I/M (Group 7). All seven groups of mice were inoculated with 1x 10^7 viable spirochetes, type strain B31 (wild type strain isolated from *I. dammini* [2]) in 0.1 ml PBS buffer solution.

(ii) A second group of C3H/HeJ mice were infected in groups of five with three different isolates of borrelia namely, B1579[32] (isolated from the tick *Amblyomma americanum*), B358 (isolated from *Ixodes scapularis*) and B532[32] (isolated from cat fleas). Organisms were injected in 0.1ml PBS buffer, (a) Intra-dermally in the chest I/D, (b) Intra-cranially (I/C) and finally, (c) In the hind footpads (F/P) separately for each isolate.

Mice in both experiments were observed daily for disease symptoms and observations were recorded. Blood samples were taken from the tail vein every ten days and serum collected for serology. At the end of five bleeds mice were terminated using CO_2 inhalation. Samples of brain, heart, spleen, liver, and rear limb joints were taken from each C3H/HeJ mouse. Half of the organs were cultured in modified Kelly's medium for spirochete recovery,[33] and the other half were fixed in 10% buffered formalin for histopathological examination.

Borrelia isolates

In all four borrelia isolates were used namely; B31,[2, 16, 17] 1579,[32] 358 and 532.[32] Table 1 shows strain designation and origin of the four isolates. Isolates maintained in Kelly's medium were received from Dr. Julie Rawlings, Dept. of Health, Austin, TX, U.S.A.

Table 1- Strain Designation And Origin Of *B.burgdorfei* Isolates

Strain	Origin
B31	First *B. burgdorferi* isolate from Deer Tick *(Ixodes dammini)*. Shelter Island, New York, U.S.A.
532	Pooled isolate from five cat fleas *(Ctenocephalides felis)*. Fort Band County, Texas, U.S.A.
358	Isolated from deer tick *(Ixodes scapularis)*.
1579	Isolated from Texas Lone Star Tick *(Amblyomma americanum)*. Dry ice collection.

Serology

Antibody to *B. burgdorferi* was measured by two methods. Serum samples were tested for *B. burgdorferi* IgG antibodies by a previously described IFA[34] test and RIA (Radio immuno assay) recently developed at Texas A&M university, TX, U.S.A. The IFA endpoint was the highest serum dilution to show distinct florescence of the spirochetes.

In the RIA, 96 well PVC flexible plates were coated with 50 µl antigen (1×10^5 spirochetes per well) in 0.01% Na azide/PBS buffer solution and incubated overnight at 20 to 25^0C to allow antigen to stick to well surface. Antigen was then removed and plates washed with 0.1% gelatin/PBS solution. Wells were filled with 0.1% gelatin/PBS and incubated for 30 minutes to block any unbound sites. The plates were then washed with the same solution followed by two washes with PBS buffer. 50 µl dilutions of mouse antisera (10^{-1} to 10^{-4}) in 0.1% gelatin were added to each well in duplicates and incubated for 2 hours at room temperature. Antiserum was removed and plates were washed 3 times with 0.1% gelatin/PBS solution and once with PBS buffer. The assay was developed by the addition of protein-A-I[125], 50 µl/well (in 0.1% gelatin/PBS containing 100,000 cpm/well). Following incubation for one hour, the labeled material was removed and wells were washed with PBS buffer. After drying the plates, wells were cut out and counted in a gamma counter.

Histological examination

Brain, Heart, Spleen, joints and skin samples of every test animal were fixed in 10% neutral-buffered formalin (pH 7.2) at necropsy. Joints were decalcified, and tissues were processed and paraffin-embedded, sectioned at 5-7 µm, and stained with hematoxylin and eosin by using standard techniques. Selected tissues were stained for spirochetes by using Steiner's or Warthin-Starry silver staining procedures.[35]

RESULTS

Response of C3H/HeJ mice to different routes of inoculation with *B. burgdorferi* B31 isolate

Group 1(S/C), Group 2 (I/V), Group 3 (I/P), and Group 4 (I/C) seroconverted following infection with *B. burgdorferi*. In Group 1, the five mice inoculated S/C, did not show any clinical signs of illness however, all five were seropositive by IFA (Immunoflourescent antibody) and RIA (Protein-A-I[125] Radio immuno assay) tests. IFA titers for serum antibodies to *B.burgdorferi* showed high positive results in all five animals tested. Titers ranged from 1:512 to ≥1:1024 (Table 2). Similarly, the antibody tests using RIA showed high antibody titers in all five animals with the peak titers ranging from 20,000 to 23,000 cpm (counts per minute) (Figure 1). At necropsy all five mice showed splenomegaly and enlarged kidneys although no internal lesions were found. Group 2, inoculated intra-venously (I/V) in the tail vein also developed high positive antibody titers to both IFA and RIA with peak antibody titers ranging from 1:256 to ≥1:1024 (Table 2)and 18,000 to 26,000 cpm (Figure 1) respectively. All test animals showed splenomegaly, although no distinct histopathological lesions were found. Similar results were found in

Group 3 mice which were inoculated with *B. burgdorferi* I/P. IFA tests in all five animals showed antibody titers ≥1:1024 (Table 2), RIA antibody titers of ~ 30,000 cpm, and splenomegaly (Figure 1). Group 4 (I/C) showed IFA antibody titers ≥1:1024 (Table 2) and RIA antibody responses greater than 40,000 cpm (Fig 1), and splenomegaly. Serum samples from pre-bleeds of test animals and serum taken from control groups showed IFA titers <1:16 and RIA titers < 500 cpm.

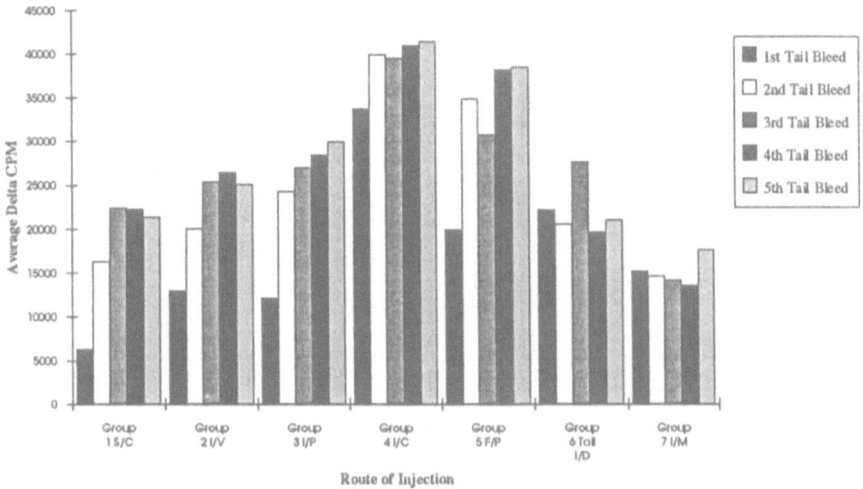

Figure 1. Anti *Borrelia burgdorferi* 31 C3H/HeJ Mouse Sera vs *Borrelia burgdorferi* 31 isolate 1:10 dilutions of 5 tailbleeds, seven differing routes of injection, 125-I protein A RIA.

Table 2. IFA Antibody Titers[1] of C3H/HeJ Mice Infected with B31 *B. burgdorferi* using different routes of inoculation.

Animal No.	Control	Group 1 S/C	Group 2 I/V	Group 3 I/P	Group 4 I/C	Group 5 F/P	Group 6 Tail I/D	Group 7 I/M
1	< 1:16	512	≥1024	≥1024	≥1024	≥1024	≥1024	256
2	< 1:16	≥1024	512	≥1024	512	≥1024	512	512
3	< 1:16	512	256	≥1024	≥1024	512	≥1024	256
4	< 1:16	≥1024	512	≥1024	≥1024	512	512	512
5	< 1:16	512	≥1024	≥1024	≥1024	≥1024	512	512

[1]Indirect Immunoflourescent antibody titer to Borrelia burgdorferi titer is given as the reciprocal of the end point dilution.

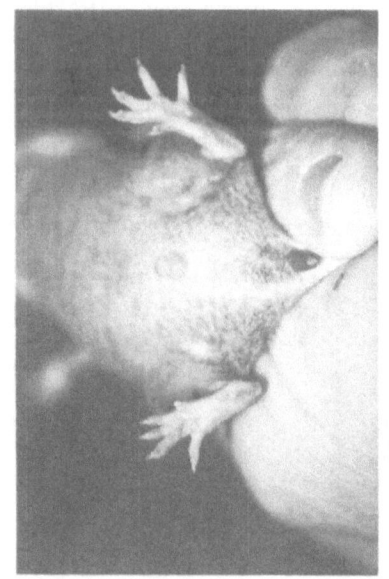

Figure 3. Footpad I/D Control mouse B31.

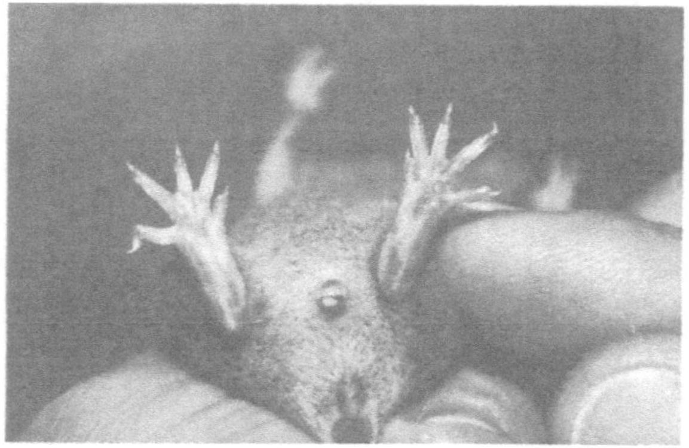

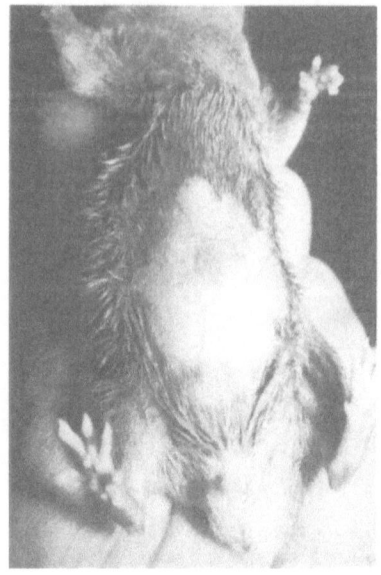

Figure 6. I/D Injection B31 C3H/HeJ Mice, 5 weeks post injection. Granulomatous skin lesion.

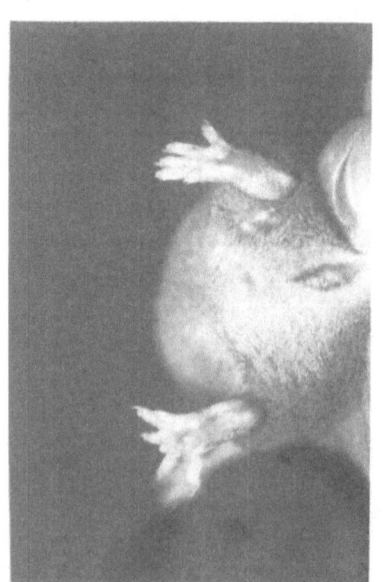

Figure 2 : Footpad I/D, B31 test mouse. 24 hours post injection. Arthus type Reaction

Figure 4. C3H/HeJ Mouse showing unilateral rear limb arthritis of the ankle. 15 days post-injection with B31 strain in left gastrocnemius muscle.

In Group 5, five mice were injected intra-dermally in both the hind footpads with a dose of ~1×10^7 antigen in 0.05 ml PBS in each of the two footpads. At 24 hours post-injection all five mice developed an arthus type reaction which showed a massive reddening and thickening of both the hind footpads (Figure 2). At 48 hours post-injection all five animals had erythema on the hind footpads whereas, control mice injected with the same volume of 0.1ml PBS without any antigen did not develop any kind of reaction (Fig 3) or erythema. All test animals showed bilateral rear-limb arthritis from 3 weeks onwards until the termination of the experiment after 50 days.. All animals showed splenomegaly and hyperplasia of the spleen. IFA titers showed titers ≥1:1024 in 3/5 animals (Table 2) and 1:512 in 2/5 animals. RIA antibody titers exceeded 35,000 cpm.

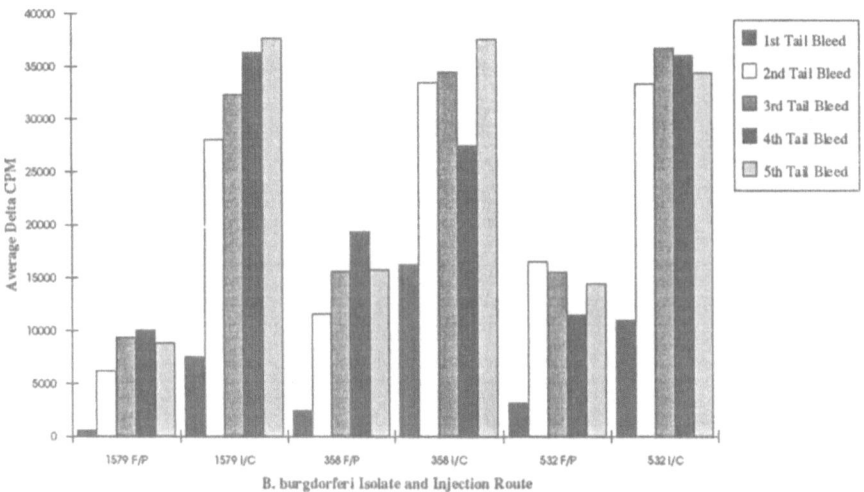

Figure 5. Anti *Borrelia burgdorferi* C3H/HeJ mouse sera vs homologous *Borrelia burgdorferi* isolates , 1:10 dilutions of 5 tailbleeds each, 125-I protein A RIA

In Group 6, mice were injected with *B. burgdorferi* in the base of tail. Mice seroconverted following B31 infection. Two out of five had IFA titers ≥1:1024 and the remaining three had titers of 1:512. Apart from splenomegaly, no other clinical signs were observed. All mice inoculated I/M in the left gastrocnemius muscle in Group 7 however, showed unilateral left rear limb arthritis (Fig 4) along with swelling in the left rear footed whereas, control animals did not show any abnormal signs. All five animals were sero-positive to *B. burgdorferi* with antibody titers in IFA (1:256 to 1:512) and RIA (14,800 to 26,300 cpm) tests.

Generally, all animals were susceptible to infection with *B. burgdorferi*, type strain B31. Intra-dermal injections to the footpads i.e., Group 5, and Intra-muscular injection to the left gastrocnemius muscle i.e., Group 7, seemed to be the most effective ways to produce symptoms of Lyme borreliosis. However, infection can be caused using all the tested routes of inoculation namely, S/C, I/V, I/P, I/C, Base of tail I/D and I/M (Table 2 and Fig 1). Control mice and serum samples from pre-bleeds of test animals were seronegative for antibodies to *B. burgdorferi*, B31 isolate.

Response of C3H/HeJ mice to infection by 1579, 358, and, 532 *Borrelia burgdorferi* isolates

It is believed that the major vector for transmitting Lyme disease in the United States are *Ixodes* ticks, particularly *I. dammini*. Although many isolates from different geographical areas exist, studies have shown that they belong to the same species,[36, 17] of *Borrelia* namely, *B. burgdorferi*, despite showing minor differences in their antigenic proteins. We conducted experiments to determine whether C3H/HeJ mice are susceptible to infection or disease with three different isolates viz, 1579,[32] 358, and 532[32] *Borrelia* strains. derived from the tick *Amblyoma americanum*, strain 358 is isolated from the tick *Ixodes scapularis*, and strain 532 is isolated from the cat flea *Ctenocephalides felis* .

All the mice developed positive antibody titers by both IFA and RIA tests (Table 3) and Fig 6 respectively). Animals injected I/C with 1579, 358, and 532 isolates produced higher antibody titers than animals injected in the footed (F/P).

Table 3. IFA Antibody Titers[1] of C3H/HeJ Mice to different isolates of *B. burgdorferi* through different routes of infection.

Animal No.	Control	1579 F/P	1579 I/C	358 F/P	358 I/C	532 F/P	532 I/C
1	< 16	< 32	256	256	512	128	256
2	< 16	64	≥1024	128	≥1024	128	512
3	< 16	32	128	32	≥1024	256	≥1024
4	< 16	128	512	32	≥1024	128	512
5	< 16	32	256	32	≥1024	128	256

[1]Indirect Immunoflourescent antibody titer to Borrelia burgdorferi titer is given as the reciprocal of the end point dilution.

Intra-dermal injections in the abdomen in all the three groups produced two forms of skin lesions. The first lesion showed erythema with vasculitis at 24 hours post injection. These lesions overtime developed into granulomatous skin lesions (Fig 6). The second form of necrotic skin lesion showed a black necrosis at the site of inoculation 24 hours post injection. This necrotic skin lesion, sloughed off after 21 days (Fig 7). A few mice were sacrificed for histopathology at day 15 post-injection. The others were kept under observation till the completion of the experiment. H&E staining of tissue sections of the granulomatous skin lesions (Fig 8) show a deep granuloma formation i.e., a central necrosis with connective tissue proliferation surrounded by inflammatory cells like macrophages, neutrophils and lymphocytes. Warthin-Starry silver staining of the lesion showed the presence of fragments of *Borrelia* and also a few intact *Borrelia* spirochetes (Fig 9). A higher magnification of the border of the border of the granulomatous lesions (Figure not shown) shows lots of *Borrelia* fragments plus phagocytosed material inside macrophages.

Staining of spleen sections using H&E stain (Fig 10), 15 days post-infection, I/D with B532, shows lymphoid depletion in the splenic follicle and only reticulo-endothelial cells remain. Many cells in the follicle show pycnosis. The follicle is surrounded by a hemorrhagic infiltration in the red pulp characterizing septicaemia. Higher magnifications of the red pulp

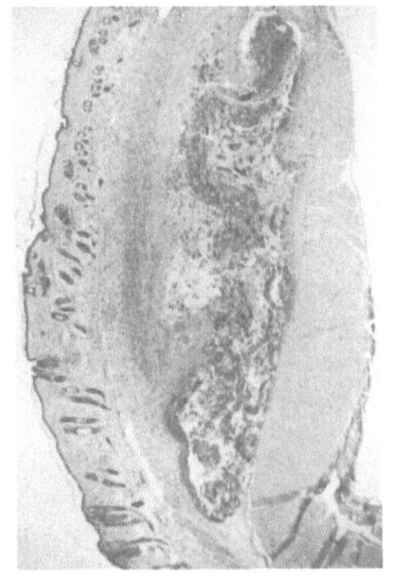

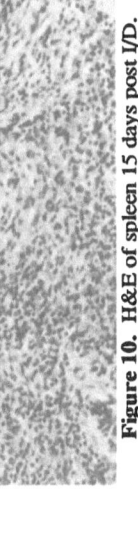

Figure 8. H&E stain . Skin 15 days post-injection with 532. Granuloma formation i.e. central necrosis, surrounded by connective tissue proliferation surrounded by inflammatory cells. (Macrophages, Neutrophils, and Lymphocytes)

Figure 10. H&E of spleen 15 days post I/D. Splenic follicle showing lymphoid deplection. Septicaemia of the red pulp is seen.

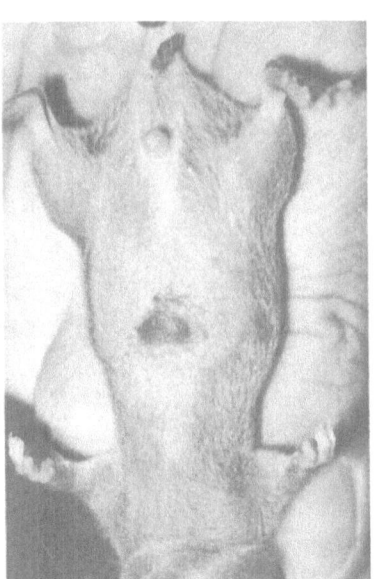

Figure 7 : I/D Injection B31. Necrotic lesion. 24 hours post injection C3H/Hej Mouse.

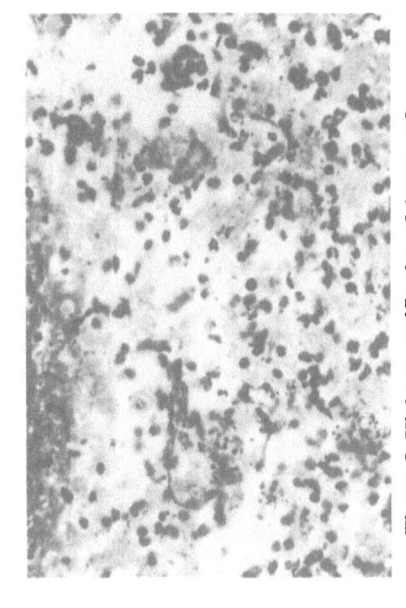

Figure 9. Higher magnification of the core of necrosis using Warthin-Starry silver stain for *Borrelia*. Antigen spotting is where fragments of *Borrelia* are, plus some intact *Borrelia*.

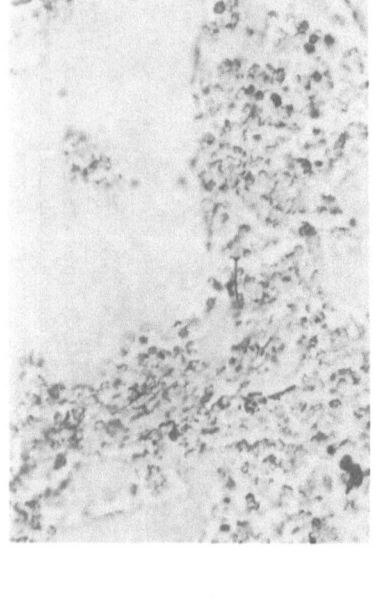

Figure 12. Venule of the heart, 15 days post infection, I/D in C3H/HeJ mouse. Intact spirochetes are seen.

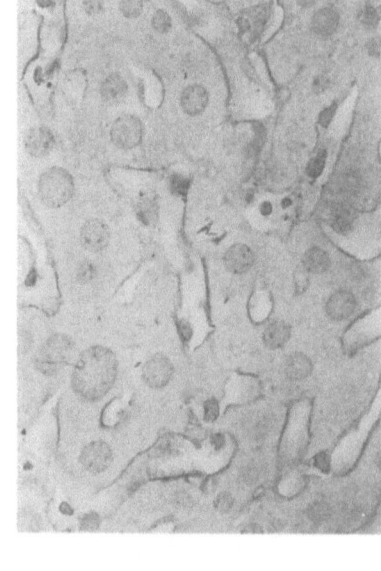

Figure 14. Silver staining of liver sections. I/D 4 weeks post injectiont injection. Liver parenchyma intact. Spirochetes intact in hepatocytes and phagocytosed by kupffer cells.

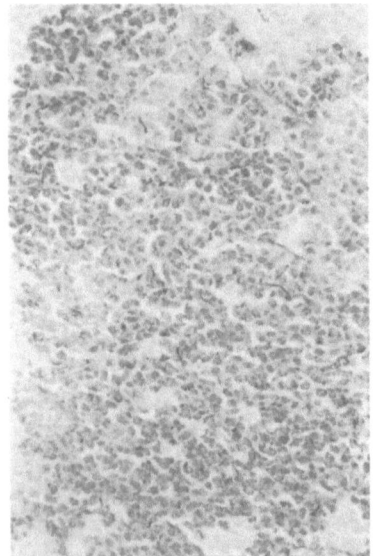

Figure 11: Silver stain of spleen, C3H/HeJ mice. 15 days post I/D injection with 532 isolate. Intact spirochetes can be seen in the red pulp.

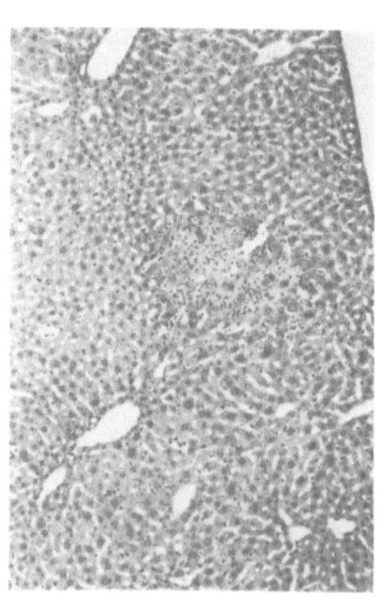

Figure 13. H&E of liver showing necrosis, loss of hepatocytes, and presence of inflammatory cells, showing coagulation necrosis.

area (figure not shown) shows an abundance of neutrophils, plasma cells, and macrophages. Warthin- Starry silver staining of the red pulp of the spleen showed intact spirochetes (Fig 11). Silver staining of sections of the heart, 15 days post-infection, I/D with B532, also showed whole spirochetes (Fig 12). H&E staining of tissue sections of liver showed multifoci necrosis. There was no predilection of necrosis for the peri-portal area thus, necrosis was random. Higher magnification of the necrotic area (Fig 13) shows a loss of hepatocytes, and the presence of lots of inflammatory cells, all characteristic of a coagulation necrosis. Silver staining of liver sections indicated (Fig 14) that the liver parenchyma was intact, with intact whole spirochetes in the hepatocytes and fragments of *Borrelia* in the kuppfer cells.

Occasional spirochetes were recovered from the brain and spleen samples of 532 I/D infected mice. Spleen, Heart, Brain and Liver sections of test animals inoculated I/D and I/C with 1579 and 358 did not show any histopathological lesions. Tissue sections of the necrotic skin lesions of the test animals injected I/D with 1579 and 358 showed the presence of spirochetes. However, all three groups of animals injected in the hind footpads with all three strains developed massive erythema of the hind footpads. Histopathological studies of organs from these mice did not reveal any significant results.

DISCUSSION

It has been reported that selected inbred strains of immunocompetent mice, especially C3H/HeJ mice, develop severe polyarthritis and carditis, while other strains such as BALB/C mice, develop only mild disease following intraperitoneal inoculation with *B. burgdorferi*.[32] Further, the infectious dose for disease-susceptible and disease resistant strains of mice is the same indicating that genetic susceptibility to disease manifestations is not spirochete dose dependent but is influenced by the route of inoculation. Studies conducted earlier in rats, hamsters, and mice confirm that skin is a preferential site for *B. burgdorferi* infection. Contact transmission and intranasal and intragastric inoculation of spirochetes are inefficient routes of infection.[32] Skin or I/D route of inoculation may therefore provide a facilitating role in the initial phase of infection via unknown mechanisms. Our experiments using different isolates of *Borrelia burgdorferi* show that infectivity of *B. burgdorferi* in 3-week-old C3H/He mice was significantly influenced by route of inoculation. Inoculation with the four isolates, B31, 1579, 358, and 532, using different routes of inoculation was successful in causing infection as indicated by the high IFA and RIA antibody titers. Nevertheless, I/D inoculations induced a broader spectrum of disease symptoms from the four isolates as compared to I/C inoculations. I/D inoculations in the abdomen using 1579, 358, and 532 strains elicited two different types of lesions namely black necrotic lesions and granulomatous lesions in all the groups. The difference, we believe is due to the depth of I/D injections. I/D inoculations with all four strains in the footpads produced erythema and arthritis. Spirochetes were observed in the heart and spleen of animals inoculated intradermally. Antigenic differences between strains may play a role in the clinical picture of Lyme borreliosis.[37]

It is not clear as to how antibody levels in the body effect disease pathogenesis. I/C and I/P routes of inoculation developed higher antibody titers as compared to S/C, F/P, and I/M inoculations with the B31 strain of *B. burgdorferi* (Fig 1). I/C and I/P inoculated mice having high antibody titers did not show any clinical symptoms of Lyme disease, on the other hand, I/D and I/M inoculated mice having comparatively low antibody titers developed dermatological, cardiac and arthritic lesions whereas, S/C inoculated mice having low antibody titers did not develop any obvious signs of infection. Route of inoculation seems to be the only differentiating factor influencing the response to *B.burgdorferi* infection.

More research needs to be conducted using genetically stable inbred mouse models such as C3H/HeJ mice. Also, the use of congenic and recombinant strains with known haplotypes would greatly facilitate in understanding the immunology and pathology of Lyme disease. Our studies conducted show that the intra-dermal route is the most effective way of inducing infection and establishing Lyme disease in C3H/HeJ mice irrespective of *Borrelia burgdorferi* strain used. It makes it an important finding that dermatological, cardiac, and arthritic lesions can be reproduced using the I/D route of infection in this animal model. In other words, Lyme disease can be experimentally reproduced in C3H/HeJ mice since, the observations described suggest that C3H/HeJ mice develop clinical and pathological Lyme borreliosis like humans and also that they are susceptible to different routes of infection and to different strains of *B. burgdorferi* . The C3H/HeJ mouse is a useful experimental model for increasing our knowledge of the interaction of *B. burgdorferi* with animal hosts. It is a relatively well studied animal and can be used to study the immunology and pathology of Lyme disease, the feasibility of a vaccine, and in vivo antibiotic susceptibility. This report outlines a promising animal model of Lyme disease that has several clinical and pathological correlations to the corresponding illness in humans.

REFERENCES

1. W. Burgdorfer, A.G. Barbour, S.F. Hayes, O.Peter, and A. Aeschlimann, Erythema chronicum migrans - a tick borne spirochetosis, *Acta Trop.* 40:79-83, (1983).
2. R.C. Johnson, G.P. Schmid, F.W. Hyde, A.G. Steigerwatt, and D.J. Brenner, *Borrelia burgdorferi* sp. nov: Etiological agent of lyme disease, *Int J Syst Bact.* 34:496-497 (1984).
3. A.C. Steere, R.L. Grodzicki, A.N. Kornblatt, J.E. Craft, A.G. Barbour, W. Burgdorfer, G.P. Schmid, E. Johnson, and S.E. Malawista, The spirochetal etiology of lyme disease, *N Eng J Med* . 308:773-740 (1983).
4. A. Buchwald, Ein fall von diffuser idiopathischer haut-atrophie, *Arch derm syph* .*(Berlin)* 15:553-556, (1983).
5. K. Herxheimer, and K. Hartman, Uber acrodermatitis chronica atrophicans, *Acta Dermatol* .*(Berlin)* 61:57-76, (1902).
6. A. Afzelius, Verhandlungen der dermatologishen gesellschaft zu Stockholm on October 28, 1909, *Arch Deramatol Syph.* 101:404, (1910).
7. A. Afzelius, Erythema chronicum migrans, *Acta Derm Venereal.* 2:120-125, (1921).
8. B. Lipschutz, Uber eine settene erytheform (erythema chronicum migrans), *Arch Dermatol.* 349-356, (1918).
9. C. Garin, and Bujadoux, Paralysie pae les Tiques, *J Med Lyon.* 71: 765-767, (1922).
10. A. Bannwarth, Chronische lymphocytare meningitis entzundliche polyneuritis and "Rheumatismus" eing beitrag zum problem "Allergic and Nervensystem", *Arch Psychiatr Nervenkr* 113:284-376, (1941).
11. A. Bannwarth, Zur klinik und pathogenese der "chronischen lymphocytaire meningitis" *Arch Psychiatr Nervenkr.* 117:682-716, (1944).
12. E. Ludwig, Erythema chronicum migrans in Fruhstadium der acrodermatitis chronica atrophicans herxheimer, *Hautartz.* 7:41-42, (1956).
13. A. C. Steere, S.E. Malawista, D.R. Snydman, et.al., Lyme arthritis: an epidemic of aligoarticular arthritis in children and adults in three Connecticut communities, *Arthritis Rheum.* 20:7, (1977).
14. A. C. Steere, S.E. Malawista, J.A. Hardin, S. Ruddy, P.W. Askenase, and W.A. Andiman, Erythema chronicum migrans and lyme arthritis: the enlarging clinical spectrum, *Ann Intern Med.* 86:685-698, (1977).
15. A. C. Steere, T.E. Broderick, and S.E. Malawista, Erythema chronicum migrans and lyme arthritis: epidemiologic evidence for a tick vector, *Am J Epidemiol.* 108:322-327, (1978).
16. W. Burgdorfer, A.G. Barbour, S.F. Hayes, J.L. Benach, E. Grunwaldt, and J.P. Davis, Lyme disease - a tick borne spirochetosis? *Science.* 216:1317-1319, 1982.
17. F.W. Hyde, and R.C. Johnson, Genetic relationship of lyme disease spirochetes to *Borrelia*, *Trepanema*, and *Leptospira* species, *J Clin Microbiol.* 20:151-154, (1984).

18. A. C. Steere, and S.E. Malawista, Cases of lyme disease in the United States: Locations correlated with distribution of *Ixodes dammini*, *Ann Intern Med*. 91:730:733, (1979).

19. W. Burgdorfer, R.S. Lane, A.G. Barbour,, R.A. Gresbrink, and J.R. Anderson, The western black-legged tick *Ixodes pacificus*: a vector of *Borrelia burgdorferi* , *Am J Trop Med Hyg*. 34:925-930, (1985).

20. H.E. Krampitz, In vito isolation and maintenance of some wild strains of European hand tick spirochetes in mammalian and arthropod hosts: a parasitologist's view.

21. E.J. Dekonenko, A. C. Steere, V.P. Beraudi, and L.N. Kravchuk, Lyme borreliosis in the Soviet Union: a co-operative US-USSR report, *J Infect Dis*. 158:748-753, (1988).

22. T.L. Schulze, G.S. Bowen, E.M. Bolser, M.F. Lakat, W.E. Parkin, R. Altman, B.G. Ormiston, and J.K. Shisler, *Amblyyomma americanum*: a potential vector of lyme disease in New Jersey, *Science*, 224:601-603. (1984).

23. L.A. Magnareli, J.F. Anderson, and A.J. Barbour, The etiological agent of lyme disease in deer flies, horse flies, and mosquitoes, *J Infect Dis*. 154:355-358, (1986).

24. E.C. Burgess, *Borrelia burgdorferi* infection in Wisconsin horses and cows, *Ann NY Acad Sci*. 539:235-243, (1988).

25. A.N. Kornblatt, P.H. Urband, and A.C. Steere, Arthritis caused by *Borrelia burgdorferi* in dogs, *J Am Vet Med Assoc*. 186:960-964, (1985).

26. S.W. Barthold, K.D. Moody, G.A. Terwilliger, P.H. Duray, R.D. Jacoby, and A.C. Steere, Experimental lyme arthritis in rats infected with *Borrelia burgdorferi* , *J Infect Dis*. 157:842-846, (1988).

27. A. Hejka, J.L. Schmitz, D.M. England, S.M. Collister, and R.F. Schell, Histopathology of lyme arthritis in LSH hamsters, *Am J Path*. 134:1113-1123, (1989).

28. S. Hedstrom, Erythema chronicum migrans Afzelii, *Acta Derm Venereol*. 11:315-321, (1930).

29. H. Salde, Erythema chronicum migrans, *Lakartidningen* 43:2381-2388, (1946)

30. T. Dalsgaard-Neilsen, Polyradiculitis after tick bite (*Ixodes ricinus*) , *Nord Med*. 35:1754-1755 (1947). (In Danish)

31. S.W. Barthold, Infectivity of *Borrelia burgdorferi* relative to route of inoculation and genotype in laboratory mice. *J Infect Dis*. 163:419-420, (1991).

32. G.J. Teltow, P.V. Fournier, and J.A. Rawhnys, Isolation of *Borrelia burgdorferi* arthropods collected in Texas, *Am J Trop Med Hyg*, 44:467-474, (1991).

33. E.C. Burgess, T.E. Amundson, J.P. Davis, R.A. Koslow, and R. Eddman, Experimental inoculation of peromyscus sp with *Borrelia burgdorferi* : Evidence of contact transmission, *Am J Trop Med Hyg*. 35:359-363, (1986).

34. J.A. Rawlings, Lyme disease in Texas, *Bakt Hyg*. A 263:483-487, (1986).

35. S.W. Thompson, Modified Warthin-Starry method for the demonstration of spirochetes, in: "Selected histochemical and histopathological methods," Charles C. Thomas, Springfield, IL. (1966).

36. G.P. Schmid, Steigerwalt, S.E. Johnson, A.G. Barbour, A.C. Steere, I.M. Robinson, and D.J. Brenner, DNA characterization of the spirochete that causes lyme disease. *J Clin Microbiol*. 20: 155-158, (1984).

SECTIONAL UPTAKE AND CYTOSOLIC PROCESSING OF

BORRELIA BURGDORFERI BY HUMAN PHAGOCYTES

Michael G. Rittig[1], Michael Kressel[2], Thomas Häupl[3], and
Gerd R. Burmester[3]

[1]Dept. of Anatomy I, University of Erlangen-Nürnberg, Germany
[2]Dept. of Anatomy, University of Zürich, Switzerland
[3]Dept. of Medicine III, University of Erlangen-Nürnberg, Germany

THE INITIAL PROBLEM

As soon there was compelling evidence that the spirochete *Borrelia burgdorferi* was the causative agent of Lyme disease (Burgdorfer et al. 1982), the first ultrastructural studies were performed on the non-specific cellular immune response to this pathogen. The examination of ultrathin sections with the electron microscope showed the presence of *B. burgdorferi* cells within phagolysosomes (Benach et al. 1984a,b; Peterson 1984).

However, considering the striking disproportion in size between the spirochetes and the phagocytes, the phagocytic mechanisms were difficult to understand. Thus, the apparent question was how the phagocytes could internalize the spirochetes in phagolysosomes, since these organisms were several times longer than the phagocytes. In order to integrate the unusual morphology of the spirochetes into the general model of phagocytosis, it was assumed that the spirochetes were somehow folded up to a package (Benach et al. 1984a). This folding would make the complete enclosure of the spirochetes within a phagolysosome possible. Indeed, in electron microscopical sections many phagocytes displayed intracellular vacuoles which contained more than one spirochete in section. This observation suggested that either several *B. burgdorferi* organisms were present in this vacuole or that one *B. burgdorferi* cell was turned back upon itself several times.

THE SECTIONAL UPTAKE

In order to obtain a better three-dimensional picture, we decided to perform an investigation using scanning electron microscopy (SEM). Adherent peripheral blood monocytes from human adults were incubated with *B. burgdorferi* cells and processed for SEM. Two findings emerged from this study (Rittig et al. 1993). Obviously, some kind of chemotaxins were released in the course of the co-incubation, because the monocytes grouped to cell clusters. Secondly, the spirochetes were not ingested in total as a package by the monocytes but rather stepwise, since segments of spirochetes were seen to project out of the phagocytosing monocytes. Regardless if there was some folding of the ingested spirochetes inside the phagosomes or not, this observation clearly demonstrated that the monocytes were not able to handle the spirochetes in a way that would allow the usual engulfment.

Lyme Borreliosis, Edited by J.S. Axford and
D.H.E. Rees, Plenum Press, New York, 1994

This was the first clue that some unusual features were evident with the uptake of the spirochetes by professional phagocytes. Much to our surprise we later were became aware that this observation had already been done on the related organism *Borrelia recurrentis*, the causative agent of louse-borne relapsing fever, more than hundred years ago. In a treatise on relapsing fever Muellendorff (1879) mentioned a feature termed "head of medusa", seen in the blood of patients suffering from this disease. Under the microscope conglomerations of leukocytes were seen, while numerous segments of spirochetes were radially projecting out of these leukocytes. Unlike some other early microscopists who had also observed this typical microscopic picture, however, Muellendorff correctly interpreted this feature as the uptake of *B. recurrentis* by the leukocytes.

In subsequent experiments we used confocal laser-scanning microscopy (CLSM) and video-enhanced microscopy (VEM) to monitor the phagocytosis of the spirochetes by adherent monocytes. With time-lapse records in VEM an interesting difference became apparent: the monocytes started to ingest viable motile spirochetes from either end of the string-shaped organisms. In contrast, they would engulf any part of non-viable spirochetes sinking as non-motile organisms onto the bottom of the well (unpublished data).

For CLSM, the spirochetal outer surface protein A (OspA) was immuno-labelled with a fluorochrome. Thereby, the extracellular segments of the spirochetes were displayed brightly. Along the intracellular segments, however, the staining was weak or totally absent. This could mean that either the OspA was altered early during the uptake of the spirochetes and lost its antigenicity - which has interesting implications concerning the specific immune response to OspA -, or the antibodies could not gain access to the antigen. To overcome this problem, the phagocyte-derived actin cortex surrounding the intracellular spirochete segments was labelled with an additional fluorochrome. Using this procedure both the extra- and intracellular routes of a given *B. burgdorferi* cell could be followed in a satisfactory way. Also, segments that were only attached to the surface of the monocytes could be distinguished from truly intracellular ones. It was evident that the same spirochete could enter and leave one phagocyte several times. In addition, different segments of one spirochete were engulfed independently by different phagocytes, an observation which could be confirmed by SEM (Rittig et al. 1993).

LEAKAGE OF LYSOSOMAL ENZYMES

There are interesting consequences of this sectional uptake. It has been known for long that under certain circumstances lysosomal enzymes may escape from the lysosomes into the extracellular space. One of these particular events is the uptake of long objects, where one end is already situated in a phagolysosome while the opposite end projects out of the phagocytes. Thus, lysosomal enzymes will leak along the projecting object into the extracellular space, an event termed "regurgitation during feeding" (Weissmann et al. 1971,1972). Both the morphology and the uptake characteristics predestine *B. burgdorferi* for that kind of leakage - together with other spirochete species, as we learned from Muellendorff (1879).

To elucidate the intracellular events following the sectional uptake of the spirochetes, different types of professional phagocytes such as peripheral blood monocytes, polymorphonuclear leukocytes, and synovial macrophages were incubated with *B. burgdorferi* organisms and processed for transmission electron microscopy (TEM). It was observed that the spirochetes were internalized by two different uptake mechanisms, conventional and coiling phagocytosis, resulting in different intracellular pathways (Rittig et al. 1992).

Investigating the conventional intracellular processing of the spirochetes, there was compelling evidence for the assumed mechanism of regurgitation during feeding. It was found that the phagolysosomes were continuous with the extracellular space via slit-like invaginations of the cell surface (Rittig et al. 1992,1993). For the uptake of smaller objects, which can be enclosed completely by the cytoplasm of the phagocytes, tubular connections comparable to these narrow clefts have been found using serial sections in TEM (Bowers 1980; Fan et al. 1982; Petersen and van Deurs 1983; Willingham and Pastan 1983).

In some TEM experiments we used the electron-dense dyes, tannic acid or ruthenium red to stain all cell membranes which were exposed to the extracellular space. Alternatively, the cell surface was labelled using gold-conjugated lectins. These experiments

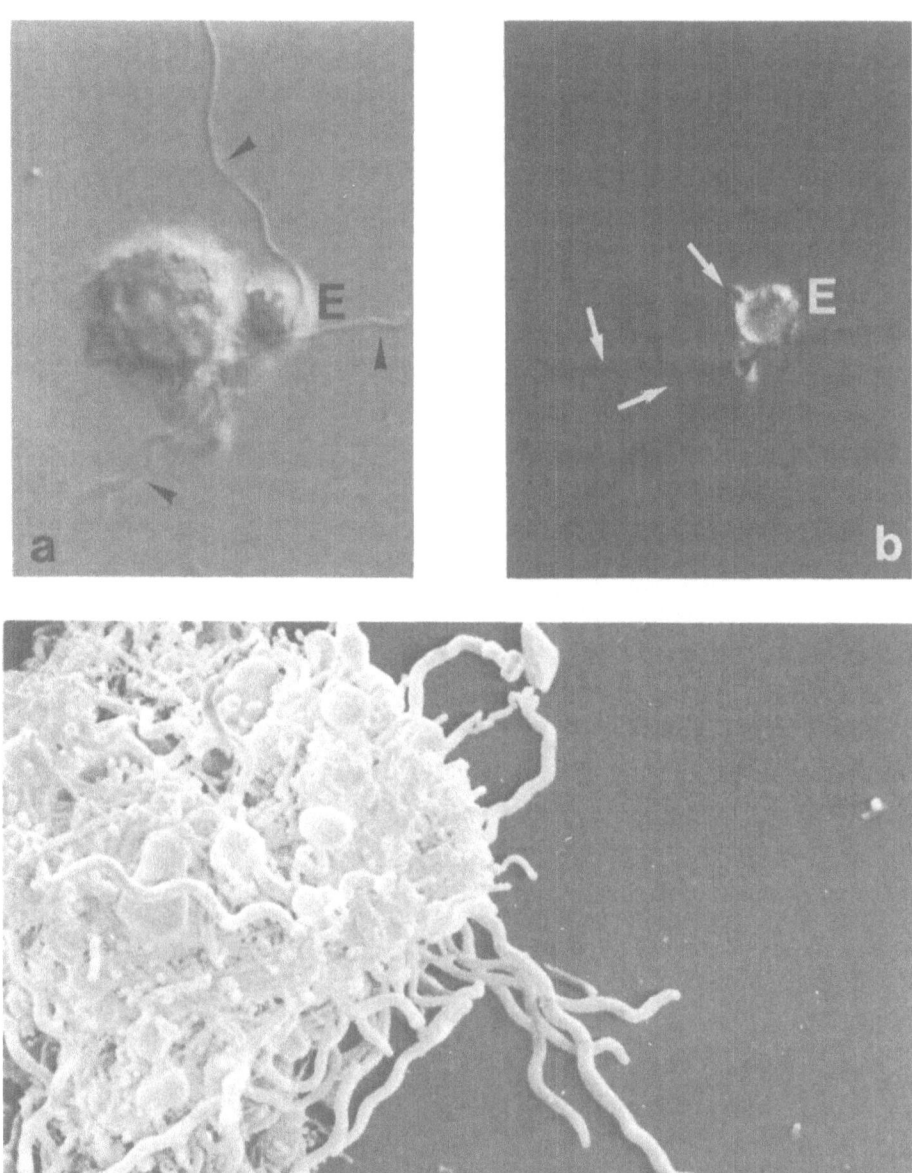

Figure 1. Sectional uptake of *B. burgdorferi*. The transmitted light DIC image [a] with the corresponding CLSM image [b] show that extracellular (arrowheads) and intracellular (arrows) segments of a spirochete are continuous (x5000). E = erythrocyte attached to the monocyte. [c] SEM micrograph of a phagocytosing monocyte. Numerous segments of spirochetes partially engulfed are projecting out of the cell (x15.000).

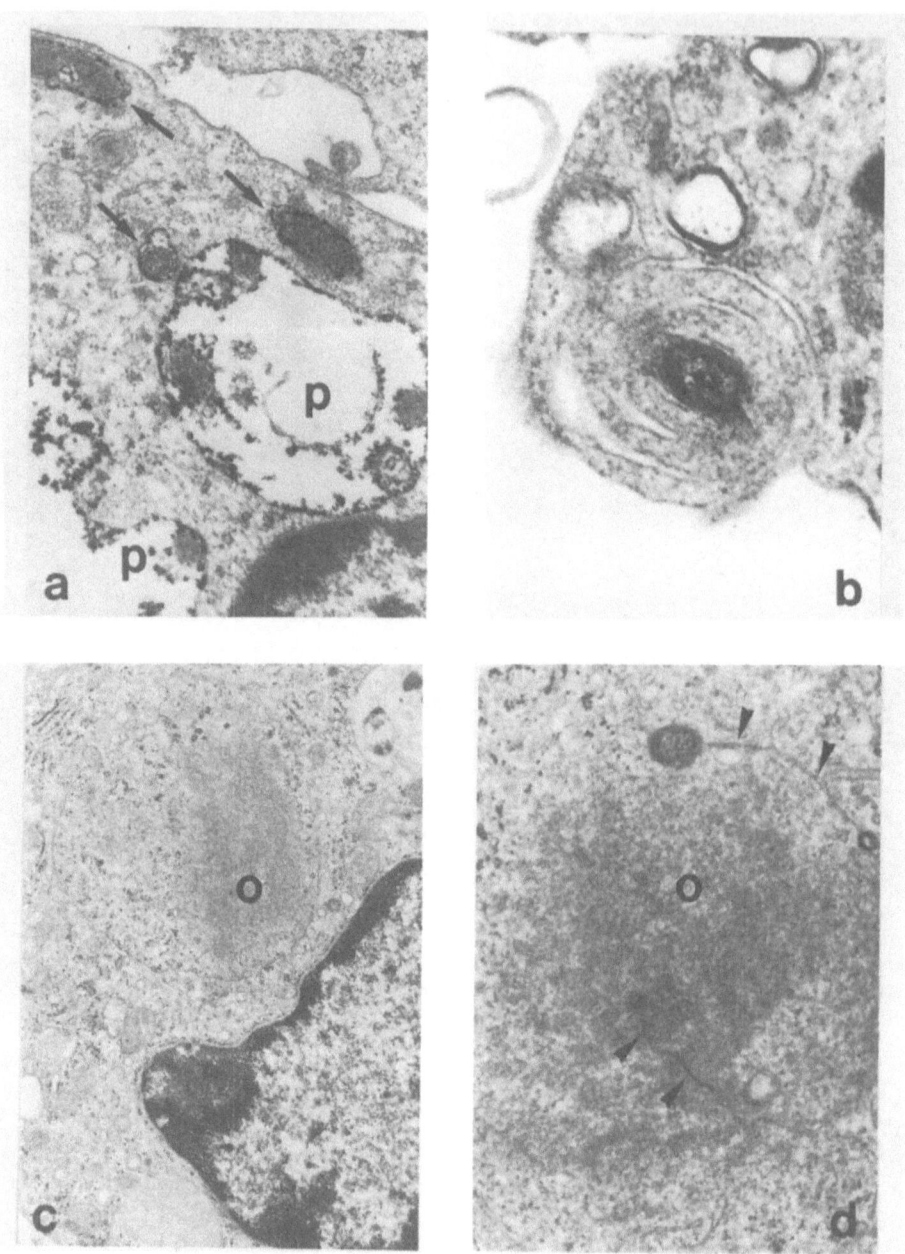

Figure 2. Leakage of lysosomal enzymes and cytosolic processing of *B. burgdorferi* as seen in TEM.
[a] Micrograph showing a monocyte with spirochetes ingested conventionally (x30.000). Reaction product of
acid phosphatase, a major degradative enzyme, is located both in the phagolysosomes (P) and along the
narrow clefts (arrows). [b] Coiling phagocytosis of a spirochete by a monocyte (x50.000), resulting in the
cytosolic degradation of the spirochetes within a specific organelle-exclusion zone (O), as seen in lower
[c; x10.000] and higher [d; x20.000] magnification. Note the remnants of the membrane clefts (arrowheads)
leading to the internalized spirochetes.

unequivocally demonstrated that the narrow clefts were accessible from the outside, and that even part of the intracellular vacuoles were stained (Rittig et al. 1992, 1993). As these vacuoles contained spirochete-derived debris in addition to the dye, it could be assumed that leakage of lysosomal enzymes was possible. Indeed, by means of histocytochemical methods the reaction products of acid phosphatase, which is a major lysosomal enzyme, was demonstrated both in the vacuoles and along the narrow clefts (Rittig et al. 1993). This leakage mechanism of regurgitation during feeding is likely to cause damage to adjacent cells and tissues, mediating the onset of inflammatory processes and immunologic reactions.

CYTOSOLIC PROCESSING

Coiling phagocytosis of *B. burgdorferi* by human neutrophils had been described occasionally as a phenomenon with unknown relevance (Szczepanski and Fleit 1988). When we performed semiquantification analysis of numerous TEM sections we surprisingly found that coiling phagocytosis was the preferential uptake mechanism for the spirochetes as compared to conventional phagocytosis (Rittig et al. 1992). The analysis of the processing sequence revealed an unusual pathway for the spirochetes engulfed by coiling phagocytosis. The spirochetes were released into the cytosol, but still were enclosed by an distinct cytoplasmic layer derived from the coiled pseudopod.

This pseudopod-derived area surrounding the enclosed spirochete created a specific cytoplasmic inclusion body (Rittig et al. 1993). Similar inclusion bodies have been found frequently in two rare variations of Alport's syndrome, the Fechtner syndrome (Peterson et al. 1985; Heynen et al. 1988; Takai et al. 1989) and the Sebastian platelet syndrome (Greinacher et al. 1990), respectively. Occasionally, such bodies have been observed in a case of a child suffering from an unclear disease (Zucker-Franklin 1975). The striking similiarity between the inclusion bodies found in a state of infection and those observed in inherited metabolic diseases shows a new aspect of phagocyte behaviour.

The *B. burgdorferi* cells in the center of this inclusion body obviously were degraded, but no lysosomal structures were seen in the vicinity of these spirochetes (Rittig et al. 1992). It can be argued that the pseudopod-derived cytoplasmic capsule surrounding the spirochetes may physically hinder the access of lysosomes. This capsule is rich in contractile filaments but poor in other cell organelles. The utilization of the host cell cytoskeleton to protect from lysosomes or to mediate cell-to-cell spread is known for some bacterial pathogens (reviewed by Falkow et al. 1992).

The mechanisms leading to the cytosolic degradation of *B. burgdorferi*, which is highly unusual for bacterial pathogens, are not quite clear. So far, the variety of non-lysosomal catabolic enzymes present in the cytosol have been thought to be involved in the turn-over of endogenous proteins, rather than in the destruction of invading organisms (reviewed by Hough et al. 1988; Orlowski 1990). Nevertheless, the ultrastructural investigations revealed that either uptake mechanism, conventional or coiling phagocytosis, resulted in the degradation of the internalized spirochetes. Thus, our results do not provide evidence of an intracellular persistance of *B. burgdorferi* in phagocytes, as has been proposed for mouse macrophages (Montgomery et al. 1993).

The non-lysosomal, cytosolic degradation of *B. burgdorferi* cells is likely to influence the subsequent antigen processing. In general, the location within the cell will determine the pathway of antigen processing (reviewed by Brodsky 1991; Harding 1991). Antigens derived from lysosomal degradation are expected to be processed for MHC class II-restricted presentation, whereas cytosolic degradation will lead preferentially to MHC class I-restricted antigen presentation. The different target cells of the MHC class I and II molecules may contribute to the apparent dissociation between the cellular and humoral immune reactivity in Lyme borreliosis (for further discussion of that topic see the article by Burmester et al. in this volume).

SUMMARY

The spirochete *Borrelia burgdorferi* is a considerably long organism, so that its uptake and processing require special means by the phagocytes. Morphological studies on these phagocytic events may give significant clues on the pathogenesis of Lyme disease.

The results of such studies show that the spirochetes were not ingested in total but rather stepwise, since segments of the spirochetes were projecting out of the phagocytes. This sectional uptake of the spirochetes strongly suggests leakage of lysosomal enzymes during conventional phagocytosis.

Following the uptake via coiling phagocytosis, the internalized sections of the *B. burgdorferi* cells were released into the cytosol, but remained enveloped by a dense cortex of cytoskeletal filaments. This envelope may hinder the access of vesicular cell organelles, as the spirochetes disintegrated without apparent participation of lysosomes. This cytosolic processing most likely results in MHC class I-restricted antigen processing.

Depending on the uptake mechanism used by the phagocytes, the internalized spirochetes were degraded either in phagolysosomes or in distinct cytoplasmic areas. There was no ultrastructural evidence of intracellular survival.

In conclusion, the morphological investigation of the uptake and processing of *B. burgdorferi* by human professional phagocytes provided several novel observations. It is concluded that both the sectional uptake and the cytosolic processing of the spirochetes may contribute to the pathogenesis of Lyme borreliosis.

REFERENCES

Benach JL, Fleit HB, Habicht GS, Coleman JL, Bosler EM, and Lane BP. 1984a. Interactions of phagocytes with the Lyme disease spirochete: role of the Fc receptor. J Infect Dis 150:497-507

Benach JL, Habicht GS, Gocinski BL, and Coleman JL. 1984b. Phagocytic cell responses to in vivo and in vitro exposure to the Lyme disease spirochete. Yale J Biol Med 57:599-605

Bowers B. 1980. A morphological study of plasma and phagosome membranes during endocytosis in *Acanthamoeba*. J Cell Biol 84:246-260

Brodsky FM. 1991. Intracellular routes for antigen processing and presentation. Res Immunol 142:453-458

Burgdorfer W, Barbour AG, Hayes SF, Benach JL, Grunwaldt E, and Davies JP. 1982. Lyme disease - a tick-borne spirochetosis? Science 216:1317-1319

Falkow S, Isberg RR, and Portnoy DA. 1992. The interaction of bacteria with mammalian cells. Annu Rev Cell Biol 8:333-363

Fan JY, Carpentier JL, Gorden P, van Obberghen E, Blackett NM, Grunfeld O, and Orci L. 1982. Receptor-mediated endocytosis of insulin: role of microvilli. PNAS (USA) 79:7788-7791

Greinacher A, Nieuwenhuis HK, and White JG. 1990. Sebastian platelet syndrome: a new variant of hereditary macrothrombocytopenia with leukocyte inclusions. Blut 61:282-288

Harding CV. 1991. Pathways of antigen processing. Curr Opin Immunol 3:3-9

Heynen MJ, Blockmans D, Verwilghen RL, and Vermylen J. 1988. Congenital macrothrombocytopenia, leucocyte inclusions, deafness and proteinuria: functional and electron microscopic observations on platelets and megakaryocytes. Brit J Haematol 70:441-448

Hough RF, Pratt GW, and Rechsteiner M. 1988. Ubiquitin/ATP-dependent protease. In: "Ubiquitin", Rechsteiner M (ed), Plenum Press, New York

Montgomery RR, Nathanson MH, and Malawista SE. 1993. The fate of *Borrelia burgdorferi*, the agent for Lyme disease, in mouse macrophages. J Immunol 150:909-915

Muellendorff J. 1879. Über Rückfallstyphus nach Beobachtungen im städtischen Krankenhaus zu Dresden 1879, Teil III. Dt Med Wschr 50:642-644

Orlowski M. 1990. The multicatalytic proteinase complex, a major extralysosomal proteolytic system. Biochem 29:10289-10297

Petersen OW, and van Deurs B. 1983. Serial-section analysis of coated pits and vesicles involved in adsorptive pinocytosis in cultured fibroblasts. J Cell Biol 96:277-281

Peterson LC, Rao KV, Crosson JT, and White JG. 1985. Fechtner syndrome - a variant of Alport's syndrome with leukocyte inclusions and macrothrombocytopenia. Blood 65:397-406

Peterson PK, Clawson CC, Lee DA, Garlich DJ, Quie PG, and Johnson RC. 1984. Human phagocyte interactions with the Lyme disease spirochete. Infect Immun 46:608-611

Rittig MG, Krause A, Häupl T, Schaible UE, Modolell M, Kramer MD, Lütjen-Drecoll E, Simon MM, and Burmester GR. 1992. Coiling phagocytosis is the preferential phagocytic mechanism for *Borrelia burgdorferi*. Infect Immun 60:4205-4212

Rittig MG, Häupl T, Krause A, and Burmester GR. 1993. Ultrastructural alterations in human professional phagocytes induced by the uptake of *Borrelia burgdorferi*. Am J Path (submitted for publication)

Szczepanski A, and Fleit HB. 1988. Interaction between *Borrelia burgdorferi* and polymorphonuclear leukocytes. Phagocytosis and the induction of the respiratory burst. Ann NYAS 539:425-428

Takai K, Sanada M, Hattori A, Koike T, and Shibata A. 1989. Fechtner syndrome: report of two families and review of the literature on the related disorders [english abstract]. Acta Haematol Jpn 52:644-654

Weissmann G, Zurier RB, and Hoffstein S. 1972. Leukocytic proteases and the immunologic release of lysosomal enzymes. Am J Pathol 68:539-559

Weissmann G, Zurier RB, Spieler PJ, and Goldstein IM. 1971. Mechanisms of lysosomal enzyme release from leukocytes exposed to immune complexes and other particles. J Exp Med 134:149s-165s

Willingham MC, and Pastan I. 1983. Formation of receptosomes from plasma membrane coated pits during endocytosis: analysis by serial sections with improved membrane labeling and preservation techniques. PNAS (USA) 80:5617-5621

Zucker-Franklin D. 1975. Physiological and pathological variations in the ultrastructure of neutrophils and monocytes. Clin Haematol 4:485-508

PHYSICAL AND GENETIC MAPS OF THE BORRELIA BURGDORFERI SENSU LATO CHROMOSOMES

I. Saint Girons[1], I.G. Old[1], C. Ojaimi[2], J. MacDougall[1] and B. E. Davidson[2]

[1]Unité de Bactériologie Moléculaire et Médicale Institut Pasteur, 75724 Paris Cedex 15, France Tel: 33 1 45 68 83 66, Fax 33 1 40 61 30 01
[2]Russell Grimwade School of Biochemistry, University of Melbourne, Parkville, Victoria 3051, Australia

ABSTRACT

Physical and genetic maps of the three species of *Borrelia burgdorferi* sensu lato associated with Lyme borreliosis were constructed. The ribosomal genes which are not organised as trancriptional units, were mapped at the centre of the linear chromosome. The region spanning *dna*A encoding the initiatior protein for chromosome replication was found to be atypical. *dna*A and genes normally associated with the origin of replication in eubacteria were found very close to the ribosomal RNA genes.

INTRODUCTION

The spirochaete, *Borrelia burgdorferi* was discovered in 1982 as the agent of Lyme borreliosis (1). Since then, three species have been delineated, namely *B. burgdorferi* sensu stricto, *B. garinii* and *B. afzelii* (2,3). The only species found in the USA is *B. burgdorferi* sensu stricto while all three species are found in Europe.

The lack of a system for genetic transfer has hindered conventional analysis of spirochaetes at the genetic level. However the analysis of large DNA fragments is possible using Pulsed Field Gel Electrophoresis (4). Eubacterial chromosomes are usually circular molecules which do not penetrate the pulsed field gels. The *Borrelia* chromosome is an exception in that it behaves as a linear molecule (5,6). In order to determine whether differences exist between the three species at the chromosomal level, we have constructed both physical and genetic maps for each to enable comparison of their chromosomes.

RESULTS

A restriction map of the chromosome of *B. burgdorferi* sensu stricto 212 was constructed (7). Digestion sites for eight different rare cutting endonucleases, *SgrAI*, *SmaI*, *SacII*, *MluI*, *BssHII*, *EagI*, *NaeI* and *ApaI* were located. The mapping and organization of the ribosomal RNA genes in *B. burgdorferi* was adressed. In procaryotes, the ribosomal RNA genes are usually organized as multiple operons with the order *rrs* (16S), *rrl* (23S), *rrf* (5S). There are however several exceptions and in *B. burgdorferi* it was shown that there was one *rrs* gene encoding 16S RNA separated from two *rrl* (23S) and two *rrf* (5S) genes which are organized in tandem (8,9). The ribosomal RNA genes have a central location within the linear chromosome (7).

Other genes isolated by us or others have been located throughout the chromosome (10). The physical maps of *B. burgdorferi*, *B. garinii* and *B. afzelii* chromosomes show significant differences in restriction enzyme patterns although the genetic maps are quite similar (Ojaimi, Old, Saint Girons and Davidson unpublished results).

Comparison between bacterial DNA replication origin sequences has allowed features essential to their function to be elucidated. This comparative approach also provided information about the genetic organization of the bacterial chromosome in the vicinity of the replication origin. The Pseudomonad origins lie in the promoter region for genes corresponding to *dnaA* (the initiator protein for chromosome replication), *dnaN* (DNA polymerase III beta subunit) and *recF* (inducer of SOS repair) while these genes are located 40 kb from *oriC* in the enterobacteria (11) (Figure 1A).

Among the clones which were randomly sequenced from a *B. burgdorferi* library, a *dnaA* gene homologue was identified (12). The organization of the *B. burgdorferi dnaA* region is atypical since *dnaA* and *dnaN* are inverted with respect to *rpmH* (ribosomal protein L34) and *gyrB* (DNA gyrase, beta-subunit) and no *recF* is found (Figure 1B). Furthermore no DnaA binding sites or DnaA boxes are found in the vicinity of *B. burgdorferi dnaA*. Another example where no DnaA boxes were found in the *dnaA* region is *Buchnera aphidicola*, a symbiont with a low GC% (13).

DISCUSSION

The dogma that eubacteria have a single circular chromosome has fallen. Different cases are found such as two circular chromosomes (14), a circular chromosome and a linear plasmid (15). *B. burgdorferi* is unique in that it has a linear chromosome and several linear and circular plasmids (5, 6, 16).

We have also shown that the arrangement of genes in the *dnaA* region is different from the consensus found in eubacteria. The *dnaA* gene is located at the centre of the linear chromosome, suggesting, by analogy with the case found in eubacteria, the close presence of the origin of replication. However, examination of the region near *gidA*, 300 kbp further away from *dnaA*, for the presence of DnaA boxes, characteristic of replication origin is under way. Further evidence such as experiments to show the functionality of an eventual replication origin would have nonetheless to be performed.

A. Eubacteria consensus

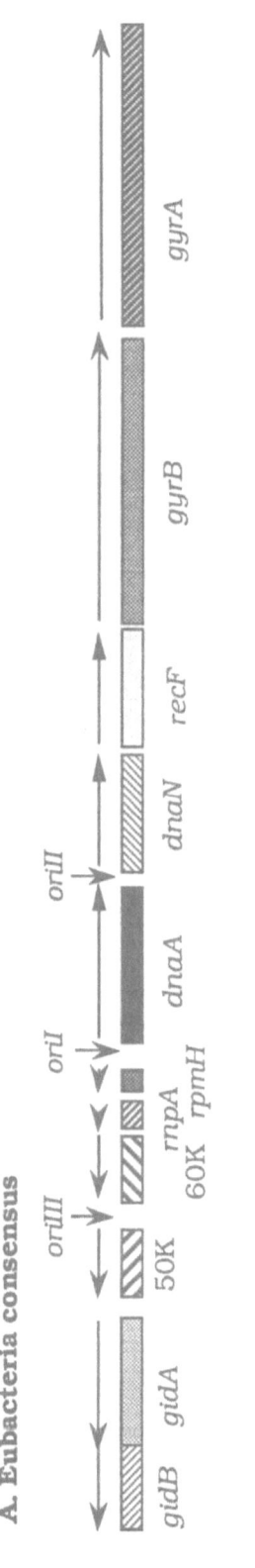

B. Borrelia burgdorferi

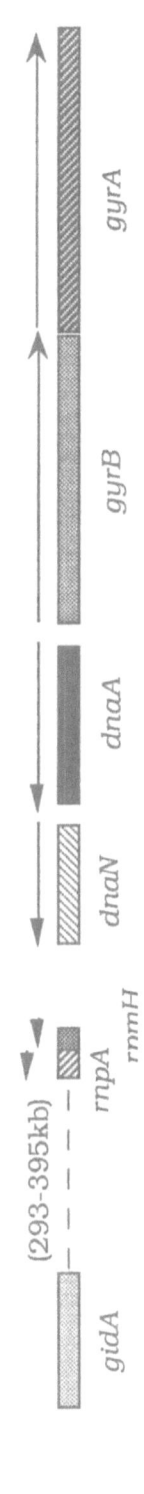

Figure 1

Consensus genetic organisation in the *dnaA* region of eubacteria (A) compared to *Borrelia burgdorferi* (B) (adapted from Smith et al. 1991). The direction of transcription is shown by single headed horizontal arrows while origin regions are marked by vertical arrows. The genes shown are: *gidA*, *gidB* (encoding glucose inhibited division proteins), *rnpA* (ribonuclease P, protein component), *rmpH* (ribosomal protein L34), *dnaA* (chromosome replication initiation protein), *dnaN* (DNA polymerase III, β-subunit), *recF* (inducer of SOS repair), *gyrB*, *gyrA* (DNA gyrase, β and α subunits). 50K and 60 K encode proteins of unknown function. In Gram negative bacteria, *gyrA* is located elsewhere on the chromosome and in *E. coli gidA*, *gidB*, and *oriC* have been translocated to 40 kb from the *dnaA* region. There is no *recF* gene in the *B. burgdorferi dnaA* region and homologues of *gidB* and genes encoding the 50 kDa and 60 kDa polypeptides have not yet been isolated.

251

REFERENCES

1. Burgdorfer W, Barbour A.G, Hayes S.F, Benach JL, Grunwaldt E, Davis J.P. Lyme Disease - A Tick-Borne Spirochetosis? *Science* 1982; **216**: 1317-1319.
2. Baranton G, Postic D, Saint Girons I, Boerlin P, Pifaretti J.C, Assous M, Grimont P.A.D. Delineation of *Borrelia burgdorferi* sensu stricto, *Borrelia garinii* sp. nov. and group VS461 associated with Lyme borreliosis. *Int J System Bacteriol* 1992; **42**: 378-383.
3. Marin Canica M, Nato F, du Merle L, Mazie J.C, Baranton G, Postic D. Monoclonal antibodies for identification of *Borrelia afzelii* sp. nov. associated with late cutaneous manifestations of Lyme borreliosis. *Scandinavian J of Infect Dis* 1993; in press.
4. Schwartz D.C, Cantor C.R. Separation of yeast chromosome sized DNAs by pulsed field gradient gel electrophoresis. *Cell* 1984; **37**: 67-75.
5. Baril C, Richaud C, Baranton G, Saint Girons I. Linear chromosome of *Borrelia burgdorferi*. *Res in Microbiol* 1989; **140**: 507-516.
6. Ferdows M.S, Barbour A.G. Megabase-sized linear DNA in the bacterium *Borrelia burgdorferi*, the Lyme disease agent. Proc Nat Acad Sci 1989; **86**: 5969-5973.
7. Davidson B, MacDougall J.H, and Saint Girons I. Physical map of the linear chromosome of the bacterium *Borrelia burgdorferi* 212, a causative agent of Lyme disease. *J Bacteriol* 1992;**174**: 3766-3774.
8. Schwartz J.J, Gazumyan A, Schwartz I. rRNA genes organization in the Lyme disease spirochete, *Borrelia burgdorferi*. *J Bacteriol* 1992; **174**: 3757-3765.
9. Fukunaga M, Sohnaka M. *Biochem Biophys Res Comm* 1992; **183**: 952-957. Tandem repeat of the 23S and 5S ribosomal RNA genes in *Borrelia burgdorferi*, the etiological agent of Lyme disease.
10. Old I, MacDougall J.H, Saint Girons I, Davidson B.E. Mapping of genes on the linear chromosome of the bacterium *Borrelia burgdorferi*: possible locations for its origin of replication. *FEMS Microbiol Lett* 1992; **99**: 245-250.
11. Smith D.W, Yee T.W, Baird C. Krishnapillai V. Pseudomonad replication origins: a paradigm for bacterial origins? *Molecular Microbiology* 1991; **5**: 2581-2587.
12. Old I, Margarita D, Saint Girons I. Unique genetic arrangement in the *dna*A region of the *Borrelia burgdorferi* chromosome: nucleotide sequence of the *dna*A gene. *FEMS Microbiol Lett* 1993; in press.
13. Lai C.Y, Baumann P. Genetic analysis of an aphid endosymbiont DNA fragment homologous to the *rnp*A-*rpm*H-*dna*A-*gyr*B region of eubacteria. *Gene* 1992; **113**: 175-181.
14. Suwanto A, Kaplan S. Physical and genetic mapping of the Rhodobacter sphaeroides 2. 4. 1. genome: presence of two unique circular chromosomes. *J Bacteriol* 1989; **171**: 5850-5859.
15. Kinashi H, Shimaji M, Sakai A. Giant linear plasmids in *Streptomyces* which code for antibiotic biosynthesis genes. *Nature* 1987; **328**: 454-456.
16. Barbour A. G, Garon C.F. Linear plasmids of the bacterium *Borrelia burgdorferi* have covalently closed ends. *Science* 1987; **237**: 409-411.

REPEATED DNA SEQUENCES ON CIRCULAR AND LINEAR PLASMIDS OF *BORRELIA BURGDORFERI* SENSU LATO

Wolfram R. Zückert,[1] Elisabeth Filipuzzi-Jenny,[1] Jennifer Meister-Turner,[2] Margaretha Stålhammar-Carlemalm,[2] and Jürg Meyer[1]

[1]Department of Preventive Dentistry and Oral Microbiology, University of Basel Dental Institute, Petersplatz 14, CH-4051 Basel, Switzerland
[2]Department of Microbiology, Biozentrum, University of Basel, Klingelbergstrasse 70, CH-4056 Basel, Switzerland

ABSTRACT

We cloned a 3.8 kb *Eco*RI fragment from the 29 kb circular plasmid of the *Borrelia burgdorferi* sensu stricto type strain B31, which hybridized to several additional restriction fragments of the 29 kb circular and the 50 kb linear plasmid of this strain (1.4 kb, 6.4 kb, 7.0 kb *Eco*RI and 4.5 kb *Hin*dIII, respectively). Three of the latter fragments were also cloned and their crosshybridizing segments identified on subclones.

Partial DNA sequence analysis of three fragments disclosed 5' regions of putative open reading frames longer than 600 bp displaying 85-95 % homology among each other.

Homologies on linear and circular plasmids of isolates of *B. burgdorferi* sensu stricto, *B. garinii* and the genospecies VS461 were detected as well, but none on *B. coriaceae*, *B. hermsii*, *B. parkeri* and *B. turicatae* genomic DNA. Thus the repeated DNA seems to be specific for *B. burgdorferi* sensu lato and therefore may be useful in nucleic acid based diagnosis of Lyme disease.

INTRODUCTION

Lyme borreliosis is a multisystemic infectious syndrome caused by the spirochete *Borrelia burgdorferi* sensu lato[1] and mimicking several disorders, infectious or non-infectious.[2] Laboratory diagnosis by cultural identification of *Borrelia* or by antibody determination are often unsatisfactory.[3] The use of DNA based diagnostics e.g. PCR[4-7] has so far been hampered by the genetic heterogeneity of isolates. Recently, the species has been partitioned into *B. burgdorferi* sensu stricto (*Bb*), *B. garinii* (*Bg*), and the genospecies VS461.[8]

We have studied the genetic variations among *B. burgdorferi* sensu lato isolates and their relationships by determining chromosomal restriction fragment length polymorphisms (RFLPs) and plasmid profiles.[9-11] Among the DNA probes used in Southern blot hybridizations, a 3.8 kilobase (kb) *Eco*RI fragment originating from the circular plasmid fraction of the *Bb* type strain B31 hybridized to several additional restriction fragments of circular and linear plasmids of this strain. In order to determine whether they contain repeated DNA, we cloned the fragments of interest using *E. coli* host-vector systems, and started to analyze the homologous regions at the DNA sequence level. We also studied the presence of homologies on circular and linear plasmids of six other strains of *B. burgdorferi* sensu lato as well as on genomes of four relapsing fever *Borrelia* species.

MATERIALS AND METHODS

The procedures followed the descriptions of Sambrook et al.[12] unless stated differently.

Bacterial strains, plasmid vectors and culture conditions

Borrelia **strains.** *Bb* strains B31, S1, NE56, *Bg* strains NE2, NE83, VS464 and strain VS461 were used.[9,11] *B. parkeri* and *B. coriaceae* strains were obtained from O. Péter (Sion), *B. hermsii* and *B. turicatae* strains from G. Gassmann (Freiburg i.Br.). All strains were grown in BSKII medium[13] at 34 °C.

E. coli **K12 strains and plasmid vectors.** JM103 or XL1-Blue (Stratagene) strains were used for propagation of pUC19,[14] pBluescript SK⁻ (Stratagene) and recombinant plasmids. Strains were grown in Luria broth or on Luria agar supplemented with ampicillin (100 µg/ml) for selection of recombinant plasmids.

DNA techniques

DNA prepariation. Chromosomal, circular and linear plasmid DNA was isolated from the seven *B. burgdorferi* sensu lato strains by CsCl-EtBr gradient centrifugation.[9] Genomic DNA was isolated from *B. coriaceae*, *B. hermsii*, *B. parkeri* and *B. turicatae* strains by a modified CTAB method.[15] Recombinant plasmids were obtained by "shot-gun" cloning of digested chromosomal, circular and linear plasmid DNA. Transformation of *E. coli* strains was achieved by electroporation (ECM 600 system, Biotechnologies & Experimental Research Inc.).

Agarose gel electrophoresis. Unrestricted circular plasmids, restriction digests of chromosomal, circular and linear plasmid *B. burgdorferi* sensu lato DNA as well as recombinant plasmids were separated by conventional electrophoresis. Separation of unrestricted linear plasmids was achieved by pulsed field electrophoresis[11] (CHEF System, Bio-Rad).

Hybridization. DNA transfer to nylon membranes (0.2 µm Pall Biodyne A) was achieved by capillary[16] or vacuum (VacuGene XL apparatus, Pharmacia LKB) Southern blotting. The inserts of recombinant plasmids used as probes were isolated from agarose gels (SeaPlaque, FMC BioProducts) and labelled with peroxidase using the ECL kit (Amersham). The same probes were used in dot blot hybridizations.

DNA Sequencing. Double-stranded DNA sequencing was performed using the T7 (Pharmacia) and Sequenase Version 2.0 (United States Biochemical) sequencing kits according to the manifacturers' instructions.

RESULTS AND DISCUSSION

Homologies on circular and linear plasmids of *B. burgdorferi* sensu lato strains

The recombinant plasmid pOMB10 contains a 3.8 kb *Eco*RI fragment originating from the circular plasmid fraction of *Bb* strain B31 which displayed homologies to the 29 kb circular and 50 kb linear plasmids of this strain. Furthermore, the insert of pOMB10 hybridized to several circular as well as linear plasmids of six other isolates (Figures 1 and 2). Interestingly, *B. burgdorferi* sensu lato strains which are closely related according to chromosomal RFLPs[9,11] e.g. NE2 and NE83 may carry the repeated sequences on plasmids of different sizes.

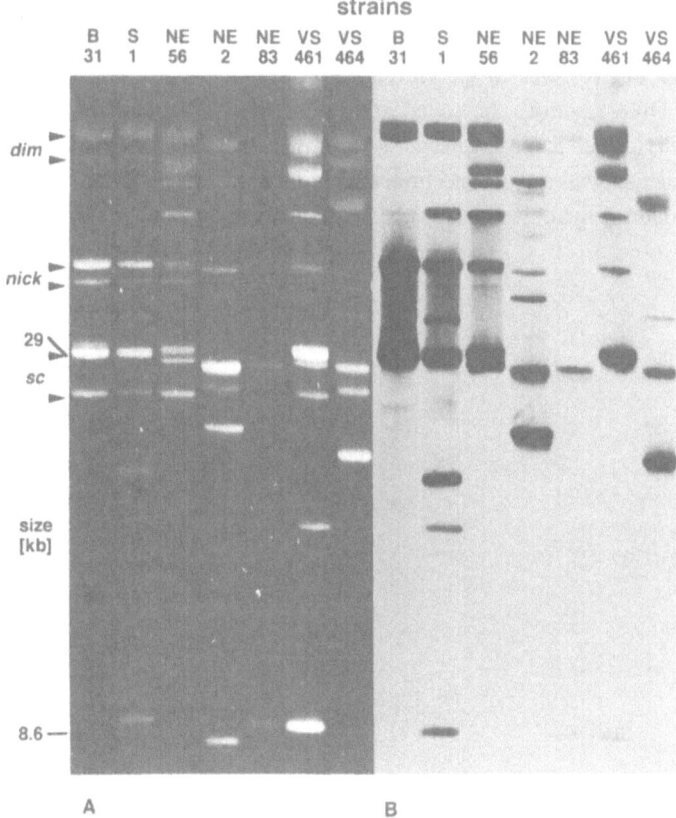

Figure 1. Homologies to pOMB10 on circular plasmids. A: Separation of the circular plasmid fractions of seven strains by electrophoresis in a 0.4% agarose gel; B: Southern blot hybridization using the insert of pOMB10 as a probe; *dim*, *nick* and *sc* indicate putative dimer, nicked and supercoiled forms of the circular plasmids of strain B31; numbers indicate plasmid sizes in kb (according to Stålhammar-Carlemalm et al.[9]).

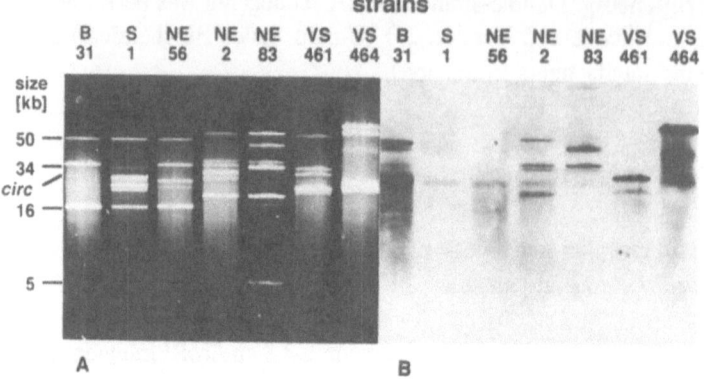

Figure 2. Homologies to pOMB10 on linear plasmids. A: Separation of the linear plasmid fractions of seven strains by pulsed field electrophoresis in a 1.0% agarose gel; B: Southern blot hybridization using the insert of pOMB10 as a probe; *circ* indicates the hybridization signal with a 29 kb fragment presumably representing contaminating linearized 29 kb circular plasmids of strain B31; numbers indicate plasmid sizes in kb (according to Stålhammar-Carlemalm et al.[9]).

The repeated DNA appears to be absent from the chromosome, because the insert of pOMB10 failed to hybridize to the chromosomal fraction of *Bb* strain B31 (Figure 3). Chromosomal DNA fractions, however, appeared to be contaminated by linear forms of circular plasmids as well as linear plasmids (Figure 3). It is therefore not surprising that the same *Eco*RI fragment was also cloned from the chromosomal DNA fraction[9] and, when used as a probe in Southern hybridizations, revealed differences between strains, presumably at the plasmid level.

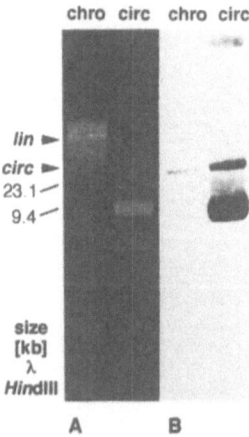

Figure 3. Absence of the repeated DNA from the chromosome. A: Separation of the chromosomal (chro) and circular plasmid (circ) fraction of strain B31 by electrophoresis in a 0.4% agarose gel; B: Southern blot hybridization using the insert of pOMB10 as a probe; *circ* and *lin* most likely indicate hybridization signals with contaminating linearized 29 kb circular plasmid and 50 kb linear plasmid of strain B31; numbers indicate sizes of λ DNA marker fragments in kb.

Cloning and characterization of the repeated DNA

DNA fragments hybridizing to pOMB10 were cloned from *Bb* strain B31 (Figure 4 and Table 1). Simpson et al.[17] recently cloned repeated DNA from *Bb* strain Sh-2-82. Two of their clones, pSPR13 and pSPR14, but not the third, pSPR9, gave weak hybridization signals with the insert of pOMB10, indicating some relationship (Figure 5).

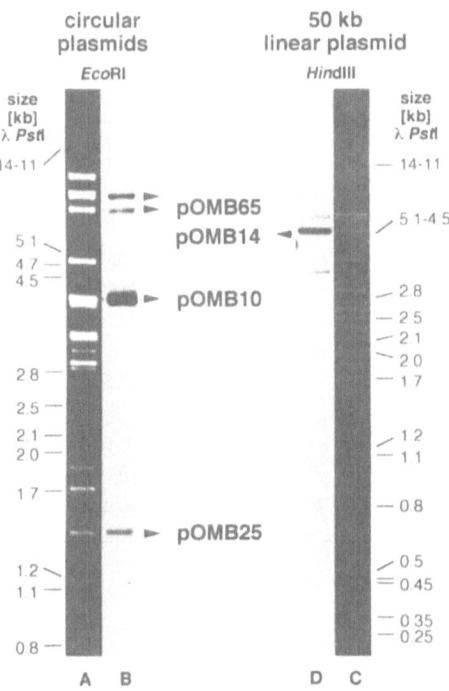

Figure 4. Restriction fragments homologous to pOMB10. A: Separation of an *Eco*RI restriction digest of the circular plasmid fraction of strain B31 by electrophoresis in a 1.0% agarose gel; B: Southern blot hybridization using the insert of pOMB10 as a probe; C: Separation of a *Hind*III restriction digest of the previously purified 50 kb linear plasmid of strain B31 by electrophoresis in a 1.0% agarose gel; D: Southern blot hybridization using the insert of pOMB10 as a probe; arrows labeled with pOMB indicate cloned fragments; the 7.0 kb *Eco*RI fragment from the circular plasmid fraction labelled with an arrow alone has not been cloned yet; numbers indicate sizes of λ DNA marker fragments in kb.

Table 1. Recombinant plasmids containing *Bb* strain B31 DNA

Plasmid	insert size	restriction fragment	origin
pOMB10	3.8 kb	*Eco*RI	29 kb circular plasmid
pOMB25	1.4 kb	*Eco*RI	29 kb circular plasmid
pOMB65	6.5 kb	*Eco*RI	29 kb circular plasmid
pOMB14	4.5 kb	*Hind*III	50 kb linear plasmid

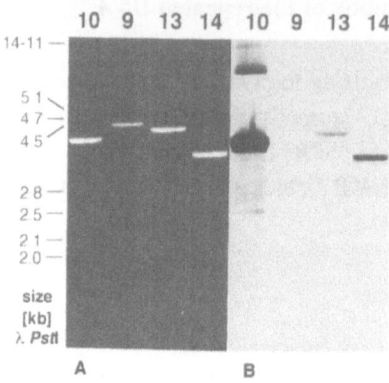

Figure 5. Homology of pOMB10 to cloned repeated sequences of *Bb* strain Sh-2-82. A: Separation of previously purified inserts of pOMB10 (10), pSPR9 (9), pSPR13 (13), and pSPR14 (14) by electrophoresis in a 1.0% agarose gel; B: Southern blot hybridization under low stringency conditions using the insert of pOMB10 as a probe; numbers on the left indicate sizes of λ DNA marker fragments in kb.

Restriction maps of the inserts of the recombinant plasmids were established (Figure 6). A region of homology surrounding an *Xba*I site common to pOMB10, pOMB25 and pOMB14 was determined by crosswise Southern hybridizations (results not shown) of the inserts of pOMB10, pOMB65, pOMB25 and pOMB14 to their respective subfragments.

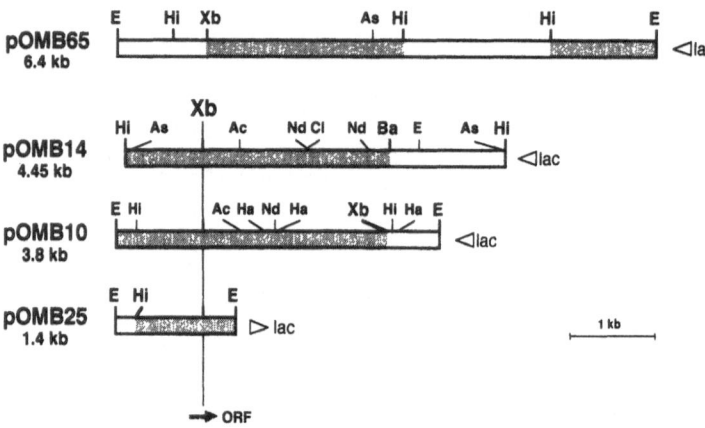

Figure 6. Restriction maps of pOMB10, pOMB25, pOMB65, and pOMB14. Restriction sites are indicated as follows: Ac (*Acc*I), As (*Asp*700), Ba (*Bam*HI), Cl (*Cla*I), E (*Eco*RI), Ha (*Hae*III), Hi (*Hin*dIII), Nd (*Nde*I), Xb (*Xba*I); open triangles indicate the direction of the lac operon of pBluescriptII SK⁻; pOMB10, pOMB25 and pOMB14 are aligned at a common *Xba*I site; shaded boxes indicate the regions thought to be mainly contributing to crosshybridization; a shaded arrow indicates the location and direction of the putative ORF.

Preliminary DNA sequencing data on this region indicated that pOMB10, pOMB25 and pOMB14, respectively, share 85-95 % sequence homology and 5' ends of putative open reading frames (ORFs) longer than 600 basepairs (bp) (results not shown). Whether these sequences are being transcribed and yield a polypeptide, is under investigation.

Species specificity of the repeated DNA

Hybridizations of pOMB10 to genomic dot blots of *B. coriaceae*, *B. hermsii*, *B. parkeri* and *B. turicatae* strains under low stringency conditions showed an estimated 100- to 500-fold decrease in signal strength with the DNA of relapsing fever *Borreliae* as compared to *Bb* strain B31 (Figure 7). Under high stringency conditions, however, no hybridization signals except to B31 DNA were observed (results not shown). This suggests that the repeated DNA is specific for *B. burgdorferi* sensu lato.

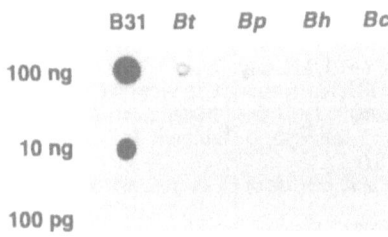

Figure 7. Species specificity of the repeated DNA. Genomic DNA of *B. coriaceae* (Bc), *B. hermsii* (Bh), *B. parkeri* (Bp) and *B. turicatae* (Bt) as well as of *Bb* strain B31 (B31) was blotted in decreasing amounts (100 ng, 10 ng and 100 pg); the dot blot was then hybridized under low stringency conditions using the insert of pOMB10 as a probe.

Repeated DNA sequences have been successfully used in nucleic acid based diagnostics of infectious diseases, either as primers for PCR[18] or as probes in hybridization assays.[19] Future experiments will evaluate the suitability of the repeated sequences of *B. burgdorferi* sensu lato for use in the diagnosis of Lyme borreliosis.

ACKNOWLEDGEMENTS

We thank O. Péter (Sion) and G. Gassmann (Freiburg i. Br.) for bacterial strains, C.F. Garon (Hamilton) for plasmids pSPR, S. Rijpkema (Bilthoven) for advice, and H. Sandmeier (Basel) for comments on the manuscript. This work was supported by grant Nr. 31-25680.88 of the Swiss National Science Foundation.

REFERENCES

1. W. Burgdorfer, A.G. Barbour, S.F. Hayes, J.L. Benach, E. Grunwaldt, and J.P. Davis. Lyme disease - a tick-borne spirochetosis? *Science* 216:1317-1319 (1982)
2. P.H. Duray and A.C. Steere. Clinical pathologic correlations of Lyme disease by stage. *Ann. N. Y. Acad. Sci.* 539:65-79 (1988)

3. A.G. Barbour. Laboratory aspects of Lyme borreliosis. *Clin. Microbiol. Rev.* 1:399-414 (1988)

4. P.A. Rosa and T.G. Schwan. A specific and sensitive assay for the Lyme disease spirochete *Borrelia burgdorferi* using the polymerase chain reaction. *J. Infect. Dis.* 160:1018-1029 (1989)

5. R.N. Picken. Polymerase chain reaction primers and probes derived from flagellin gene sequences for specific detection of the agents of Lyme disease and North American relapsing fever. *J. Clin. Microbiol.* 30:99-114 (1992)

6. J.L. Goodman, P. Jurkovich, J.M. Kramber, and R.C. Johnson. Molecular detection of persistent *Borrelia burgdorferi* in the urine of patients with active Lyme disease. *Infect. Immun.* 59:269-278 (1991)

7. W.H. Krüger and M. Pulz. Detection of *Borrelia burgdorferi* in cerebrospinal fluid by the polymerase chain reaction. *J. Med. Microbiol.* 35:98-102 (1991)

8. G. Baranton, D. Postic, I. Saint Girons, P. Boerlin, J.C. Piffaretti, M. Assous, and P.A.D. Grimont. Delineation of *Borrelia burgdorferi* sensu stricto, *Borrelia garinii* sp. nov., and group VS461 associated with Lyme borreliosis. *Int. J. Syst. Bact.* 42:378-383 (1992)

9. M. Stålhammar-Carlemalm, E. Jenny, L. Gern, A. Aeschlimann, and J. Meyer. Plasmid analysis and restriction fragment length polymorphisms of chromosomal DNA allow a distinction between *Borrelia burgdorferi* strains. *Zbl. Bakt.* 274:28-39 (1990)

10. J. Meister-Turner, E. Filipuzzi-Jenny, O. Péter, A.G. Bretz, M. Stålhammar-Carlemalm, and J. Meyer. Genotypic and phenotypic diversity among nine Swiss isolates of *Borrelia burgdorferi*. *Zbl. Bakt.* in press (1993)

11. E. Filipuzzi-Jenny, M. Blot, N. Schmid-Berger, J. Meister-Turner, and J. Meyer. Genetic diversity among *Borrelia burgdorferi* isolates - more than three genospecies? *Res. Microbiol.* in press (1993)

12. J. Sambrook, E.F. Fritsch, and T. Maniatis. "Molecular Cloning - A Laboratory Manual," Cold Spring Harbor Laboratory Press, Cold Spring Harbor N.Y. (1989)

13. A.G. Barbour. Isolation and cultivation of Lyme disease spirochetes. *Yale J. Biol. Med.* 57:521-525 (1984)

14. C. Yanisch-Perron, J. Vieira, and J. Messing. Improved M13 phage cloning vectors and host strains - nucleotide sequences of the M13mp18 and pUC19 vectors. *Gene* 33:103-119 (1985)

15. K. Wilson, Preparation of genomic DNA from bacteria, *in*: "Current Protocols in Molecular Biology," F.M. Ausubel, R. Brent, R.E. Kingston, D.D. Moore, J.G. Seidman, J.A. Smith, and K. Struhl, eds., Wiley, New York N.Y. (1987)

16. K.C. Reed and D.A. Mann. Rapid transfer of DNA from agarose gels to nylon membranes. *Nucl. Acids Res.* 13:7207-7221 (1985)

17. W.J. Simpson, C.F. Garon, and T.G. Schwan. *Borrelia burgdorferi* contains repeated DNA sequences that are species specific and plasmid associated. *Infect. Immun.* 58:847-853 (1990)

18. K.D. Eisenach, M.D. Cave, J.H. Bates, and J.T. Crawford. Polymerase chain reaction amplification of a repetitive DNA sequence specific for *Mycobacterium tuberculosis*. *J. Infect. Dis.* 161:977-981 (1990)

19. R.L. Zuerner and C.A. Bolin. Repetitive sequence element cloned from *Leptospira interrogans* serovar hardjo type hardjo-bovis provides a sensitive diagnostic probe for bovine leptospirosis. *J. Clin. Microbiol.* 26:2495-2500 (1988)

BIOCHEMICAL AND IMMUNOLOGICAL ANALYSIS OF A POLYMORPHIC

LOW-MOLECULAR-WEIGHT LIPOPROTEIN OF *BORRELIA BURGDORFERI*

Reinhard Wallich,[1] Claudia Helmes,[1] Ulrich E. Schaible,[2] Michael D. Kramer,[3] and Markus M. Simon[2]

[1]Applied Immunology, FS0440, dkfz, INF280, 69120 Heidelberg, Germany
[2]Max-Planck-Institut für Immunbiologie, Freiburg, Germany
[3]Institut für Immunologie und Serologie, Universität Heidelberg, Germany

ABSTRACT

The cloning, expression and molecular characterization of a novel polymorphic *B. burgdorferi* lipoprotein recognized by monoclonal antibody LA7 is described. lpLA7 shows sequence similarity with other prokaryotic lipoproteins, including the immunodominant *B. burgdorferi* outer surface proteins OspA, OspB, OspC and OspD.

Upon solubilization of intact *B. burgdorferi* with the non-ionic detergent Triton X-114, lpLA7 was extracted together with other lipoproteins into the detergent phase. Indirect immuno-labeling studies indicated that the epitope recognized by mAb LA7 is mainly localized to the periplasmic space. Two dimensional gel electrophoresis and immunoblotting confirmed the calculated acidic pI of 5.7 for lpLA-7. The *LA7* gene is shown to be species specific and to be located on the linear chromosome of *B. burgdorferi*. The analysis of 40 individual spirochetal isolates by restriction fragment length polymorphism analysis revealed a considerable genotypic heterogeneity of *LA7* corresponding to that previously found for *OspA*.

Native and recombinant lpLA-7 were recognized by immune sera from infected mice as well as some human sera and may also prove useful as an additional marker for serodiagnosis of Lyme disease.

Lyme Borreliosis, Edited by J.S. Axford and
D.H.E. Rees, Plenum Press, New York, 1994

INTRODUCTION

Lyme disease is the most common vector borne infectious disease of the temperate climate. The etiological agent, the spirochete *Borrelia burgdorferi*, causes a multisystemic illness in humans which may affect skin, nervous system, joints and heart (Steere, 1990). *B. burgdorferi* strains isolated from different biological sources and geographic areas are heterogeneous and it is assumed, that the patterns of disease manifestations are influenced by antigenic differences of the spirochetal strains (Barbour et al., 1983; Wallich et al., 1992; Wilske et al., 1988). Although several attempts to classify *B. burgdorferi* by immunological or molecular criteria have recently been reported, the taxonomy of *B. burgdorferi* is still a matter of controversy and active research (Wilske et al., 1992).

To date, a variety of *B. burgdorferi* antigens such as outer surface protein A (OspA), OspB, pC and p100, are employed for serological diagnosis and as putative candidates for vaccine development (Simon et al., 1991). However, because of their obvious heterogeneity unambiguous criteria for the generation of a species-specific diagnostic standard and for the development of a polypeptide vaccine that would guarantee protection against any subspecies are still lacking. Therefore, more detailed information on the polymorphism of immunologically relevant *B. burgdorferi* component(s) is urgently required.

We have described previously a series of monoclonal antibodies (mAb) that recognize distinct *B. burgdorferi* associated antigens (Kramer et al., 1990). In the present paper we describe the genetics and further biochemical characterization of a novel polymorphic *B. burgdorferi* lipoprotein recognized by mAb LA7.

RESULTS AND DISCUSSION

A pUEX1 expression library of *B. burgdorferi* ZS7 genomic DNA was screened using mAb LA7. Three independent recombinant clones were identified. The physical maps of these three plasmid inserts (pLA7-2, pLA7-7, and pLA7-8) and some restriction sites are shown in Fig. 1.

The molecular weight of the protein predicted from the amino acid sequence of 194 residues is 21.865 Da. Molecular analysis and sequence comparison of lpLA7 with other proteins reveals sequence similarity to the signal peptides of murein lipoproteins (Wu and Tokunaga, 1986; Wu, 1987). Cleavage at the cysteine residue by signal peptidase II would yield a polypeptide with a predicted mass of 19.341 Da.

The RFLP analysis for *LA7* using endonuclease BamHI or HindIII revealed at least six distinct hybridization patterns among 40 *B. burgdorferi* isolates tested (Table 1).

The distinct patterns of the *LA7*-specific RFLP correlated with the differential expression of the epitope defined by mAb LA7. Only *B. burgdorferi* strains of the OspA genotypes I, III and V but not those of the OspA genotype II, IV and VI reacted with mAb LA7 (Table 1).

In contrast to the *OspA* gene which is located on the 49-kilobase extrachromosomal linear plasmid, lpLA7 is encoded by a chromosomal gene as elucidated by pulse field gel electrophoresis (data not shown). DNAs isolated from members of other species of *Borrelia* did not hybridize to the *LA7*-specific probe indicating specificity for *B. burgdorferi*.

To analyse the expression of lpLA7 in *B. burgdorferi* organisms, whole-cell lysates were separated on 2D gels and subsequently either silver stained (Fig. 2A) or immuno-blotted using mAb LA7 and as controls mAbs directed against OspA, OspB and flagellin (Fig. 2B). lpLA7 is clearly separated from other low molecular weight proteins. Another 20-kDa protein migrated at a charge position similar to that of OspB and reacted with OspB mAb LA32 (Kramer et al., 1990).

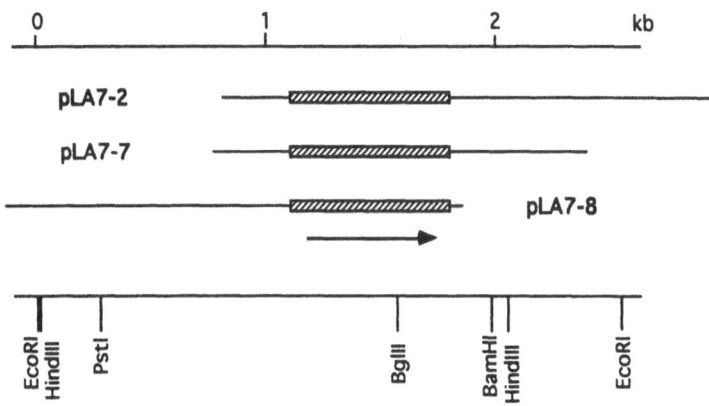

Figure 1. Restriction map of three overlapping B. burgdorferi DNA fragments cloned in pUEX1. The region corresponding to the protein encoding part of the LA7 gene is shown as a hatched box. Arrow indicates the direction of transcription.

ZS7 *B. burgdorferi* organisms were labeled with [^{3}H]palmitate and subsequently extracted by Triton X-114. Radiolabeled proteins were separated by 2D gel electrophoresis. The proteins identified included lpLA7, OspA and other as yet unidentified molecules (Fig. 3). Due to the limited pI range from 4 to 8 the OspB molecule is not seen, which exhibits a calculated pI of 9.7.

To explore the localization of the LA7 epitope in intact spirochetes, mAb LA7 as well as a selected range of mAbs directed against different *B. burgdorferi* structures were tested for staining the bacteria by immunofluorescence (data not shown). To further elucidate the localization of lpLA7 protein, intact *B. burgdorferi* ZS7 organisms were treated with proteinase K and subsequently analysed by SDS-PAGE (Fig. 4). The bands corresponding to OspA, OspB and a 24-kDa protein showed a decreased intensity compared to the untreated control. In contrast, lpLA7 was resistent to degradation under these conditions, as revealed by Western blot analysis using mAb LA7.

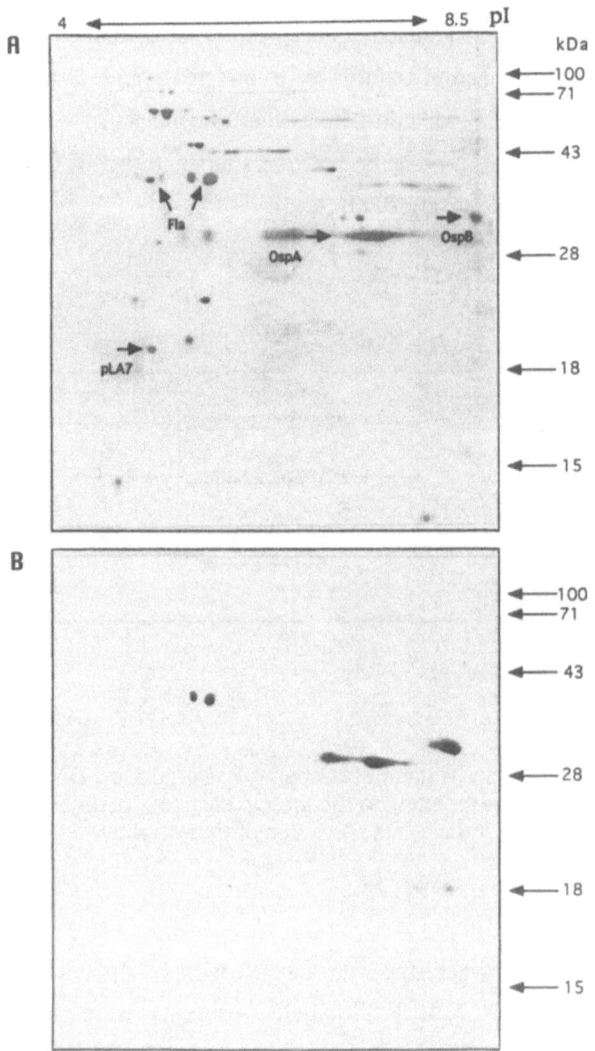

Figure 2. 2D gels of *B. burgdorferi* whole cell lysates (first dimension, pH 4 to 8.5; second dimension, 13% SDS-PAGE). Gels were either silver stained (A) or electrophoretically transferred to a nylon membrane and reacted with a mixture of mAbs LA7, LA21 (flagellin), LA2 (OspA) and LA32 (OspB) (B). Molecular masses are indicated on the right and the pI values on top. The arrowhead indicates proteins subsequently identified by immunoblotting.

Table 1: *B. burgdorferi* isolates used in this study.

Borrelia burgdorferi isolate	Biological origin*	Geographical origin	Genotype (fla,HSP60/Hsp70)	Genotype (OspA)	Reactivity mAb LA7 (Western)	Genotype LA7
B31	Tick (I. dammini)	USA	AAA	I	+	2.0
ZS7	Tick (I. ricinus)	Germany	AAA	I	+	2.0
Z37	Tick (I. ricinus)	Germany	AAA	I	+	n.d.
GeHo	Skin (ECM)	Germany	AAA	I	+	n.d.
B1	Skin (ECM)	Germany	AAA	I	+	n.d.
B2	Skin (ECM)	Germany	AAA	I	+	n.d.
B3	Skin (AD)	Germany	AAA	I	+	n.d.
20004	Tick (I. ricinus)	France	AAA	I	+	n.d.
19535	Mouse (Peromyscus)	USA	AAA	I	+	2.0
26816	Vole (Microtus)	USA	AAA	I	+	n.d.
28691	Tick (I. dammini)	USA	AAA	I	+	2.0
21305	Mouse (Peromyscus)	USA	AAA	I	+	2.0
21343	Mouse (Peromyscus)	USA	AAA	I	+	n.d.
26815	Chipmunk	USA	AAA	I	+	2.0
297	Cerebrospinal fluid	USA	AAA	I	+	n.d.
Mac3	Skin	USA	AAA	I	+	n.d.
20001	Tick (I. ricinus)	France	AAA	I	+	n.d.
CTiP7	Dog tick	USA	AAA	I	+	n.d.
SH-2-82	Tick (I. dammini)	USA	AAA	I	+	n.d.
CA-2-87	Tick (I. pacificus)	USA	AAA	I	+	n.d.
S12/14	Tick (I. ricinus)	Germany	AAA	I	+	n.d.
IP1	Cerebrospinal fluid	France	AAA	I	+	2.0
NE2	Tick (I. ricinus)	Switzerland	AAA	I	+	n.d.
R7NE4	Tick (I. ricinus)	Switzerland	AAA	I	+	n.d.
LW2	Skin	Germany	AAA	I	+	n.d.
LW2.4	Skin	Germany	AAA	I	+	n.d.
						n.d.
ZQ1	Tick (I. ricinus)	Germany	BBB	II	-	0.8
NE4	Tick (I. ricinus)	Switzerland	BBB	II	-	0.8
NE58	Tick (I. ricinus)	Switzerland	BBB	II	-	0.8
NE11H	Tick (I. ricinus)	Switzerland	BBB	II	-	0.8
IP3	CSF	France	BBB	II	n.d.	0.8
R3NE2	Tick (I. ricinus)	Switzerland	BBB	II	-	n.d.
N34	Tick (I. ricinus)	Germany	BBB	II	-	n.d.
20047	Tick (I. ricinus)	France	BBB	V	+	0.6
S90	Tick (I. ricinus)	Germany	BBB	VI	-	>0.8
19857	Rabbit kidney	USA	B(A/B)A	III	+	0.75
21038	Larva (I. dentatus)	USA	B(A/B)A	III	+	0.75
ACA-1	Skin (ACA)	Sweden	BBA	IV	-	1.9
Bo23	Skin (ECM)	Germany	BBA	IV	-	1.9
So2	Tick (I. ricinus)	England	BBA	IV	n.d.	1.9

* ECM, erythema chronicum migrans; ACA, acrodermatitis chronica atrophicans; AD, atrophodermia.

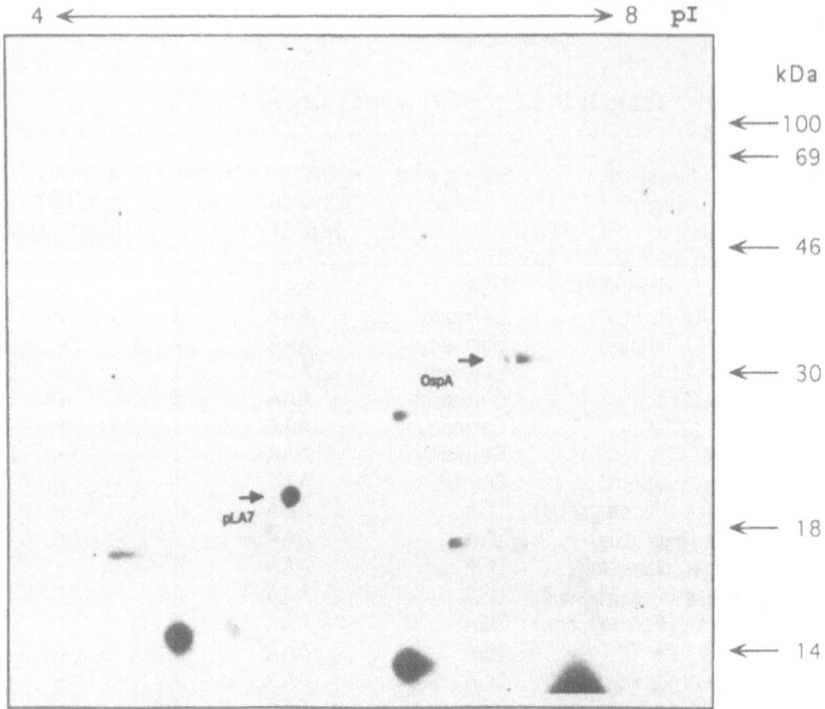

Figure 3. 2D gel analysis of *B. burgdorferi* strain ZS7 lipoproteins, as identified by biosynthetic labeling with [^{3}H]palmitate and Triton-X114 fractionation.

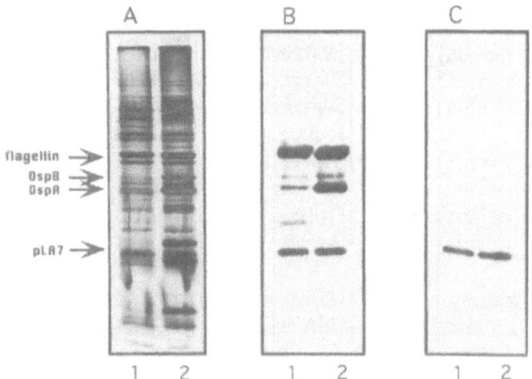

Figure 4. Surface proteolysis of *B. burgdorferi* strain ZS7 indicating degradation of OspA/OspB and additional other proteins but not lpLA7. *Borreliae* were washed and exposed to proteinase K (0.5mg/ml) for 40 min. at 25°C (lanes 1) or left untreated (lanes 2). Proteins were separated by SDS-PAGE and visualized by Coomassie blue staining (A), or electrophoretically transferred to nylon membrane and reacted with a mixture of mAbs LA7, LA21, LA2 (B), or with LA7 alone (C).

Sera derived from mice previously inoculated with 10^8 *B. burgdorferi* (ZS7) reacted with native lpLA7 antigen as well as with the two recombinant proteins encoded by pLA7-2 or pLA7-8 but not with lysates of *E.coli* carrying only the cloning vector (data not shown).

On the other hand when human sera from either uninfected individuals, those from a high risk group (forest workers) or from patients with ECM and ACA were tested it was found that only sera from seropositive but healthy forest workers contained antibodies reactive with lpLA7 (data not shown). At present there is no explanation for this discrepancy. Whether the coincidence of sero-conversion and reactivity to lpLA7 with the asymptomatic status in the three infected forest workers is of significance and a marker for protection against the development of Lyme disease has to await further experimentations. However it should be noted that adoptive transfer experiments in mice clearly demonstrate that immune sera to lpLA7 unlike those to OspA and OspB, are unable to protect SCID mice against the development of arthritis (U. Schaible et al., unpublished). In any case the data clearly indicate that lpLA7 is also immunogenic, at least in part, in humans and suggest that this lipoprotein is a further useful marker for diagnosis of Lyme disease.

REFERENCES

Barbour,A.G., Tessier,S.L., and Todd,W.J. (1983). Lyme disease spirochetes and ixodid tick spirochetes share a common surface antigenic determinant defined by monoclonal antibody. Infect.Immun. **41**:795.

Kramer,M.D., Schaible,U.E., Wallich,R., Moter,S.E., Petzoldt,D., and Simon,M.M. (1990). Characterization of Borrelia burgdorferi associated antigens by monoclonal antibodies. Immunobiol. **181**:357.

Simon,M.M., Schaible,U.E., Wallich,R., and Kramer,M.D. (1991). A mouse model for *Borrelia burgdorferi* infection: approach to a vaccine against Lyme disease. Immunology Today **12**:11.

Steere,A.C. (1990). Lyme disease. N.Engl.J.Med. **321**:586.

Wallich, R., Helmes,C., Schaible,U.E., Lobet,Y., Moter,S.E., Kramer,M.D. and Simon,M.M. (1992). Evaluation of genetic divergence among *Borrelia burgdorferi* isolates by use of OspA, fla, HSP60, and HSP70 gene probes. Infect.Immun. **60**:4856.

Wilske,B., Preac-Mursic,V., Schierz,G., Kühbeck,R., Barbour,A.G., and Kramer,M.D. (1988). Antigenic variability of *Borrelia burgdorferi*. Ann.NY Acad.Sci. **539**:126.

Wilske,B., Barbour,A.G., Bergström,S., Burman,N., Restrepo,B.I., Rosa,P., Schwan,T., Soutschek,E. and Wallich,R. (1992). Antigenic variation and strain heterogeneity in *Borrelia* spp. Res.Microbiol. **143**:583.

Wu,H.C., and Tokunaga,M. (1986). Biogenesis of lipoproteins in bacteria. *In* Current Topics in Microbiology and Immunology. Vol. **125**:127.

Wu,H.C. (1987). Posttranslational modification and processing of membrane proteins in bacteria, p.37-74. *In* M. Inouye (ed.), Bacterial outer membranes as model systems. John Wiley & Sons, Inc., New York.

DETECTION OF LYME DISEASE SPIROCHAETE
DNA IN CLINICAL SAMPLES

KJ Cann*#, ML Wilson#, C Akintunde *, L Archard#, DJM Wright*

Departments of Medical Microbiology* and Biochemistry#, Charing Cross and Westminster Medical School, St. Dunstan's Road, W6 8RF

ABSTRACT

Objective- Serology has traditionally been the mainstay of diagnosis in Lyme disease (LD). This approach has limitations because of the delayed humoral response to infection, the presence of cross-reactive antibodies arising from commensal and pathogenic spirochaetes and the heterogeneity of the species of Borrelia causing Lyme disease. The polymerase chain reaction (PCR) is a highly sensitive and specific technique which offers an alternative approach to laboratory diagnosis.

Methods- We have developed a nested PCR method which employs unique primers complementary to conserved regions of the *osp* A gene to detect the DNA of LD spirochaetes in clinical specimens. We have applied a simple and quick method of processing fluid samples for PCR, which incorporates diatoms to adsorb nucleic acids. The necessity of the preservation of DNA in samples in transit by the addition of alcohol has been established.

Results- Using this test we can detect 1fg of DNA in a sample without using radiolabelled probes. The reaction is specific for all three of the LD spirochaetes currently defined - *B. burgdorferi* sensu stricto, *B. garinii* and VS461 while closely related spirochaetes such as *B. duttonii* and *Treponema denticola* are not amplified. We have successfully detected LD spirochaete DNA in serum, urine, CSF and skin, retrospectively and prospectively, from patients with serologically confirmed LD as well as one patient who was sero-negative.

Conclusions- Further prospective studies are in progress to determine the level of sensitivity required for detection in various types of sample taken at different stages of infection and following treatment.

INTRODUCTION

Lyme disease (LD) is a multi-system infection caused by the spirochaetes *Borrelia burgdorferi sensu stricto, Borrelia garinii* and VS461[1,2]. Transmission of the disease is mediated by *Ixodes* spp. ticks. The diverse and delayed nature of presenting signs and symptoms hamper clinical diagnosis. Erythema migrans (EM) is one of the most common early clinical manifestations following tick bite but later the disease presents with neurological, cardiac and arthritic symptoms.

Although these spirochaetes are cultivable, this is an insensitive, time-consuming and expensive technique so that in the past serology has been the key to confirming the diagnosis of infection. Initially, immunofluorescent antibody tests were developed but these have been largely superceded by enzyme linked immunosorbent assays (ELISA) and western blotting. There are intrinsic limitations to the use of serology in the diagnosis of this infection. Cross-reacting antibodies, arising from immunisation with commensal and occasionally pathogenic spirochaetes, are present in a significant proportion of patients sera[3] so that the positive predictive value of the test is diminished. All of these problems are compounded by the recent description of three separate species of LD spirochaete: *B. burgdorferi sensu stricto*, *B. garinii* and VS461 so that many patients sera may not have been tested against the appropriate antigen.

Polymerase chain reaction (PCR)[4] has been used to detect *B. burgdorferi* DNA in ticks[5], urine [6,7], CSF[8,9,10], serum[11] and skin[12,13]. Targets for amplification include *osp* operon, the rRNA operon, the clone Ly-1 and the flagellin gene. Comparable specificity is achieved by use of these sequences and sensitivities of 1fg DNA are achieved by nested PCR or single round PCR followed by hybridisation of the product with labelled probes.

Our aim was to develop a PCR assay which detects the DNA of all currently defined LD spirochaetes, in a range of clinical samples and to establish a standard method of sample transport and processing. Diagnostic accuracy can then be improved in early disease before the development of the humoral response, in late disease to differentiate active and previous infection and in previous infection in clinically plausible cases where serology is negative or equivocal.

MATERIALS AND METHODS

POLYMERASE CHAIN REACTION

To ensure the amplification of DNA from different Borrelia species PCR primers were constructed complementary to the conserved regions of the gene encoding the Osp A protein of *B. burgdorferi* strains B31, ACA1 and Ip90[14]. A nested PCR was designed to obtain maximal sensitivity while avoiding the use radiolabelled probes. The first round primer pair, *Osp A* 1 (5'- CAG TAG ACA AGC TTG AGC -3') *Osp A* 6 (5'- TTG AGT CGT ATT GTT GTA C -3') and second round primer pair, *Osp A* 3 (5'- GCG ATG GAT CTG GAA AAG CT -3') with *Osp A* 6, again, correspond to nucleotide nos. 207-224,770-788 and 492-511 respectively of the *osp A* gene. Amplification was performed in 50ml reactions as follows: first round amplification mix contained 0.025mM *Osp A* 1 and O*sp A* 6 primers, 120mM (each) dATP, dGTP, dCTP and dTTP, 1.5mM MgCl2, 10mM Tris-HCl (pH 9.0), 50mM KCl, 0.1% (v/v) Triton X-100 and 2.5U *Taq* polymerase (Promega). The reaction mix was warmed to 94°C for 5 mins before the enzyme was added, to improve specificity[15] . The first cycle parameters were: denaturation at 94°C for 45s, annealing at 50°C for 1 min 30s and extending at 72° C for 3 min. This was followed by a further 19 cycles employing the same temperatures but denaturing and annealing for 45s and extending for 1 min 30s. The second round amplification mix differed from the first in the primer concentration only, incorporating 0.25mM *Osp A* 3 and *Osp A* 6. A hot start was not used in the second round and the first cycle parameters were: denaturation at 94°C for 4 min, annealing at 50°C for 1 min 30s and extension at 72°C for 3 min. This was followed by a further 29 cycles employing the same temperatures but denaturing and annealing for 45s and extending for 1 min 30s. A volume of 20ml of processed urine, CSF, serum or synovial fluid was added to the first round reaction; with tissue samples 500ng of DNA was added. Twenty microlitres of the products of the first round reaction was added to the second.

DETECTION

Products (20ml) were resolved on 2% agarose gels, run for 1h at 100V in TBE buffer (89mM Tris, 89mM boric acid, 2mM EDTA) followed by staining with ethidium bromide (5.0mg/ml) for 1 hour.

CONTROLS

Negative controls were run between each sample. A positive control (100pg) of *B. burgdorferi* (strain B31) DNA was included at the end of each run. Human primers directed against the anti-thrombin gene were used as a reaction control for each sample [16].

SAMPLE PROCESSING

Clinical material contains undefined constituents which vary from patient to patient and may inhibit PCR. We have modified the method of Boom *et al.*[17], to produce a quick method of processing fluid samples which preserves and concentrates nucleic acids while removing inhibitory constituents. Lysis buffer - 900ml(24g guanidium isothyocyanate (Fluka), 20ml 0.1M Tris-HCl, pH6.4, 4.4ml 0.2M EDTA, pH8.0 and 0.52g Triton X -100) is mixed with 50ml of diatom slurry (4g diatomaceous earth (Sigma), 20ml sterile dH_2O and 200ml HCL (32%w/v)) and 50 ml of the fluid sample. After vortex mixing and standing for 10 min at room temperature the diatoms are pelleted in a micro centrifuge. The supernatant is removed and the diatoms are washed twice with wash buffer (24g guanidium isothyocyanate, 20ml 0.1M Tris-HCl,pH 6.4), twice with 70% (v/v) ethanol and once with acetone. The diatoms are vacuum dried for 2 min before the addition of 100ml TE (10mM Tris, 1mM EDTA pH 8.0) and incubated at $65^{o}C$ for 10 mins to elute the nucleic acids. The diatoms are pelleted again and the supernatant transferred to a fresh tube, ready for use in the PCR. Tissue samples were processed by a standard methodology using digestion with proteinase K (Sigma) and phenol extraction[18].

PRESERVATION OF DNA IN SAMPLES

Many samples sent to the laboratory by post have been in transit at room temperatures for several days before receipt. We have therefore studied the effect of storage at room temperature for one week on the detection of DNA in "spiked" samples, with and without the addition of ethanol.

RESULTS

SPECIFICITY

We have successfully amplified the type strains *B. burgdorferi sensu stricto*, *B. garinii and* VS461 (kindly provided by G. Baranton, Institute Pasteur). We have also amplified a UK isolate of *B. burgdorferi sensu lato* as well as Japanese isolates HP3 and HO14 and Chinese isolates C8 and C23. *B. duttonii, T. denticola, Staphylococcus aureus, Staphylococcus epidermidis, Escherichia coli, Klebsiella aerogenes* and *Pseudomonas aeruginosa* were not amplified (data not shown).

SENSITIVITY

The sensitivity of detection was between 10 - 1.0 fmol of DNA.

B. *burgdorferi* DNA has been successfully detected in "spiked" specimens of serum, CSF, urine and synovial fluid as well as culture positive mouse bladder tissue (data not shown).

SAMPLE STORAGE

Urine samples stored at room-temperature for up to a week showed a decrease in the PCR products detected whereas samples preserved in alcohol showed no decrease (Figure 1.). All samples collected specifically for PCR are now transported in the presence of ethanol (35%v/v), end concentration.

CLINICAL SAMPLES

PCR products have been detected in serum, urine, CSF and skin biopsies from patients with positive LD serology.

Here we present three clinical cases that illustrate our experience to date:

Case 1: A 55 year old lady with an eight month history of erythematous plaques; initially on the right buttock and progressing to both knees and dorsal aspect of the feet. For the previous eight years the patient had regularly visited her mother in the New Forest where she walked and fed the deer. She also spent annual holidays in Normandy.
Diagnosis : Clinically A.C.A.
L.D. serology: IgG ELISA > 100 units, IgG and IgM Western blots - positive.
PCR of skin biopsy - positive, urine - negative, serum - negative. (Figure 2).
The patient was initially treated with doxycycline for 28 days but relapsed three months later and was subsequently treated with ceftriaxone for 21 days.

Case 2: A 59 year old male, who presented with polyradiculopathy. Clinically there was no meningism but a CSF sample showed a lymphocytic infiltrate. Three months previously he had taken a sailing holiday in the Norwegian Fjords when he camped and walked in the forests.
Diagnosis: L.D. serology: IgG ELISA > 100 units, IgG Western blots - positive.
PCR of urine - positive, serum - positive, CSF - positive.
The patient was treated with cefotaxime for 28 days. He has made a good clinical recovery with some mild residual weakness in his foot. We have tested this patients urine samples by PCR for five months following treatment - they remain positive although the patient continues to do well.

Case 3: A 64 year old lady with 10 year history of presumptive morphea with lichen sclerosis and atrophicans.
Diagnosis: L.D. serology: IgG ELISA negative, IgG Western blot - negative.
PCR of urine - positive, serum - positive, skin - positive. A second urine sample was negative (Figure 3).
This patient has been treated with doxycycline for 28 days. There has been marked improvement of her vaginal erosions although less improvement has been seen in the morphea lesions. We continue to follow this patient.

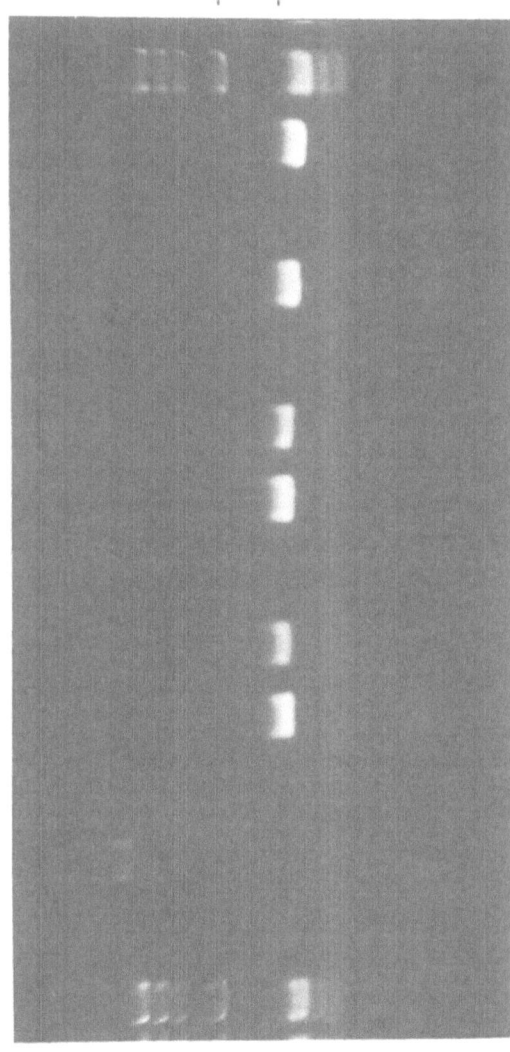

Figure 1. Preservation of DNA in clinical samples by the addition of ethanol. Urine samples were spiked with B31 *B. burgdorferi* DNA 50ng/ml, stored at room temperatures for one week, pelleted and 50ml used in diatom processing. Lanes 2, 3, and 4:- 10ml, 5ml and 1ml pellets, respectively, no ethanol added. Lanes 5, 6 & 7:- 10ml, 5ml and 1 ml pellets, respectively, stored in 35% ethanol (end concentration). Lanes 8, 9 & 10:- 10 ml, 5ml and 1ml pellets, respectively, stored in 50% ethanol (end concentration). Lane 11 3ml pellet stored in 70% ethanol (end concentration). Lane 12 negative control. Lane 13 positive control. Lanes 1 & 14 φX 174/ Hae III markers.

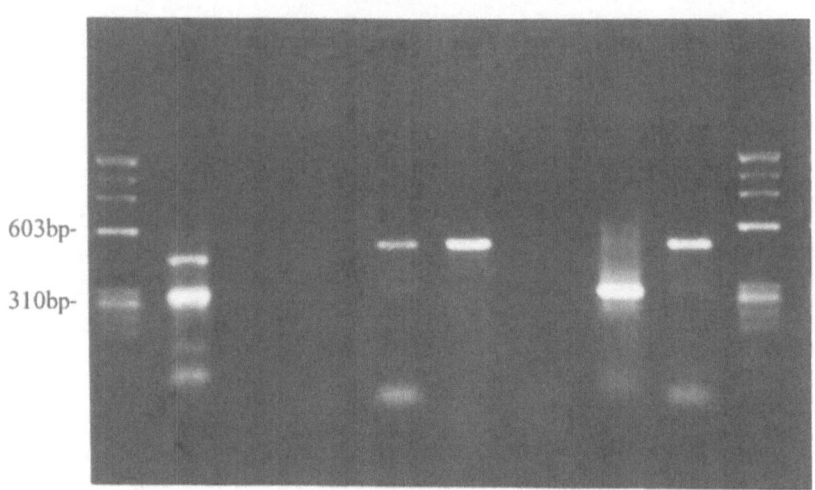

Figure 2. Detection of *B. burgdorferi* DNA in skin. Lane 1 & 10:- φX 174/Hae III markers. Lanes 2, 3 & 4 patient 5:- skin biopsy, urine and negative control using *B. burgdorferi* primers, lanes 5, 6 & 7:- patient 5 skin biopsy, urine and negative control using anti-thrombin gene primers. Lanes 8 & 9:- positive controls - *B. burgdorferi* and anti-thrombin gene, respectively.

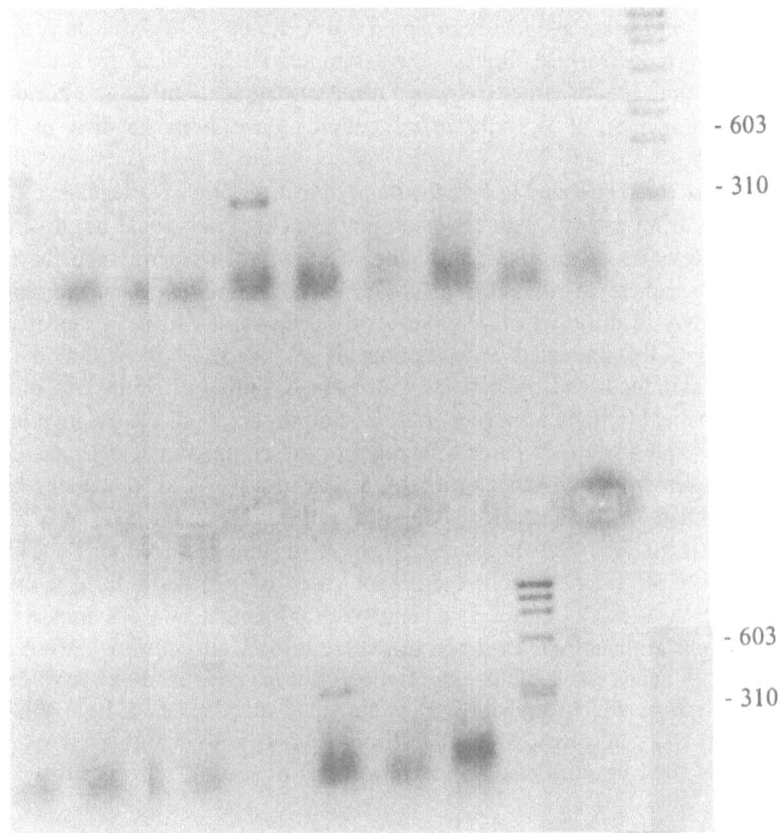

Figure 3. Lanes 1, 2 & 3 urine, negative control, urine amplified with anti-thrombin primers. Lanes 4, 5 & 6 serum, negative control, serum amplified with anti-thrombin primers. Lanes 7, 8 & 9 urine, negative control, urine amplified with anti-thrombin primers. Lane 10 φX174/HaeIII markers. Lanes 11,12 & 13 skin biopsy, negative control, skin amplified with anti-thrombin primers. Lane 14 blank. Lane 15 *B.burgdorferi* positive control. Lane 16 & 17 human DNA positive and negative controls. Lane 16 φX174 HaeIII markers.

TICKS

PCR has been used to detect LD spirochaete DNA in tick pools collected from New Forest, Richmond Park and the Salisbury area. LD spirochaete DNA was detected in approx. 36%, 38% and 3.6% of ticks respectively.

DISCUSSION

The results show that PCR can be used to detect *B. burgdorferi* DNA in samples from patients with confirmed LD. These samples had not been transported in alcohol to prevent DNA degradation and had been stored at 4°C for up to 14 days while serological tests for LD were performed. Retrospective samples are not ideal for study not only because of possible loss of target DNA as a result of degradation but also because of the risks of contamination if *B. burgdorferi* cultures have been handled in the same laboratory area .

Serum samples were not found to be consistently positive, other workers have found that serum samples from patients with erythema migrans (EM) may yield negative results[11]. This may be because of the transient nature of the spirochaetaemia and the ability of spirochaetes to adhere to all blood cells .We have demonstrated the presence of *B. burgdorferi* DNA in urine which has previously been shown to be a site of persistent infection in half the untreated cases examined[7]. Schwartz *et al.*[13], found that PCR performed on skin biopsies from patients with EM was only positive in 59% of untreated patients. A smaller study by Melchers *et al.* [12] showed that PCR was positive in 3 out of 4 cases of EM and 4 out of 5 cases of acrodermatitis chronicum atrophicans (ACA). A skin biopsy taken from our patient with ACA was positive. CSF specimens have been shown to contain *B. burgdorferi* DNA in 2 out of 2[9] and 5 out of 6 cases[8] of neuroborreliosis, respectively. In a large retrospective study Keller *et al.*[10] , showed that 10 out of 11 CSF samples from confirmed cases of neuroborreliosis contained *B. burgdorferi* DNA as did 28 out of 38 patients with inflammatory CNS disease with only peripheral blood immunologic evidence of exposure to *B. burgdorferi.* Seven out of 7 patients with clinically plausible neuroborreliosis but negative serology were also found to have *B. burgdorferi* DNA in their CSF. In our small study we have been able to detect *B. burgdorferi* DNA in some CSF samples from patients with neuroborreliosis. The lack of consistency may be attributed to the limitations of testing retrospectively collected samples.

The clinical validation of this test is currently being assessed by studying prospective and retrospective samples from different patient groups - including those with negative serology but clinically plausible LD. These studies will be undertaken in a 3 room suite of laboratories designated for diagnostic PCR work.

ACKNOWLEDGEMENTS

This work is supported by The Charing Cross and Westminster Medical School Special Trustees.

REFERENCES

1. Johnson, R.C., Schmid,G.P., Hyde ,F.W., Steigerwalt, A.G. & Brenner, D.J. 1984 *Borrelia burgdorferi* sp. nov.: etiological agent of Lyme disease. *International Journal of Systematic Bacteriology* 34, 496-497.

2. Baranton, G., Postic, D., Saint Girons, I., Boerlin, P., Piffaretti, J-C., Assous, M. & Grimont, P.A.D. 1992 Delineation of *Borrelia burgdorferi* Sensu Stricto, *Borrelia garinii* sp. nov., and Group VS461 associated with Lyme Borreliosis. *International Journal of Systemic Bacteriology* 42, 378-383.

3. Magnarelli, L.A., Anderson, J.F. & Johnson, R.C. 1987 Cross-reactivity in serological tests for Lyme disease and other spirochaetal infections. *Journal of Infectious Diseases* 156,183-188.

4. Saiki, R.F., Gelfand, D.H., Stoffel, S. et al. 1988 Primer-directed enzymatic amplification of DNA with a thermostable DNA polymerase. *Science* 239, 487-491.

5. Persing, D.H., Telford, S.R. III, Spieman, A. & Barthold, S.W. 1990 Detection of *Borrelia burgdorferi* infection in *Ixodes dammini* ticks with the polymerase chain reaction. *Journal of Clinical Microbiology* 28, 566-572.

6. Malloy, D.C., Nauman R.K. & Paxton, H. 1990 Detection of *Borrelia burgdorferi* using the polymerase chain reaction. *Journal of Clinical Microbiology* 28, 1089-1093.

7. Goodman, J.L., Jurkovich, P., Kramber, JM & Johnson, R.C. 1991 Molecular detection of persistent *Borrelia burgdorferi* in the urine of patients with active Lyme disease. *Infection and Immunity* 59, 269-278.

8. Benoit,G., Nicolini, P., Piedmont, Y. & Monteil H. 1991 Detection of *Borrelia burgdorferi* in cerebrospinal fluid in patients with Lyme borreliosis. *Lancet* i,1440.

9. Kruger, W.H. & Pulz, M. 1991 Detection of *Borrelia burgdorferi* in cerebrospinal fluid by the polymerase chain reaction. *Journal of Medical Microbiology* 35, 98-102.

10. Keller, T.L., Halperin, J.J. & Whitman, M. 1992 PCR detection of *Borrelia burgdorferi* DNA in cerebrospinal fluid of Lyme neuroborreliosis patients. *Neurology* 42,32-42.

11. Guy, E.C. & Stanek, G. 1991 Detection of *Borrelia burgdorferi* in patients with Lyme disease by the polymerase chain reaction. *Journal of Clinical Pathology* 44, 610-611.

12. Melchers, W., Meis, J., Rosa, P., Claas, E., Nohlmans, L., Koopman, R., Horrevorts, A. & Galama, J. 1991 Amplification of *Borrelia burgdorferi* DNA in skin biopsies from patients with Lyme disease. *Journal of Clinical Microbiology* 29, 2401-2406.

13. Schwartz, I., Wormser, G.P.,Schwartz, J.J., Cooper, D., Weissensee, P., Gazumayan, A., Zimmerman, E., Goldberg, N.S., Bittker, S., Campbell, G.L. & Pavia, C.S. 1992 Diagnosis of early Lyme disease by polymerase chain reaction amplification and culture of skin biopsies from Erythema Migrans lesions. *Journal of Clinical Microbiology* 30, 3082-3088.

14. Jonsson, M., Noppa,L., Barbour, A.G. & Bergstrom, S. 1992 Heterogeneity of outer membrane proteins in *Borrelia burgdorferi:* comparison of *osp* operons of three isolates of different geographic origins. *Infection and Immunity* 60,1845-1853.

15. D'Acquila, R.T.D., Bechtel,L.J., Videler, J.A., Eron, J.J., Gorczyca,P. & Kaplan, J.C. 1991 Maximising sensitivity and specificity of PCR by pre-amplification heating. *Nucleic Acids Research,* 19, 3749.

16. Wakefield, A.E., Pixley, F.J., Banerji, S., Sinclair, K., Miller , R.F., Moxon, R, & Hopkin, J.M. 1990 Amplification of mitochondrial ribosomal RNA sequences from *Pneumocystis carinii* DNA of rat and human origin. *Molecular and Biochemical Parasitology* 43, 69-76.

17. Boom, R , Sol, C.J.A., Salimans, M.M.M., Jansen, C.L., Wertheim-Van-Dillen, P.M.E. & Van Der Noordaa, J. 1990 Rapid and simple method for purification of nucleic acids. *Journal of Clinical Microbiology* 28,495-503.

18. Sambrook, Fritsch and Manniatis. 1989 Molecular cloning: A laboratory manual. *Cold spring Laboratory Press.*

CLINICAL AND SEROLOGICAL STUDY OF LYME BORRELIOSIS

IN A POPULATION OF NEUROLOGICAL PATIENTS

Elisabetta Capello, Gerolamo Bianchi*, Patrizia Monteforte*,
Laura Buffrini*,Angelo Schenone, Sandro Ratto**,
Nicola Dagnino***, Guido Rovetta*, Gian Luigi. Mancardi

Italian Group for the Study of Lyme Borreliosis (IGSLB)
Institute of Clinical Neurology and * Institute E. Bruzzone
Rheumatological Center, University of Genova, Italy
** Division of Neurology, Hospital of Lavagna, Italy
*** Division of Neurology, Hospital of Sestri Ponente, Italy

INTRODUCTION

Lyme Borreliosis (LB) is a multisystem infection caused by *Borrelia burgdorferi*
(Bb) that is associated with a variety of neurological abnormalities. Peripheral nervous
system (PNS) involvement is prevalent, while central nervous system (CNS)
manifestations are less frequent and generally signs of CNS impairment are observed only
in cases with a long lasting course (1-4).

In Liguria, an Italian region where Lyme disease is endemic (5) exposure to Bb, and
consequentely specific immunoreactivity are widespread, making it difficult to determine
if in the individual patient neurologic symptoms are causally related to Bb infection.

This study reports our experience gathered to detect the incidence of sieropositivity
against Bb in a neurological population resident in an Italian endemic area, and to
differentiate patients with occasional association of neurological disease and previous
spirochetal exposure from patients with definite neuroborreliosis (6),

MATERIAL AND METHODS

Neurological in-patients admitted in the years 1989 through 1992 at the Institute of
Clinical Neurology, University of Genoa, and at the Divisions of Neurology of Sestri
Ponente and of Lavagna were tested for antibodies against Bb by immunofluorescence
(IFA), Globally 109 sera and 50 CSF samples were tested.

Patients were divide according to diagnoses: 35 cases of cranial neuritis of which 3 cases of multiple neuritis, 21 multiple sclerosis patients, 16 polyneuropathies, 8 optic neuritis, 8 cases of Guillain Barre' Syndrome (GBS), 5cases of multiple motor neuropathies (MMN), 4 motor neuron disease cases (MND), 4 meningoradiculoneuritis, 4 encephalitis, 2 cases of myeloradiculophaties, and finally 2 patients affected by mononeuropathies.

Some patients had a positive clinical history for previous tick bite, or signs of multisystem involvement, as indicated in table 1, while others were only characterized by vague neurological symptoms with no other recognizable cause.

Antibodies titers against Bb greater than 1:256 in the serum, and greater than 1:5 in the CSF were considered as positive.

RESULTS

Antibodies against Bb were detected in 36 out of 109 patients (31%) . In 21 of them the titer was 1:256, in 7 was 1:512 and in 8 was attested at 1:1024 as showed in table 2.

Table 1. Data of neurological patients.

	tick bite	skin lesions	cardiac or joint involvement	systemic symptoms	rural addresses
Cranial neuritis n 35	2 (2)	3 (3)	1 (1)	3	12 (1)
Multiple sclerosis n 21				3	7
Polyneuropathies n 16		2	1 (1)		2
Optic neuritis n 8				1	3
Guillain.Barre' syndrome n 8				3	
Multifocal motor neurophaties n 5					2
Motor neuron diseases n 4	1				1
Meningo-radiculopathies n 4	1 (1)	1 (1)	1 (1)	1 (1)	2 (2)
Encephalitis n 4		1			2
Myelo-radiculopathies n 2					1
Mononeuropathies n 2				1	1

( ): number of cases with a diagnosis of definite LB.

Table 2. Seric antibodies against Bb.

	1:256		1:512		1:1024	
	IgG	IgM	IgG	IgM	IgG	IgM
Single cranial neuritis (n 32)	5	1 (1)	3 (3)		3 (3)	
Multiple cranial neuritis (n 3)			1 (1)		1 (1)	
Multiple sclerosis (n 21)	3					
Polyneuropathies (n 16)	1				1 (1)	
Optic neuritis (n 8)	2					
Guillain Barre' Sindrome (n 8)	1		1		1	
Multifocal Motor Neuropathies (n 5)	2					
Motor Neuron.Disease (n 4)	1					
Meningo-radiculopathies (n 2)	2	1 (1)	1 (1)		2 (2)	
Encephalitis (n 4)			1			
Myelo-radiculopathies (n 2)	2					
Mononeuropathies (n 2)	1					

( ) : number of cases with a diagnosis of definite LB.

In 2 patients low titers of anti-Bb antibodies of IgM class were initially present, with the following detection of high antibody titer of IgG class. In 3 out of 40 patients CSF anti-Bb IgG antibodies were present. In other 2 patients, a case of recurrent cranial neuropathy and a case of polyneuropathy, there was evidence at isoelectrofocusing (IEF) of muliple oligoclonal bands at alkaline pH, but no CSF antibodies against Bb were detected by IFA.

According to Reik criteria (1991) the diagnosis of neuroborreliosis was confirmed in 11 out of 36 patients : 5 cases of facial nerve palsy, 2 cases of multiple cranial neuropathy (VII and V nerves), 1 case of oculomotor nerve palsy, 3 cases of lymphocytic meningoradiculitis (Bannwarth syndrome), 1 case of sensory-motor polyneuropathy with vasculitic changes.

A total of 8 cases of cranial neuropathy were observed. All patients with cranial neuropathies had supportive clinical history and/or signs of multisystem involvement. In 2 of them, together with high titer of seric anti-Bb antibodies of IgG class, CSF specific immunoreactivity was present. In another 2 patients the functional CNS evaluation revealed optic nerve impairment with normal CSF parameters and no MRI evidence of white matter lesions. In half of cranial neuropathies cases clinical improvement was observed after antibiotic therapy.

Three cases of lymphocytic meningoradiculitis (Bannwarth syndrome) were observed. In the first case the recovery was spontaneous, in a second was complete after antibiotic therapy. The last case, characterized by the presence of IgM and IgG antibodies in the serum and IgG antibodies in CSF was treated with two cicle (2 wks) of ten millions units of intravenous penicillin with slow regression of neurologic signs and persistence of weakness, polyarthritis and erythematous skin lesions.

In the first two cases of meningoradiculitis no subsequent signs of multisystem involvement were observed.

A peculiar case of neuroborreliosis was observed in a 64 years old man, resident in the country-side, who developed at 62 years severe septic arthritis, with periarticular ulcers and anchilosis, and a slow progressing motor-sensory polyneuropathy. In this patient seric anti-Bb antibodies of IgG class were detected at 1:1024 dilution, with presence at immunoblotting of a 31 Kd band. CSF analysis showed an increase of protein concentration with intrathecal IgG synthesis, and evidence at IEF of four bands between 8.6-8.9 pH. Sural nerve biopsy was performed. A mild loss of myelinated fibres, more evident for larger ones, was observed, together with marked tickening of the vessels wall, duplication of the basement membranes, and inflammatory infiltrates around small epineurial vessels. Treatment with i.v. penicillin (ten millions units for 2 wks) determined a satisfactory clinical recovery.

Weak positivities were observed in 25 cases (23%) and were all characterized by absence of CSF specific immunoreactivity (8).

In detail seric anti-Bb antibodies were present in 14% of multiple sclerosis , in 20% of optic neurites and in 37% of GBS patients. In two cases of GBS high seric positivity was detected.

In all the 25 cases only weak supportive clinical history was referred, and in no patien signs of skin or joint involvement were recognized.

CONCLUSIONS

We found significant titers of anti-borrelia antibodies in 36 out of 109 patients with a neurological disease compatible with LB. However only in 11 out of these 36 cases a final definite diagnosis of neuroborreliosis was carried out when we used the strict diagnostic criteria of Reik.

These cases were characterized by an exclusive involvement of the peripheral nervous system, with evidence in two cases of subclinical optic neuritis. According to the criteria proposed by Reik (6) a diagnosis of definite neuroborreliosis was performed when a compatible neurologic disorder was associated to a history of tick bite or to the presence of skin lesions, when there was immunoreactivity against Bb in the serum and in the CSF, or when serum immunoreactivity cohexisted with either symptoms and signs of involvement of other organs.

We did not observe a single case of CNS disorder certainly due to Bb infection. In only 14% of our MS cases a slight positivity in the serum was found, but in all MS no immunoreactivity was detected in CSF, thus confirming recent reports that there is no relationship between MS and Borrelia infection (9).

All cases of GBS positive for seric anti-Bb antibodies were negative for CSF immunoreactivity, suggesting that in these cases the presence of seric anti Bb antibodies is only part of the wide humoral activation directed against various antigens, often detected in autoimmune inflammatory poliradiculoneuropathies.

Unspecific positivities were characterized by low antibody titer without cohexisting CSF antibody positivity.

Our data suggest that in Liguria, an Italian region endemic for LB, peripheral nervous system impairment is mainly observed and cranial nerves are more frequently involved in patients with LB.

REFERENCES

1. Krüger H, Reuss K, Pulz M. Meningoradiculitis and encephalomyelitis due to Borrelia Burgdorferi: a follow-up study of 72 patients over 27 years. J Neurol 1989; 236:322-328.
2. Weder B, Wiedersheim P, Matter L. Chronic progressive neurological involvement in Borrelia Burgdorferi infection. J Neurol 1987; 234:40-43.
3. Halperin JJ, Luft BJ, Anand AK. Lyme neuroborreliosis: central nervous system manifestations Neurol 1989; 39:753-759.

4. Logigian EG, Kaplan RF, Steere AC, Chronic neurologic manifestations of Lyme Disease. N Engl. J Med 1990; 323:1438-44.

5. G.Bianchi , G. Rovetta, P. Monteforte, Fumarola D, Trevisan G, Crovato F, Cimmino MA.. Articular Involvement in european patients with Lyme disease. A report of 32 italian patients. Br.J.Rheum. 1990; 29:178-80.

6. L.Reik. "Lyme Disease and The Nervous System", Thieme Medical Publisher, Inc. New York (1991)

7. Halperin J, Luft BJ, Volkman DJ. Lyme neuroborreliosis. Peripheral nervous system manifestations Brain 1990; 113:1207-1221.

8. Baig S, Olsson T, Link H, Predominance of Borrelia Burgdorferi specific B cells in cerebrospinal fluid in neuroborreliosis. Lancet, 1989; July 8, 71-74.

9. Heller J, Holzer G, Schimrigk K. Immunological differentiation between neuroborreliosis and multiple sclerosis. J. Neurol. 1990; 237:465-470.

PITFALLS IN THE LABORATORY DIAGNOSIS OF LYME BORRELIOSIS

Sally J. Cutler and David J. M. Wright

Department of Medical Microbiology
Charing Cross Hospital
Fulham Palace Road
London, W6 8RF
England

INTRODUCTION

Over a decade has passed since the identification of *Borrelia burgdorferi* as the causative agent of Lyme borreliosis, yet laboratory diagnosis of Lyme borreliosis remains problematic. Conventional bacteriological methods of microscopy and culture have been used with only limited success. Numbers of spirochaetes are too low for diagnosis by direct microscopic examination and pseudospirochaetes may falsely be reported as positive. Results of culture for *Borrelia* in the United Kingdom have been disappointing, with only one isolate obtained to date in our laboratory.

SERODIAGNOSIS

Serology is currently the method of choice for the laboratory diagnosis of Lyme borreliosis. Indirect immunofluorescence (IFA) and enzyme-linked immunosorbent assay (ELISA) have been used for serological screening, however, both techniques have their limitations. In a comparative trial we showed these two methods had similar specificity, but ELISA was more sensitive detecting 96.6% of 146 seropositive samples, whereas IFA only found 66.4% [1]. The ELISA was less subjective to interpret and better suited for screening

Lyme Borreliosis, Edited by J.S. Axford and
D.H.E. Rees, Plenum Press, New York, 1994

large numbers of samples[1]. Improvements in the specificity of tests has been explored using either purified antigen such as the 41 kDa flagellin protein[2] or semi purified preparations of *B. burgdorferi*[3]. While attempts have been made to increase sensitivity by utilising antibody capture techniques[4]. These techniques however are still plagued by problems of specificity and sensitivity, especially in early disease when patients frequently present with erythema migrans but insignificant levels of antibodies. This is illustrated by the finding of only 16.4% of 258 patients with possible erythema migrans, investigated by our laboratory, giving positive serology. When individual serological tests were evaluated with known cases of erythema migrans confirmed by a dermatologist, serology at its best could only detect half of these patients (table 1). Lowering the diagnostic cut off levels of the ELISA would not resolve this problem as no statistically significant difference could be detected when absorbance values from 51 erythema migrans cases were compared with 120 healthy blood donors (Mann-Whitney U test, p=0.498). Negative results have even been found when serum was tested against a patient's own isolate [5]. The humoral response, unlike the cellular immune response, may be selectively suppressed during early infection.

Table 1. Comparison of serological tests using patients with erythema migrans (n=53).

Serological test	Percent seropositive
Positive by IFA	9.6%
Positive by ELISA	28.3%
Positive by Immunoblotting	50.9%

Confirmation of serological reactivity can be achieved using immunoblotting. As the disease progresses, antibodies are produced against a range of borrelial proteins, both specific and non-specific. The interpretation of immunoblot patterns is an area of controversy with many different schemes available. Some studies have reported poor specificity of immunoblotting[6], while others maintain that the technique is highly specific[7]. These discrepancies largely result from differing methods used for interpretation. The principal methods used for the interpretation of immunoblots are those based on band counting, or those which require possession of selected bands. We prefer to use a weighted numerical system for interpretation, with greater value being attached to bands considered to be specific, while other bands considered less important, are awarded lesser values[8]. Using this method, features of both the band counting and possession of specific bands can be combined to form a simplified method for interpreting immunoblots.

When ELISA and immunoblot results were compared, only 16.3% of 1222 ELISA positive sera were found to be positive by immunoblotting (table 2). When weakly reactive sera are removed, this proportion of ELISA results positive by immunoblotting only rose to 24.4%. This leaves the question, what do the remaining sera with positive ELISA results but negative immunoblot findings represent? These may illustrate a specificity problem of the ELISA, with individuals producing antibody against non-specific proteins of *Borrelia*. Alternatively, these may be patients with insufficient disease duration to produce a positive immunoblot pattern. Not all individuals may be able to produce an immune response to diagnostic bands and early antibiotic treatment may prevent a positive immunoblot pattern developing. In addition, the diversity of borrelial strains causing Lyme borreliosis may result in falsely negative immunoblots if a different strain or species is employed as antigen.

Table 2. IgG Immunoblot results for sera positive by ELISA.

Immunoblot result	Negative	Positive
Weakly positive ELISA	559 (91.9%)	49 (8.1%)
Strongly positive ELISA	464 (75.6%)	150 (24.4%)
Total	1023 (83.7%)	199 (16.3%)

Geographical differences can be seen in the United Kingdom, with the proportion of sera from forestry workers in Scotland which could be confirmed by immunoblotting being less than that found in similar groups tested from England (table 3).

Table 3. ELISA and immunoblot findings for British forestry workers.

Group	No.	ELISA positive		Immunoblot positive	
New Forest Rangers	42	12	(28.6%)	5	(11.9%)
Thetford Forest	78	16	(20.5%)	11	(14.1%)
Scottish Foresters	142	23	(16.2%)	2	(1.4%)
Scottish Deer workers	43	7	(16.3%)	2	(4.6%)
Total	305	58	(19%)	20	(6.6%)

A major problem encountered with serological tests for Lyme borreliosis is the lack of specificity. Patients with a variety of other spirochaetal diseases and unrelated clinical conditions have been reported to give false positive serological results. Sera may cross-react by binding to nonspecific antigens possessed by *B. burgdorferi* such as the 60 kDa common antigen or heat shock related proteins[9,10].

Considerable diversity exists between strains of *B. burgdorferi,* especially among European isolates, resulting in the recent proposition of new genomic species[11]. This may account for differences in seropositivity reported, depending upon which strain of borrelia was used as antigen, and highlights the problem of which isolate should be used for diagnostic purposes. Possibly a panel of different isolates should be used for preparing antigens for serodiagnosis?

Problems common to all serological tests are that the IgG response to *B. burgdorferi* rises slowly, often not achieving significant levels until the sixth week following infection. Titres of IgG can remain elevated years after clinical remission, or following asymptomatic exposure[12], complicating diagnosis in patients with unrelated clinical presentation Additionally, there are patients with clinical Lyme borreliosis who fail to mount a significantly elevated antibody response[13]. This may result from prompt treatment which has been reported to abrogate the antibody response[13]. Alternatively, antibody production may be restricted to the site of active disease, for example, in neuroborreliosis the CSF may be positive but the serum negative[14]. Antibody may remain sequestered in immune complexes and thus not detectable by commonly used assay methods. Cases have been reported with no significant B cell response, however, specific T cell proliferative responses could be demonstrated[15].

ANTIGEN DETECTION

Although antigen can be detected using dot blot and membrane capture techniques, polymerase chain reaction (PCR) has become the most widely used method for detecting borrelial DNA in patient samples. PCR has been used to diagnose Lyme borreliosis with mixed success. Primers used may fail to detect all isolates of *B. burgdorferi*, or are not totally specific for *B. burgdorferi*. Inhibitors of the PCR may be present in test samples, especially blood. Degradation of target DNA may lead to false negatives, while, if there is poor laboratory practice, contamination problems may result in false positive findings. One of the problems with this technique is deciding which sites and how many samples to test before accepting a negative result. Although PCR analysis can provide valuable diagnostic information, and even if made semiquantitative, is unlikely to be valuable for clinical follow up. The role of PCR in the diagnosis of Lyme borreliosis has yet to be fully determined.

CONCLUSION

Until the problems of sensitivity and specificity can be overcome, results of laboratory findings must be interpreted with caution. The results of diagnostic tests must be considered in conjunction with clinical manifestations before attributing *B. burgdorferi* infection as the cause of the patients' complaints.

REFERENCES

1 S.J. Cutler, and D.J.M. Wright, Comparison of immunofluorescence and enzyme linked immunosorbent assays for diagnosing Lyme disease, *J Clin Pathol.* 42:869-71 (1989).

2 K.Hansen, P. Hindersson, and N.S. Pedersen, Measurement of antibodies to the *Borrelia burgdorferi* flagellum improves serodiagnosis in Lyme disease, *J Clin Microbiol.* 26:338-46 (1988).

3 L.A. Magnarelli, J.F. Anderson, and A.G. Barbour, Enzyme-linked immunosorbent assays for Lyme disease: reactivity of subunits of *Borrelia burgdorferi*, *J Infect Dis.* 159:43-9 (1989).

4. V.P. Berardi, K.E. Weeks, and A.C. Steere, Serodiagnosis of early Lyme disease: analysis of IgM and IgG responses using an antibody-capture enzyme immunoassay, *J Infect Dis.* 158:754-60 (1988).

5 B.W. Berger, A.B. MacDonald, and J.L.Benach, Use of an autologous antigen in the serologic testing of patients with erythema migrans of Lyme disease, *J Am Acad Dermatol.* 18:1243-6 (1988).

6 M.G. Golightly, and A.L. Viciana, ELISA and immunoblots in the diagnosis of Lyme borreliosis: sensitivities and sources of false-positive results, *in:* Lyme disease molecular and immunologic approaches. Current Communications in Cell & Molecular Biology, 6:283-97, S.E. Schutzer, ed., Cold Spring Harbor Laboratory Press (1992).

7 R.L. Grodzicki, and A.C. Steere, Comparison of immunoblotting and indirect enzyme-linked immunosorbent assay using different antigen preparations for diagnosing early Lyme disease, *J Infect Dis.* 157:790-7 (1988).

8 S.J.Cutler, D.J.M. Wright, and V. H. Luckhurst, Simplified method for the interpretation of immunoblots for Lyme borreliosis, *FEMS Immunol & Med Microbiol.* 6:281-5 (1993).

9 K. Hansen, J.M. Bangsborg, H. Fjorduang ,N.S. Pedersen, and P. Hindersson Immunochemical characterization of and isolation of the gene for a *Borrelia burgdorferi* immunodominant 60 kilodalton antigen common to a wide range of bacteria, *Infect Immun.* 56:2047-53 (1988).

10 M.M. Carreiro, D.C. Laux, and D.R. Nelson, Characterization of the heat shock response and identification of heat shock protein antigens of *Borrelia burgdorferi*, *Infect Immun*, 58:2186-91 (1990).

11 G. Baranton, D. Postic, I. Saint Girons *et al.* Delineation of *Borrelia burgdorferi* sensu stricto, *Borrelia garinii* sp. nov., and group VS461 associated with Lyme borreliosis, *Int J Syst Bacteriol.* 42:378-83 (1992).

12 H. Fahrer, S.M. van der Linden, M.J. Sauvain *et al.*, The prevalence and incidence of clinical and asymptomatic Lyme borreliosis in a population at risk, *J Infect Dis.* 163:305-10 (1991).

13 V. Preac-Mursic, K. Weber, H.-W. Pfister *et al.*, Survival of *Borrelia burgdorferi* in antibiotically treated patients with Lyme borreliosis, *Infection* 17:355-9 (1989).

14 S. Baig, T. Olsson, and H. Link Predominance of *Borrelia burgdorferi* specific B cells in cerebrospinal fluid in neuroborreliosis, *Lancet* 334:71-4 (1989).

15 R.J. Dattwyler, D.J. Volkman, B.L. Luft *et al.*, Seronegative Lyme disease. Dissociation of specific T- and B-lymphocyte responses to *Borrelia burgdorferi*, *New Engl J Med.* 319:1441-6 (1988).

SERODIAGNOSIS OF LYME DISEASE IN THE UK

Edward Guy[1], Ian Ferguson[2], Rahim Sorouri-Zanjani[3], Susan O'Connell[3]

1. Southampton Public Health Laboratory
2. Hereford Public Health Laboratory
3. Department of Microbiology, University of Southampton

INTRODUCTION

Erythema migrans (EM), the early skin manifestation of Lyme disease, was first observed in the UK in the mid-1970s in Scotland[1] and East Anglia[2]. However, it was not until 1985 that the disease itself was recognised in the UK, in the New Forest, Hampshire[3]. Since that time, studies of the incidence of the disease in man have identified a number of endemic areas including much of rural central-southern and southwestern England, East Anglia, parts of northeastern England, and coastal areas of Scotland. Subsequent studies aimed at the detection of *Borrelia burgdorferi*, the causative organism of Lyme disease, in the tick vector *Ixodes ricinus* have confirmed the presence of potential foci of infection throughout the UK. The Public Health Laboratory Service Lyme Disease reference facilities at Hereford and Southampton have provided a national serodiagnostic service for England and Wales since 1987. All samples are tested initially by ELISA with subsequent confirmatory testing where appropriate by immunoblot.

Serological testing for Lyme disease is difficult due to the often poor immune response in early infection and the relatively high prevalence of cross-reacting antibody in the normal population. The IgM response often does not peak until 4-6 weeks after infection but at six weeks, most patients are seropositive both for IgM and IgG by ELISA[4]. However, greater than 50% of the normal population were shown to possess detectable levels of antibody reacting with components of *Borrelia burgdorferi* by immunoblot[5]. One recent approach to improving the specificity of ELISA has been the use of purified borrelial antigens or antigen enriched fractions, and several versions of such tests are now available commercially. The most successful to date rely on flagellin or flagellin enriched antigens where the protein has been extracted under essentially non-denaturing conditions that appear to conserve some *B. burgdorferi*-specific epitopes.

Lyme Borreliosis, Edited by J.S. Axford and
D.H.E. Rees, Plenum Press, New York, 1994

RESULTS

In order to confirm the specificity of the ELISA kit currently in use in our laboratories, samples were tested from blood donors living in various inner city locations in the UK (n=192) and in the New Forest (NF), a known Lyme-endemic area (n=300). The prevalence of *B. burgdorferi* positive results among inner-city sera was found to be 0-2% which is consistent with the claimed specificity for the kit of 98% (Table 1). In addition, a higher prevalence of *B.burgdorferi*-specific antibody in New Forest inhabitants was clearly demonstrated. Where the diagnostic cut-off of the ELISA was decreased according to the manufacturer's protocol in order to aid detection of 'borderline' sera that might contain the low levels of antibody typical in early Lyme disease, there was no clear difference between endemic and non-endemic areas implying that this would not be helpful in diagnosis in our laboratory.

Table 1. Prevalence of *B.burgdorferi*-specific antibody in UK Blood Donor sera.

Location	Positive (%)	'Borderline' (%)
Birmingham	2	4
London	0	10
Sheffield	0	4
Leeds	2	3
NF: Bransgore	2	6
NF: Fordingbridge	9	1
NF: Hythe	11	2
NF: Verwood	10	4

A particular problem in the serodiagnosis of Lyme disease, particular to areas such as the UK, is the relatively low incidence of the disease here (0.7/100,000 *p.a.* in central Southern England) compared to elsewhere in Europe and the US. In a 12 month period a total of 6,354 samples were submitted for testing to Hereford and Southampton PHL from throughout the UK, of which 43 (0.68%) were confirmed as having active Lyme disease. Assuming the reported specificity of 98%, the ELISA would be expected to generate 127 positive results from an equivalent number of healthy adult sera. In order to eliminate a proportion of ELISA false-positive results, and hence to reduce the potential for the problem of overdiagnosis[6], confirmatory testing by immunoblot is also carried out. Despite the subjective nature of this test, in the hands of the experienced laboratory immunoblot can be helpful in improving both sensitivity and specificity of serodiagnosis. Immunoblot has been found to be particularly helpful in:

    i)   **Elimination of non-specific antibody (false positives)** - by excluding ELISA positive results from patients with suspected chronic Lyme disease where the immune response can be shown to be restricted only to highly cross-reactive antigens such as stress (heat shock) proteins.

ii)   **Exclusion of specific antibody due to past *B.burgdorferi* infection**
A well-developed immune response to a wide variety of antigens, characteristic of several months/years duration of infection, would be inconsistent with suspected early Lyme disease. This is a particular problem in endemic areas where the prevalence of *B.burgdorferi*-specific IgG in the healthy adult population can be greater than 10%.

iii) **Assisting in the diagnosis of acute Lyme disease** - by detecting the low levels of IgM and IgG reacting with the *B.burgdorferi* antigens recognised by the early immune response.

CONCLUSIONS

Despite continuing improvements in serological testing for Lyme disease, interpretation of results is still often difficult in the absence of an appropriate clinical history. Exposure to the risk of tick bite in an endemic area is also an important factor and, since many such areas are in attractive rural and woodland settings, a travel history can also be helpful. For example, in the 12 month period cited earlier, 5/43 (12%) cases of Lyme disease occurred after vacation to woodland areas of mainland Europe or to the U.S. Raised awareness of Lyme disease in the New Forest and increased education among local clinicians as to the recognition of the early symptoms of Lyme disease has led to greater numbers of cases being treated sooner. As a result, based on the epidemiological data collected in its Lyme Disease reference facilities, the Public Health Laboratory Service, together with the Medical School of the University of Southampton and New Forest District Council Environmental Health Officers, is currently involved in a public health education effort targetting other appropriate areas of the UK.

References

1. Obasi OE. Erythema chronicum migrans. Br J Dermatol 1977;**97**:459

2. Goldin D, Champion RH, Rook A, Roberts SOB. Erythema chronicum migrans in Britain. Br Med J 1978;**2**(6144):1087

3. Williams D, Rolles CJ, White JE. Lyme disease in a Hampshire child - Medical curiosity or the beginning of an epidemic. Br Med J 1986;**292**:1560-1561

4. Shrestha M, Grodzicki RL, Steere AC. Diagnosing early Lyme disease. Am J Med 1985;**78**:235-240

5. Grodzicki RL, Steere AC. Comparison of immunoblotting and indirect enzyme-linked immunosorbent assay using different antigen preparations for diagnosing early Lyme disease. J. Inf. Dis. 1988;**157**:790-797

6. Steere A.C., Taylor E., McHugh G.L., Logigian E.L. The Overdiagnosis of Lyme Disease. JAMA 1993;**269**:1812-1816

PCR-BASED DETECTION OF CSF *BORRELIA BURGDORFERI* AS A PREDICTOR OF TREATMENT RESPONSE IN CNS LYME BORRELIOSIS

J. Halperin[1], T. Keller[2], M. Whitman[2]

[1]North Shore University Hospital/Cornell University
  Medical College, Manhasset, NY, 11050
[2]Harvard Medical School, Boston, MA 02115

ABSTRACT

Objective: To use PCR based detection of *B. burgdorferi* to predict treatment response in Lyme neuroborreliosis.
Background: Accurate diagnosis of active Lyme neuroborreliosis has been difficult, contributing to difficulty judging treatment efficacy.
Design/Methods: Forty-nine patients with immunologic evidence of exposure to *B. burgdorferi*, and nervous system symptoms consistent with this diagnosis, underwent lumbar puncture and were treated with ceftriaxone. Using previously described techniques, intrathecal antibody production (ITAb) was measured, and CSF was probed for *B. burgdorferi*-specific DNA. Response to treatment was judged clinically, with 1-4 year follow-up. Response was assessed prior to performing PCR assays, and was judged as complete, partial, or failure.
Results: (i) 19 patients failed to respond; 11 were PCR -, 8 PCR+. (ii) 22 patients responded partially; 19 were PCR+. (iii) 8 patients recovered completely; 5 were PCR+, 3 PCR-. The association between PCR positivity and treatment response was highly statistically significant (p=0.005, Chi-squared test).
Conclusion: PCR based detection of *B. burgdorferi* genetic material predicts treatment responsiveness in patients with appropriate nervous system symptoms. False negatives may occur; treatment failures may reflect inadequate treatment or false positive results.

INTRODUCTION

Lyme borreliosis, the tick-borne spirochetosis caused by infection with *B. burgdorferi*, is frequently associated with abnormalities of nervous system function[1-3]. Accurate diagnosis of central nervous system (CNS) infection is often difficult because of the difficulty culturing organisms from infected patients[4-6]. Consequently, diagnosis has relied on indirect methods, such as the demonstration of production of anti-*B.burgdorferi* antibodies within the CNS[7-9].

While this method is highly specific, sensitivity is difficult to estimate, since there is no alternative "gold standard" diagnostic method - other than clinical diagnosis - to which it can be compared. Moreover, apparent intrathecal production of specific antibody (ITAb) may persist for years after successful treatment[3], making assessment of treatment response - or disease recrudescence - difficult if not impossible.

Because of these difficulties, many groups have tried to develop polymerase chain reaction-based detection methods for *B. burgdorferi* DNA to improve diagnostic accuracy[10-16]. This technique also suffers from inherent technical limitations. Its tremendous sensitivity makes contamination a major concern. It also must be remembered that detecting bacterial DNA is not synonymous with detection of viable or pathophysiologically relevant organisms. Just as this technique can characterize DNA from mummies that have been dead for millennia, a positive result in an assay for *B. burgdorferi* DNA may merely indicate the presence of dead organisms that are slowly being shed into the cerebrospinal fluid (CSF). Alternatively, it is conceivable that small numbers of organisms might persist, yet not cause disease in an immunologically intact host, but still slowly leak small amounts of genomic material into the CSF[14].

To assess the validity of our PCR-based assay, we reviewed data on patients who were felt to have CNS Lyme borreliosis on other grounds, and evaluated PCR-positivity as a predictor of response to antimicrobial therapy.

METHODS

1. Clinical

Patients referred for possible Lyme neuroborreliosis were evaluated clinically. During the 4 year period 1988-1992, 239 individuals with peripheral blood immunoreactivity against *B. burgdorferi* (positive ELISA or evidence of cell mediated immunoreactivity against this organism) underwent lumbar puncture. All were from Lyme-endemic areas. Demographic information is summarized in Table 1; neurologic syndromes are summarized in Table 2. At the time of lumbar puncture, CSF was stored on all. In forty-nine of these patients, there appeared to be sufficient clinical and laboratory evidence of disease activity to warrant intravenous ceftriaxone treatment. CSF from these 49 patients was probed using polymerase chain reaction techniques. With 1 to 4 year follow-up, patients were judged as:

Treatment failure: no significant improvement in objective neurologic signs, and usually continued deterioration;

Treatment success: significant improvement in objective signs; no new neurologic deficits;

Partial success: significant improvement in objective signs but development of new neurologic deficits.

2. Cerebrospinal fluid

At the time of lumbar puncture, CSF was divided into 1 cc aliquots using disposable sterile microbiologic pipettes. Water and other neurologic disease (OND) controls were processed and frozen in parallel. Aliquots were frozen at

TABLE 1. Patient demographic information and CSF findings[a].

	Range		Mean ± SD		# Increased
Demographics:					
Age	14 -	86	45.1 ±	16.8	
M/F		19/29			
Duration (mos)	1 -	264	36. ±	55	
Rx Duration (d)	14 -	56	18. ±	8	
CSF					
WBC	0 -	54	6.4 ±	11.2	15
Protein	16 -	106	43.9 ±	23.0	21
OCB					7
IgG Index					10
ITAb +					20
PCR +					31

[a]Demographic information includes age, male/female ratio, disease duration prior to treatment in months and length of treatment course in days. CSF data includes number of leukocytes/mm^3, protein (mg/100 ml), number of patients with increased numbers of leukocytes, increased protein concentration, oligoclonal bands, increased IgG index, evidence of intrathecal production of anti-*B. burgdorferi* antibody, and positive CSF PCR.

TABLE 2: Neurologic disorders in included patients.

Meningitis	2
Radiculoneuritis	4
Cranial neuritis	4
Encephalomyelitis	26
Encephalopathy	13

-70º until needed. For PCR analysis, previously unopened vials were subdivided into 250 µl aliquots, coded and sent blind from Long Island, NY to Boston MA, where they were handled in a laboratory in which PCR-amplified borrelial material had never been handled. Multiple aliquots were sent blind on seventeen patients - the laboratory performing PCR analysis did not know if material consisted of patient CSF, control CSF, water, or a replicate sample on a previously studied patient. Controls consisted of water samples, and CSF samples from patients with other neurologic disorders thought to be unlikely to be causally related to *B. burgdorferi* infection.

3. PCR Methods

Extraction of DNA from patient CSF's and PCR analyses were performed as previously described[14]. Briefly, coded samples were boiled, then added to an equal volume of SDS/proteinase K lysis buffer. Glycogen and *E. Coli* were added to each sample as carriers. Samples were then incubated at 37ºC for 10', boiled and extracted with phenol: chloroform (1:1), precipitated with ethanol, dried and re- dissolved in 15 - 30 µl 10 mM Tris pH 8.3, 0.5 mM EDTA (TE).

5 µl of DNA extract from CSF was added to a PCR reaction mix containing standard concentrations of reaction buffer, nucleotide and Taq polymerase, and a concentration of 0.25 µM of primers A and B. Samples were amplified through 40 cycles in a thermal cycler using the previously described automated *Hot Start* technique[17] with Ampliwax PCR Gems (Perkin Elmer Cetus) at an annealing temperature of 40º or 48º C. Product was diluted 20-fold in TE buffer, and 5 µl of diluted material was added to 50 µl fresh reaction mix identical to that described above except for the inclusion of radio-labeled dATP and the substitution of primers C and D for primers A and B. Samples were then amplified through 20 cycles. After amplification, reaction products were analyzed by gel electrophoresis using a Stratagene Separation Matrix and reaction products were visualized by autoradiography.

Nested oligonucleotide primers were designed to recognize the C terminal region of OspA, from the published sequence of *B. burgdorferi* strain B31 OspA. Primer sequences were:

A.   5' GGCTATGTTCTTGAAGGAACTCT 3'
B.   5' CGTATTGTTGTACTGTAATTGTGTT 3'

C.   5' CTTTTGTAAACACAAGGTCTT 3'
D.   5' GGTGGTTAAAGAAGGAACTGTTACTT 3'

OBSERVATIONS

In the course of processing the samples involved in this study, approximately one water or non-patient CSF (Sigma) control and one OND control were used for every two patient samples. No contamination of these controls was observed. For each patient, samples were processed on at least two separate occasions, in blind fashion. For seventeen patients, multiple aliquots were sent coded on at least two separate occasions; results were concordant in fourteen. Maximal assay sensitivity was established to be on the order 10 fg (5 genomic copies) of *B. burgdorferi* DNA. It was not uncommon, however, for one or more low copy number (20 fg) positive controls to fail to be amplified during PCR runs in which higher copy number controls (200 fg) were efficiently amplified. These observations indicate that detection of very low numbers of OspA molecules is sporadic under our amplification conditions. The technical focus in analyzing these samples was the re-assay and confirmation of positives as well as the rigorous control for false positives. Increasing the number of times that initially negative samples were assayed would almost certainly reduce the number of false negatives.

PCR results (Figure 1) in patient material (Table 3) were found to be predictive of treatment response. Of the 30 patients who clearly improved after antimicrobial therapy, 24 were PCR positive. The association between PCR-positivity and treatment response was highly statistically significant. Interestingly, 19 of these 24 responders subsequently developed additional CNS difficulties, raising the possibility that the usually prescribed 2 weeks of intravenous ceftriaxone may be inadequate for the eradication of established parenchymal CNS *B. burgdorferi* infection. The sample size and distribution of treatment durations was such that no statistically significant correlation could be detected between treatment duration and likelihood of treatment success.

Table 3: Treatment response.

	PCR -	PCR +
No response	11	8
Partial	3	19
Response	3	5

$$p=0.005 \ (\text{Chi}^2 = 11.0)$$

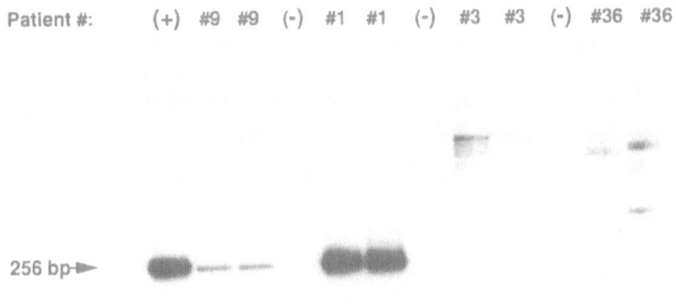

Figure 1. Duplicate PCR analysis of CSF's for *B. burgdorferi* DNA. The positive control is 500 fg of *B. burgdorferi* strain JD-1. PCR analyses of patients #9 and #3 are duplicate amplifications of a single CSF aliquot. PCR analses of patients #1 and #36 are amplifications from replicate CSF aliquots. The negative control lanes are reagent controls in which the addition of 5 µl of TE replaces the addition of target material. Amplified material from patient #9 is six weeks older than the other samples shown; consequently the signal is weaker due to radioactive decay.

Although 6 of the 17 PCR negative patients did respond to antimicrobial therapy, those who failed to respond were slightly more likely to be PCR negative (11 negative vs. 8 positive). Similarly, those who responded completely were more likely to be PCR positive than negative (5 vs 3).

DISCUSSION

The PCR procedure used in this study appears to be accurate and reliable. No contamination was observed, and patient results were internally consistent among independently processed aliquots analyzed in blind fashion. It seems unlikely that observed positive results were due to blood contamination of CSF, since all lumbar punctures included in this study were atraumatic (<3 RBC/cu. mm.). Moreover, others have repeatedly demonstrated that it is difficult to detect *B. burgdorferi* in blood of infected individuals, either by culture or PCR techniques[18-20].

The results indicate that PCR-based detection of *B. burgdorferi* genomic material in CSF is a useful predictor of treatment responsiveness - and raise the possibility that this technique might provide a useful means of assessing treatment response when applied to post-treatment samples. Clearly false negatives occur. Six patients responded to ceftriaxone, but were PCR negative. Given the objective nature of the neurologic deficits observed in these patients a placebo response seems relatively unlikely; however coincidental recovery at approximately the same time as treatment is always a possibility. Alternatively, it is possible that these patients did indeed have CNS *B. burgdorferi* infection, but insufficient genomic material was being shed into the CSF at the time of lumbar puncture to permit detection.

Conversely, PCR-positive patients who failed to respond to ceftriaxone may have represented technical false positives, biologic false positives, or treatment failures. The first possibility seems unlikely given the absence of contamination in internal controls. The second possibility is conceivable, with the positivity representing genomic material of organisms that had previously been in the CNS, but were no longer viable or pathophysiologically relevant. The third possibility is also plausible, and may be supported by treatment response in similar individuals receiving longer courses of treatment. Further work will clearly be needed to establish the true sensitivity and specificity of this technique. However this study indicates that this method provides information that accurately predicts response to antimicrobial therapy, and therefore is potentially of considerable clinical utility.

ACKNOWLEDGEMENT

This work was supported in part by the EEL Association.

REFERENCES

1. Stiernstedt G, Gustafsson R, Karlsson M, Svenungsson B, Skoldenberg B. Clinical manifestations and diagnosis of neuroborreliosis. Ann N Y Acad Sci 1988; 539:46-55.

2. Finkel MJ, Halperin JJ. Nervous System Lyme Borreliosis-Revisited. Arch Neurol 1992; 49:102.

3. Hansen K, Lebech A-M. The clinical and epidemiological profile of Lyme neuroborreliosis in Denmark 1985-1990. Brain 1992; 115:399-423.

4. Preac MV, Weber K, Pfister HW, et al. Survival of Borrelia burgdorferi in antibiotically treated patients with Lyme borreliosis. Infection 1989; 17:355-9.

5. Pfister HW, Einhaupl K, Preac MV, Wilske B, Schierz G. The spirochetal etiology of lymphocytic meningoradiculitis of Bannwarth (Bannwarth's syndrome). J Neurol 1984; 231:141-4.

6. Karlsson M, Hovind HK, Svenungsson B, Stiernstedt G. Cultivation and characterization of spirochetes from cerebrospinal fluid of patients with Lyme borreliosis. J Clin Microbiol 1990; 28:473-9.

7. Stiernstedt GT, Granstrom M, Hederstedt B, Skoldenberg B. Diagnosis of spirochetal meningitis by enzyme linked immunosorbent assay and indirect immunofluorescence assay in serum and cerebrospinal fluid. J Clin Microbiol 1985; 21:819-825.

8. Wilske B, Schierz G, Preac-Mursic V, et al. Intrathecal production of specific antibodies against Borrelia burgdorferi in patients with lymphocytic meningoradiculitis. J Inf Dis 1986; 153:304-314.

9. Halperin JJ, Luft BJ, Anand AK, et al. Lyme neuroborreliosis: central nervous system manifestations. Neurology 1989; 39:753-759.

10. Rosa PA, Schwan TG. A specific and sensitive assay for the Lyme disease spirochete Borrelia burgdorferi using the polymerase chain reaction. J Infect Dis 1989; 160:1018-29.

11. Goodman JL, Jurkovich P, Kramber JM, Johnson RC. Molecular detection of persistent Borrelia burgdorferi in the urine of patients with active Lyme disease. Infect Immun 1991; 59:269-78.

12. Persing DH, Telford SR3, Spielman A, Barthold SW. Detection of Borrelia burgdorferi infection in Ixodes dammini ticks with the polymerase chain reaction. J Clin Microbiol 1990; 28:566-72.

13. Malloy DC, Nauman RK, Paxton H. Detection of Borrelia burgdorferi using the polymerase chain reaction. J Clin Microbiol 1990; 28:1089-93.

14. Keller TJ, Halperin JJ, Whitman M. PCR detection of Borrelia burgdorferi DNA in cerebrospinal fluid of Lyme neuroborreliosis patients. Neurology 1992; 42:32.

15. Luft BJ, Steinman CR, Neimark HC, et al. Invasion of the Central Nervous System by Borrelia burgdorferi in Acute Disseminated Infection. JAMA 1992; 267:1364-7.

16. Lebech A-M, Hinderson P, J. V, Hansen K. Comparison of In Vitro culture and polymerase chain reaction for detection of Borrelia burgdorferi in tissue from experimentally infected animals. J Clin Microbiol 1991; 29:731-737.

17. D'Aquilla RT, Bechtel LJ, Videler JA, Eron JJ, Gorczyca P, Kaplan JC. Maximizing sensitivity and specificity of PCR by pre-amplification heating. Nucleic Acids Res. 1991; 19:3749.

18. Benach JL, Bosler EM, Hanrahan JP, et al. Spirochetes isolated from the blood of two patients with Lyme Disease. . NEJM 1983; 308:740-742.

19. Steere AC, Grodzicki RL, Kornblatt AN, et al. The spirochetal etiology of Lyme Disease. NEJM 1983; 308:733-740.

20. Guy EC, Stanek G. Detection of Borrelia burgdorferi in patients with Lyme disease by the polymerase chain reaction. J Clin Pathol 1991; 44:610-1.

DIAGNOSTIC DETECTION OF *BORRELIA BURGDORFERI* DNA BY THE POLYMERASE CHAIN REACTION (PCR)

Anne-Mette Lebech

Department of Infection-Immunology
Borrelia laboratory
Statens Seruminstitut
Copenhagen, Denmark

INTRODUCTION

The fundamental problem in all attempts at direct detection of *Borrelia burgdorferi* is the extremely low number of organisms present in pathological lesions or body fluids from patients with Lyme borreliosis. Therefore the recent introduction of the powerful in vitro amplification of DNA by polymerase chain reaction (PCR)[1] seemed a promising solution. PCR enables a selected DNA sequence to be copied in vitro. With optimal conditions a single strand of DNA can be copied over a million fold and then subsequently detected by conventional methods such as Southern blot hybridization.

PCR assays for detection of purified *B.burgdorferi* DNA achieve a high analytical sensitivity with a lower detection limit of about 2 femtogram DNA corresponding to approximately one genome copy[2,3,4]. Testing clinical specimens artificially seeded with in vitro cultivated spirochetes a sensitivity of 20–50 spirochetes per ml may be obtained[5].
From a diagnostic view point the optimal target DNA sequence to be amplified should be *B.burgdorferi* specific but also highly conserved, so that DNA from all *B.burgdorferi* strains would be detected. This demand are meet by parts of the *B.burgdorferi* flagellin gene[5], or the 23S- and 16S RNA gene[6,7] which allow both species specific and species wide amplification.

PCR is as effective as immunofluorescence for the demonstration of *B.burgdorferi* in infected Ixodes ticks[3]. However, due to the much higher density of spirochetes in the midgut of ticks compared with that in clinical specimens from patients with active Lyme borreliosis, the conditions are not comparable.

Regarding the diagnostic performance of PCR on clinical specimens from patients with Lyme borreliosis the data available are so far sparse and less conclusive. In tissue from experimentally infected animals[4] as well as in skin biopsies from patients with erythema migrans and acrodermatitis chronica atrophicans[6,8], PCR seems to achieve a diagnostic sensitivity comparable to culture. Thus PCR seems promising for dermatological specimens. The reported success rates using other sample sources vary considerably. *B. burgdorferi* DNA has been amplified from serum[9], joint fluid[10], cerebrospinal fluid (CSF)[5,7,10,11] and urine[5,12] however only from a limited number of patients with Lyme borreliosis. At present it seems that CSF and especially urine are the most appropriate sample source for PCR. Using CSF samples as sample source the diagnostic sensitivities reported vary from 20 to 90%[5,7,10,11]. Although only few data are available on urine[5,12] the diagnostic sensitivity on pretreatment samples are approximately 50% in cases with early disseminated disease. An interesting observation is, that the diagnostic sensitivity on urine seems to increase considerably when urine sampled 3–6 days after antibiotic treatment start is tested[5]. This phenomena may be explained by treatment induced spirochetal killing and then increased excretion of *B. burgdorferi* DNA. However, the results in the different reports are not comparable especially due to significant differences in the selection of patients. This is probably more important than differences in PCR methodology. The variable success rate of diagnostic PCR in Lyme borreliosis may be due to that the low number of spirochetes in body fluids from patients a prior is close to the detection limit of PCR thus making the PCR assay much more sensitive to Taq polymerase inhibition by tissue components.

CONCLUSION

The present perspective indicates that PCR on skin biopsy material could become a useful diagnostic procedure. Using PCR *B. burgdorferi* DNA has reliably been detected in CSF and urine from patients with Lyme borreliosis but the diagnostic sensitivity remains to be defined. Future studies need to evaluate the precis performance of PCR as a diagnostic tool on large series of patients with unequivocal Lyme borreliosis. For the time being PCR should not be accepted as the golden standard, substituting in vitro cultivation.

REFERENCES

1. Saiki R K, Gelfand D H, Stoffel S, Scharf S J, Higuchi R, Horn G T, Mullis K B, Erlich H A. Primer directed enxymatic amplification of DNA with a thermostable DNA polymerase. Science. 1988; **239**: 487–491.
2. Rosa P A, Schwan T G. A specific and sensitive assay for the Lyme disease spirochete *Borrelia burgdorferi* using the polymerase chain reaction. J Infect Dis 1989; **160**: 1018–1029.

3. Persing D H, Telford III S R, Spielman A, Barthold S W. Detection of *Borrelia burgdorferi* infection in Ixodes dammini ticks with the polymerase chain reaction. J Clin Microbiol. 1990; **28**: 566–572.

4. Lebech A M, Hindersson P, Vuust J, Hansen K. Comparison of in vitro culture and polymerase chain reaction for detection of *Borrelia burgdorferi* in tissue from experimentally infected animals. J Clin Microbiol. 1991; **29**: 731–737.

5. Lebech A M, Hansen K. Detection of Borrelia burgdorferi DNA in urine samples and cerebrospinal fluid samples from patients with early and late Lyme neuroborreliosis by polymerase chain reaction. J Clin Microbiol. 1992; **30**: 1646–1653.

6. Schwartz I, Wormser G, Schwartz J J, et al. Diagnosis of early Lyme disease by polymerase chain reaction amplification and culture of skin biopsies from erythema migrans lesions. J Clin Microbiol. 1992; **30**: 3082–3088.

7. Luft B J, Steinman C R, Neimark H C, et al. Invasion of the central nervous system by Borrelia burgdorferi in acute disseminated infection. JAMA. 1992; **267**: 1364–1367.

8. Melchers W, Meis J, Rosa R, et al. Amplification of *Borrelia burgdorferi* DNA in skin biopsies from patients with Lyme disease. J Clin Microbiol. 1991; **29**: 2401–2406.

9. Guy E C, Stanek, G. Detection of *Borrelia burgdorferi* in patients with Lyme disease by the polymerase chain reaction. J Clin Pathol. 1991; **44**: 610–611.

10. Debue M, Gautier P, Hackel C, et al. Detection of *Borrelia burgdorferi* in biological samples using the polymerase chain reaction assay. Res Microbiol. 1991; **142**: 565–572.

11. Keller, T L, Halperin J J, Whitman M. PCR detection of Borrelia burgdorferi DNA in cerebrospinal fluid of Lyme neuroborreliosis patients. Neurology. 1992; **42**: 32–42.

12. Goodman J L, Jurkovich P, Kramber J M, Johnson R C. Molecular detection of persistent *Borrelia burgdorferi* in the urine of patients with active Lyme Disease. Infect. Immun. 1991; **59**: 269–278.

IMPROVED DETECTION OF IMMUNOGLOBULIN M IN SERA OF ERYTHEMA MIGRANS PATIENTS BY WESTERN BLOTTING WITH A LOCAL *Borrelia burgdorferi* SKIN ISOLATE

Sjoerd Rijpkema[1], Herman Kuiper[2,3], Marc Molkenboer[1], Alje van Dam[2] and Joop Schellekens[1]

[1]Laboratory of Bacteriology and Antimicrobial Agents
  National Institute of Public Health and Environmental
  Protection, PO Box 1
  3720BA Bilthoven (tel. +31-30742107/fax +3130292957)
[2]Department of Medical Microbiology
[3]Department of Neurology, Academic Medical Centre
  Meiberglaan 9, 1105AZ Amsterdam, The Netherlands

INTRODUCTION

Lyme borreliosis (LB) is a multi-system disorder, which occurs in temperate zones and is caused by the tick-borne spirochaete *Borrelia burgdorferi*.[1] The clinical symptoms of LB can be differentiated in manifestations of early and late infection. The early infection comprises two stages, the localized stage which is marked by the development of an Erythema Migrans (EM) around the site of the tick bite, and the disseminated stage which is commonly characterized by meningopolyneuritis, multiple skin lesions, arthralgias, bouts of arthritis or carditis. Months to years after onset, persisting infection can result in symptoms such as chronic arthritis, encephalomyelitis or acrodermatitis chronica atrophicans.[2]

A whole cell sonicate ELISA or IgG western blot can be used for the serodiagnosis of disseminated and persistent infection.[2,3,4] Serodiagnosis of patients with an EM or acute neuroborreliosis, has proven cumbersome because serologic tests often fail to detect an antibody response. Seronegativity in early disease may be explained by poor sensitivity of the serodiagnostic tests or by the assumption that the antibody response is retarded e.g., through modulation by biologically active substances in saliva of the ticks. Indeed, studies in rodents and dogs have demonstrated that the onset of the antibody response after tick mediated infection is delayed compared to the antibody response provoked by intraperitoneal or subcutaneous injection of *B burgdorferi*.[5,6,7]

In humans, the 41-kilodalton (kD) flagellar antigen and the protoplasmic cylinder (pC) protein (22 kD) of *B burgdorferi* appear to be the main antigens recognized by the antibody response in early LB.[8,9] ELISAs based on the flagellum or a combination of recombinant flagellin and pC protein proved to be superior to the whole cell sonicate ELISAs for the serodiagnosis of early LB.[10,11,12] Seropositivity among patients with EM was significantly enhanced when a flagellum μ-capture ELISA instead of an indirect IgM flagellum ELISA was used.[13] Similar results were obtained with a μ-Capture ELISA which used the B burgdorferi whole cell sonicate as a marker.[14] Western blotting has also been used for the detection of IgM. Karlsson[15] noted an IgM seropositivity of 68% among neuroborreliosis patients in Westernblot. Only 34% of this group were IgM positive when an indirect whole cell sonicate ELISA was used. Recently, application of Western blot revealed a seropositivity for IgM of 81% among patients with EM opposed to 33% by indirect IgM flagellum ELISA.[16] This makes Western blotting the most sensitive assay for the detection of specific IgM. In the study of Lange, IgM reacted consistently with the flagellin.[16] However, Karlsson[15] and Wilske[9] demonstrated an additional reaction with the pC protein in patient sera. This discrepancy is possibly due to a low level of pC expression by the tick isolate which was used as antigen by Lange.

B burgdorferi isolates are not genetically homogeneous[17] and 7 serotypes have been identified for the outer surface protein (Osp) A.[18] Recently it has become clear that strains of OspA serotype 2 are prevailing amongst european skin isolates,[19] and that some of these strains have a high expression of the pC protein.[18] B burgdorferi OspA serotype 2 strains have been allocated to a genetically distinct group named VS461.[19,20]

In this study we determined the influence of the genospecies on the sensitivity of the IgM Westernblot. B burgdorferi sensu stricto, OspA serotype 1, and B burgdorferi VS461 were used as antigen in a Western blot for the detection of IgM in sera of patients with EM.

PATIENTS AND METHODS

PATIENTS AND SERUM SPECIMENS

Sera of 55 untreated EM patients were used. An EM was defined as an expanding lesion at least 5 centimetres in size. At the presentation of the EM there were no obvious clinical signs of disseminated infection such as secondary skin lesions, arthritis, migratory musculoskeletal pain, severe headache, neurological symptoms except for sensory manifestations limited to the skin lesion, bradycardia and collapse, severe malaise and conjunctivitis. Skin biopsies were taken 45 patients, and from 38 (84%) specimens B.burgdorferi was isolated.

Sera from 24 healthy blood donors were used to determine the specificity of the IgM westernblot. Additionally we tested the reactivity of sera of patients who had an infectious disease not related to LB: Thirteen sera from syphilitic patients, who had a positive result in the 19S IgM fluorescent treponemal antibody absorption test; fifteen sera from patients with mononucleosis infectiosa (A gift from Dr W. van Dijk, Slotervaart Hospital Amsterdam). Eight sera from patients with toxoplasmosis; five sera from patients with respiratory syncytial virus infection; four sera from patients with herpes simplex virus infection and seven sera from patients with a cytomegalovirus infection (A gift from Dr P. Herbrink, Diagnostic Centre, SSDZ Delft).

Monoclonal antibodies (moabs) LF22 1F8 (A gift from Dr B. Wilske, München, Germany), H9724 and H5332 (A gift from Dr A. Barbour, San Antonio, Texas) were used for the identification of pC, flagellin and OspA respectively.

STRAINS AND ANTIGEN PREPARATION

B burgdorferi isolates used in this study are A39S and B31 (ATCC 35210). A39S was isolated from the skin of a dutch patient with EM, and belongs to genospecies VS 461 and is OspA serotype 2 (van Dam, submitted for publication). The american tick isolate B31 belongs to *B.burgdorferi* sensu stricto and is OspA serotype 1.[19,20] Strains were cultured in BSK-II medium at 34°C and harvested when colour of the medium became distinctly orange.[21] *B burgdorferi* cells were washed three times in phosphate buffered saline (PBS, pH 7.2) containing 5mM $MgCl_2$, suspended in sterile distilled water, aliquotted and kept at -70°C until further use.

IgM WESTERN BLOTTING

The IgM Western blot procedure was performed with modifications as described.[16] Forty micrograms of *B burgdorferi* cells were solubilized in sample buffer containing 2 % sodium dodecyl sulphate (SDS), 2 % dithioerythreitol, 10 % glycerol and 0.01% bromophenol blue. *B burgdorferi* antigens were separated by discontinuous SDS-polyacrylamide gel electrophoresis,[22] loaded on a 4% stacking gel and electrophorated on a 13% separating gel at 150 mA. Antigens were transferred from the gel to nitrocellulose sheets (0.2 μm, Schleicher and Schuell, Dassel, Germany) under semidry conditions for 1 hour with a current of 250 mA. The blot was used immediately or stored at -20°C until further use.

The blot was blocked overnight by submersion in PBS containing 0.05 % Tween-20 (PBST) and 1 % non-fat dry milk powder (Nutricia, Zoetermeer, The Netherlands) at 4°C on a rotating platform. Prior to their use in Western blot, sera were treated with an anti-human IgG sorbens as described by the manufacturer (Gull Laboratories, Salt Lake City, Utah) and finally diluted 1:100 with blocking buffer. A lane of the blotting chamber was filled with 1 ml diluted serum and incubated overnight at 4°C on a rotating platform. Blots were washed two times with PBST and incubated for 90 mins with peroxidase conjugated goat anti-human immunoglobulin M (Dakopatts, Glostrup, Denmark) diluted 1:1000 in blocking buffer, which contained 2% normal goat serum (Organon, Oss, The Netherlands), on a rotating platform.

Blots were washed three times with PBST and bound conjugate was visualized by reaction of 3-3'-5-5' tetra-methylbenzidine (Sigma Chemical Company, St Louis, Missouri) and hydrogenperoxide which gives a green precipitation.[23]

RESULTS

A typical IgM Western blot result is presented in figure 1. Sera were tested on strains B31 and A39S. A sample was considered positive when a reaction was detected either with the flagellin (41 kD) or the pC (22 kD) antigen or both. Results of IgM Western blotting in relation to duration of EM are summarized in table 1. IgM reactivity of EM patient sera, donor sera, and the sera of patients with an infection not related to LB are presented in table 2.

When B31 was used as antigen, IgM was detected in sera of 28 EM patients and directed solely against the flagellin. Serum IgM reactivity of this group was extended to 42 patients when *B burgdorferi* strain A39S was used as antigen. In those 42 samples, reaction of IgM with both the flagellin of A39S and the pC protein was found in 20 sera. In 2 sera a unique reaction with the pC protein was found. The

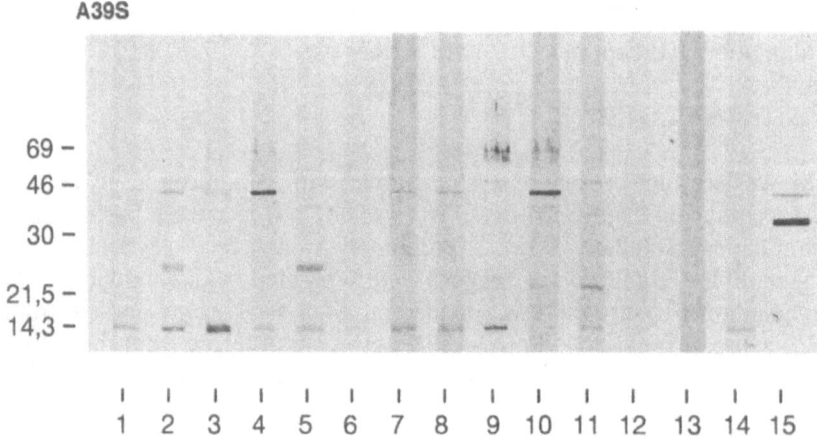

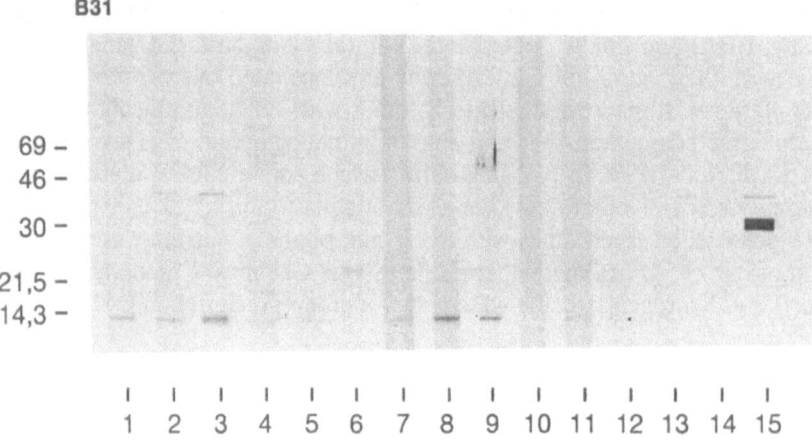

Figure 1: Reactivity of human sera and mouse monoclonal antibodies in a Western blot with *B burgdorferi* isolate A39S (A39S) or *B burgdorferi* isolate B31 (B31). Lane 1: Negative control serum, lanes 2-5: Sera from patients with EM, lanes 6 and 7: Sera from patients with syphilis, lanes 8-10: Sera from patients with mononucleosis infectiosa, lanes 11-14: Sera from blood donors, lane 15: Monoclonal antibodies H9724 and H5332. Molecular weight markers (in kD) are indicated left.

Table 1. Effect of the *B burgdorferi* strain variation and duration of EM on reactivity of patient sera with the flagellin (Fla) and the pC protein in IgM Western blot.

Duration of EM in weeks	No. of sera	No. or (%) of sera positive by IgM westernblot					
		B31			A39S		
		Fla	pC	Fla or pC	Fla	pC	Fla or pC
< 2	7	4	0	4	5	3	5
2-5	8	4	0	4	4	2	5
> 5	40	20	0	20	31	15	32
Total	55	28	0	28 (51)	40	20	42[a](76)

[a]: *A unique reaction with the pC protein was observed in 2 sera.*

discrepancy in IgM reactivity of antigen B31 and A39S is visualized in lanes 2, 4 and 5 of figure 1. Although, *B burgdorferi* strain B31 apparently lacked the 22 kD antigen by means of IgM Western blotting, the pC protein could be detected in this strain with moab LF22 1F8 (results not shown). The highest gain in sensitivity was seen in patients who had an EM with a duration of more than 5 weeks. In this group, one of 8 patients, who had a negative IgM Western blotting result, had specific IgG antibodies (results not shown). Banding in Western blotting was more pronounced when strain A39S was used as antigen.

Six out of 24 donor sera showed reactivity in the Western blot when A39S was used as antigen (table 2, figure 1 lanes 11-14). This reactivity included the flagellin in all cases. With the flagellin of B31 reactivity was found in only two donor sera. Sera of patients with syphilis showed a clear reaction with the flagellin. A high degree of reactivity was also found in sera of patients suffering from mononucleosis infectiosa or toxoplasmosis.

Both high and low intensity 41 kD bands were found for EM patients and controls. Additional reactivity against the pC protein was found in only one serum from a patient with mononucleosis infectiosa and in one serum from a blood donor.

DISCUSSION

Seronegativity is commonly found among patients with an EM or an acute neuroborreliosis. The absence of specific antibodies is an obstacle for reliable serodiagnosis of LB. The advent of ELISAs based on purified immunodominant antigens, the μ-capture flagellum ELISA and the IgM westernblot has improved the performance of serodiagnostic tests but seronegative LB still exists. Western blotting proved to be the most sensitive method to detect IgM in serum of patients with EM.[16]

In our study we used two strains as antigen for IgM Western blotting. We found a higher percentage of IgM positive sera among EM patients when skin isolate A39S as opposed to laboratory strain B31 was used as antigen. Immunoglobulin M reacted mainly with the flagellin and additional IgM reactivity with the pC protein could be detected in 18 of 40 sera of EM patients when *B burgdorferi* isolate A39S was used. However, a substantial number of sera from blood donors and patients with non related infections contained IgM directed against the flagellin. Especially the high

Table 2. Reactivity of donor sera, sera of patients with EM or infections other then LB with the flagellin (Fla) and the pC protein in IgM Western blot.

Clinical diagnosis	No. sera	B31		A39S	
		Fla	pC	Fla	pC
EM patients	55	28	0	40	20
Healthy Donors	24	2	0	6	1
Syphilis	13	3	0	5	0
Mononucleosis infectiosa	15	11	0	13	1
Toxoplasmosis	8	6	0	5	0
Respiratory syncytial virus	5	0	0	0	0
Herpes simplex virus	4	1	0	0	0
Cytomegalovirus	7	1	0	1	0

levels of reactivity among blood donors pose a problem for the specificity of the IgM Westernblot. We could not demonstrate a quantitative difference in IgM reactivity with the flagellin among EM patient sera and control sera. Our findings are in contrast with results obtained by Lange, who did not find any IgM reactivity in donor sera but did not test the reactivity of IgM in sera of patients with other infectious diseases. The discrepancies in reactivity of control sera with the flagellin can be attributed to differences in antigen concentration or blocking agent. For a limited number of sera, we compared the blocking capacity of non-fat dry milk powder to ovalbumin and did not find any differences in background reactivity at a concentration of 1 μg antigen per lane. Other studies have demonstrated IgG or IgM reactivity to the flagellin in sera obtained from non Lyme disease patients or healthy donors in Western blot.[3,4,15] This has led to a cautious interpretation of Western blotting results. A single 41 kD band in Western blot is not considered of diagnostic value. Karlsson[15] and Wilske[9] proposed that the presence of both the flagellin band and the pC band should be considered to confirm an early LB. Our results indicate that IgM reactivity to the pC protein is rarely found in the control group but can be detected in 20 of 55 sera of EM patients. These findings confirm the high immunogenicity of this antigen in early LB.[12] Reactivity to the pC antigen has been found in sera of american Lyme disease patients[24] and the presence of a pC gene has been demonstrated in american *B burgdorferi* isolates.[25] Although a weak reaction of isolate B31 with an anti-pC monoclonal antibody was noted, we did not find any pC reactivity in Western blotting. This could be due to a low level of expression of the pC protein or to a difference in serotype of pC. Heterogeneity of the pC protein has been noted for european isolates.[18]

Experiments are in progress to enhance specificity of the IgM reactivity to the flagellin by absorption of sera with *Treponema phagedenis*. The absorption might remove IgM which reacts with the common epitopes of the flagellin but will not affect the reactivity with the central *B burgdorferi* specific region of the flagellin. This region is recognized by the immune system of patients with LB.[26,27] Preliminary results show a diminished reactivity of IgM with flagellin after absorption of donor sera with *T phagedenis*.

In our opinion IgM Western blotting can be used as a confirmation test when reactivity to the pC protein is adapted as criterium. In our study, this resulted in a decline of sensitivity of the IgM Western blot from 76% to 36% if strain A39S was used as an antigen. With B31, all sera should be considered negative since anti-pC antibodies could not be detected. Therefore, IgM Western blotting should be performed with an *B burgdorferi* isolate which has a proper expression of the pC protein. Because of the heterogeneity of the pC protein the isolate should be of local origin. Our results emphasise the importance of the pC protein as antigen for the serodiagnosis of EM with IgM Western blotting.

REFERENCES

1. W. Burgdorfer, A.G. Barbour, S.F. Hayes, et al., Lyme disease—a tick borne spirochetosis?, *Science* 216:1317 (1982).
2. A.C. Steere, Lyme disease, *New Engl. J. Med.* 321:586 (1989).
3. L. Zöller, S. Burkard, and H. Schäfer, Validity of Western immunoblot band pattern in the serodiagnosis of Lyme borreliosis, *J. Clin. Microbiol.* 29:174 (1991).
4. B. Ma, B. Christen, D. Leung, and C. Vigo-Perlfrey, Serodiagnosis of Lyme borreliosis by Western immunoblot: reactivity of various significant antibodies against *Borrelia burgdorferi*, *J. Clin. Microbiol.* 30:370 (1992).
5. R.T. Greene, R.L. Walker, W.L. Nicholson, H.W. Heidner, J.F. Levine, E.C. Burgess, M. Wyand, E.B. Breitschwerdt, and H.A. Berkhoff, Immunoblot analysis of immonoglobulin G response to the Lyme disease agent (*Borrelia burgdorferi*), in experimentally and naturally exposed dogs, *J. Clin. Microbiol.* 26:648 (1988).
6. M.J.G. Appel, S. Allan, R.H. Jacobson, T.L. Lauderdale, Y.F. Chang, S.J. Shin, J.W. Thomford, R.J. Todhunter, and B.A. Summers, Experimental Lyme disease in dogs produces arthritis and persistent infection, *J. Infect. Dis.* 167:651 (1993).
7. J.T. Roehrig, J. Piesman, A.R. Hunt, M.G. Keen, C.M. Happ, and B.J.B. Johnson, The hamster antibody response to tick-transmitted *Borrelia burgdorferi* differs from the response to needle-inoculated,cultured organisms, *J. Immunol.* 149:3648 (1992).
8. J.E. Craft, D.K. Fischer, G.T. Shimamoto, and A.C. Steere, Antigens of *Borrelia burgdorferi* recognized during Lyme disease. Appearance of a new immunoglobulin M response and expansion of the immunoglobulin G response late in the illness, *J. Clin. Invest.* 78:934 (1986).
9. B.A. Wilske, V. Preac-Mursic, G. Schierz, and K.V. Busch, Immunochemical and immunological analysis of european *Borrelia burgdorferi* strains, *Zbl. Bakt. Hyg. A* 263:92 (1986).
10. K. Hansen, P. Hindersson, and N. Strandberg Pedersen, Measurement of antibodies to the *Borrelia burgdorferi* flagellum improves serodiagnosis in Lyme disease, *J. Clin. Microbiol.* 26:338 (1988).
11. T.-M. Lin, C.M. Schubert, F.F. Shih, P. Ahmad, M. Lopez, and H. Horst, Use of flagellin-enriched antigens in a rapid, simple and specific quantitative enzyme immunoassay for Lyme disease antibodies in human serum samples, *J. Immunoassay* 12:325 (1991).
12. B.A. Wilske, V. Preac-Mursic, R. Fuchs, and E. Soutschek, Immunodominant proteins of *Borrelia burgdorferi*, the etiological agent of lyme borreliosis, *World J. Microbiol. Biotechnol.* 7:130 (1991).
13. K. Hansen, K. Pii, and A.-M. Lebech, Improved immunoglobulin M serodiagnosis in Lyme borreliosis by using a µ-capture enzyme-linked immunosorbent assay with biotinylated *Borrelia burgdorferi* flagella, *J. Clin. Microbiol.* 29:166 (1991).
14. M. Karlsson, and M. Granström, An IgM-antibody capture enzyme immunoassay for serodiagnosis of Lyme borreliosis, *Serodiagnosis Immunotherapy Infect. Dis.* 3:413 (1989).
15. M. Karlsson, I. Möllegard, G. Stiernstedt, and B. Wretlind, Comparison of western blot and enzyme-linked immunosorbent assay for diagnosis of Lyme borreliosis, *Eur. J. Clin. Microbiol. Infect. Dis.* 8:871 (1989).
16. R. Lange, H. Bocklage, T. Schneider, H.W. Kölmel, J. Heesemann, and H. Karch, Ovalbumin blocking improves sensitivity and specificity of immunoglobulin M immunoblotting for serodiagnosis of patients with Erythema Migrans, *J. Clin. Microbiol.* 30:229 (1992).
17. T. Adam, G.S. Gassmann, C. Rasiah, and U.B. Göbel, Phenotypic and genotypic analysis of *Borrelia burgdorferi* isolates from various sources, *Infect. Immun.* 59:2579 (1991).
18. B.A. Wilske, J.F. Anderson, G. Baranton, A.G. Barbour, K. Hovind-Hougen R.C. Johnson, V. Preac-Mursic, Taxonomy of *Borrelia* spp., *Scand. J. Infect. Dis.* S77:108 (1991).

19. B.A. Wilske, V. Preac-Mursic, U.B. Göbel, B. Graf, S. Jauris, E. Soutschek, E. Schwab, and G. Zumstein, An ospA serotyping system for *Borrelia burgdorferi* based on the reactivity with monoclonal antibodies and OspA sequence analysis, *J. Clin. Microbiol.* 31:340 (1993).

20. G. Baranton, D. Postic, I. Saint Girons, P. Boerlin, J.-C. Piffaretti, M. Assous, and P.A.D. Grimont, Delineation of *Borrelia burgdorferi* sensu stricto, *Borrelia garinii* sp. nov., and group VS461 associated with Lyme borreliosis, *Int. J. Syst. Bacteriol.* 42:378 (1992).

21. A.G. Barbour, Isolation and cultivation of Lyme disease spirochetes, *Yale J. Biol. Med.* 57:521 (1984).

22. U.K. Laemmli, Cleavage of structural proteins during the assembly of the head of bacteriophage T4, *Nature (London)* 227:680 (1970).

23. P. Buckel, and E. Zehelein, Expression of *Pseudomonas fluorescens* D-galactose in *Escherichia coli*, *Gene* 16:149 (1981).

24. F.N. Dressler, J.A. Whalen, B.N. Reinhardt, and A.C. Steere, Western blotting in the serodiagnosis of Lyme disease, *J. Infect. Dis.* 167:392 (1993).

25. R.T. Marconi, D.S. Samuels, and C.F. Garon, Transcriptional analysis and mapping of the ospC gene in Lyme disease spirochetes, *J. Bacteriol.* 175:926 (1993).

26. R. Berland, E. Fikrig, D. Rahn, J. Hardin, R.A. Flavell, Molecular characterization of the humoral immune response to the 41-kilodalton flagellar antigen of *Borrelia burgdorferi* the Lyme disease agent, *Infect. Immun.* 59:3531 (1991).

27. J.M. Robinson, T.J. Pilot-Matias, S.D. Pratt, C.D. Patel, T.S. Bevirt, and J.C. Hunt, Analysis of the humoral immune response to the flagellin protein of *Borrelia burgdorferi*: Cloning of the regions capable of differentiating between Lyme disease from syphilis, *J. Clin. Microbiol.* 31:629 (1993).

DIFFERENTIATION OF BORRELIA BURGDORFERI ISOLATES FROM TICKS AND HUMANS BY DIFFERENT MONOCLONAL ANTIBODIES IN IMMUNO-FLUORESCENCE

Arno Schönberg and Christine Loser

Institute for Veterinary Medicine
Federal Health Office
Diedersdorfer Weg 1
1000 Berlin 48
Germany

INTRODUCTION

The clinical symptoms of Lyme borreliosis in Europe and USA show similarities. In contrast, immunological analysis of strains of <u>Borrelia burgdorferi</u>, the etiologic agent of the disease provided evidence of polymorphisms of the protein patterns. Wilske et al. (1986) suggested that owing to the considerable heterogeneity of antigens, serological tests should be carried out with local Borrelia strains. Even if this opinion has changed in the meantime, the serological differentiation of B. burgdorferi isolates by different monoclonal antibodies (MAB) is important for two reasons:

1. Using MABs in the immunofluorescence test (IFT) of B. burgdorferi, isolates from Europe and USA have shown differences in the presentation of the proteins OspA and OspB (Barbour and Schrumpf, 1986). All 24 American strains reacted with the MABs H 5332 and H 3TS, and of 21 European strains, 20 reacted with H 5332 and only three with H3TS.

2. The identification of 57 <u>B. burgdorferi</u> strains, isolated from ticks of the woodlands in Berlin, provided results which differ from those of Barbour and Schrumpf (1986): 56 strains showed reaction with H 5332 and 38 with H 3TS; one was negative with both MABs (Schmidt and Schönberg, 1990).

Lyme Borreliosis, Edited by J.S. Axford and
D.H.E. Rees, Plenum Press, New York, 1994

In the latter study, an IgM-MAB was tested additionally to get more information about the antibody reactivities of the protein OspA. The IgG-MABs as well as the IgM-MAB are directed against the OspA protein.

MATERIALS AND METHODS

366 nymphs and 80 adult Ixodes ricinus were collected in three new "Bundesländer" (table 1). Samples of 3 to 5 nymphs and single individuals of adults were examined for Borreliae by cultivation in modified Barbour-Stoenner-Kelly's medium (Schönberg et al., 1988). To reduce growth of other bacteriae neomycin (4µl/ml) was added to the medium.

From patients with the clinical sign of erythema migrans skin biopsies were cultured in BSK-medium with neomycin.

Table 1. Nymphs and adult ticks examined by cultivation

	Nymphs	Females	Males
Mecklenburg-Vorpommern	20 samples per 3 11 samples per 5	12	8
Brandenburg	10 samples per 3 14 samples per 5	15	15
Thuringia	27 samples per 3 14 samples per 5	18	12

2 ml of every cultured strain (from ticks and from skin) were pipetted into reaction tubes and centrifugated 20 minutes at 14000 g. The sediment was three times suspended and washed in PBS-MgCl$_2$ buffer (0.15 M PBS with 5 mM MgCl$_2$) and re-centrifugated. Thereafter the sediment was again suspended to a suitable density. Six fields of a marked slide were covered by 12 µl of each strain. After air drying the antigen preparations were fixed by aceton at -20°C for five minutes (Bark, 1986). Finally they were washed with the buffer and distilled water. Antigen preparation of the strain 1 B 29 (strain of Ixodes ricinus) and of B 31 served as positive controls. Of each strain two of the antigenic fields were used as negative control.

On the antigenic fields 10 µl of each monoclonal antibody (table 2) were incubated at 33°C for 30 minutes in a wet chamber. The slides were washed in PBS-MgCl$_2$ buffer and all antigen fields were covered by 10 µl of FITC-conjugate (antibodies marked with fluorescein isothiocyanate). After a further wash with buffer and distilled water the slides were covered by glycerine buffer (9 glycerine: 1 PBS) and examined by fluorescence microscopy at the magnification of 400.

Table 2. Identification with monoclonal antibodies (Barbour et al., 1983; Barbour et al., 1986; Benach et al., 1988)

Name	class of Ig	against	Infection of mouse with
H 9724	IgG	flagellin	B. hermsii, strain HS 1 serotype C
H 5332	IgG	OspA	B. burgdorferi, strain B 31 (isolate of Ixodes dammini)
H 3TS	IgG	OspA	B. burgdorferi., strain HB 19 (isolate of human blood)
11 G1	IgM	OspA	B. burgdorferi, strain B 31 (isolate of Ixodes dammini)

RESULTS

Spirochetes could be isolated by cultivation in 9 samples of nymphs, in 9 samples of adult ticks and 3 biopsies of skin. The 21 strains were identified as Borrelia resp. Borrelia burgdorferi by the indirect immunofluorescence using 3 IgG monoclonal antibodies and one IgM monoclonal antibody.

All strains (100%) reacted with H 9724, a monoclonal antibody which is specific for the genus Borrelia. Apart from one strain (out of a nymph) they also reacted with the species-specific monoclonal antibody H 5332.

By the monoclonal antibody H 3TS it was possible to identify 6 strains as the species Borrelia burgdorferi. From the total collection, only 6' (5 from ticks, 1 from a patient) reacted with the IgM-MAB.

According to the IgG-MAB reaction pattern of Barbour resulting in a formation of 3 serogroups, all strains tested were classified as follows: Type I 6 strains, Type II 14 strains and Type III 1 strain.

DISCUSSION AND CONCLUSION

This study demonstrated considerable heterogeneity of a major protein of B. burgdorferi isolated from ticks and humans. It is known that OspA protein with a molecular weight of 31 kilo dalton (kD) reacts with the MABs H 5332 and H 3TS and OspA protein with molecular weights of 32 kD to 33 kD only with H 5332 (Barbour, 1986; 1987). The result of this study with 3 MABs which are specific to B. burgdorferi

Table 3. Identification of the isolates of Borreliae by indirect immunofluorescence

Strain isolation	location	reaction with MABs			
		H 9724	H 5332	H 3TS	11 G1
I. Ticks					
24 nymph	Mecklenburg	+	+	+	-
6 nymph	Brandenburg	+	+	-	-
12 nymph	Brandenburg	+	+	-	-
24 nymph	Brandenburg	+	+	-	-
13 female	Brandenburg	+	+	+	-
15 female	Brandenburg	+	+	-	-
17 male	Brandenburg	+	+	-	-
20 male	Brandenburg	+	+	-	-
22 female	Brandenburg	+	+	-	-
24 female	Brandenburg	+	+	+	+
25 female	Brandenburg	+	+	-	-
27 male	Brandenburg	+	+	-	+
3 nymph	Thuringia	+	+	-	-
15 nymph	Thuringia	+	+	-	+
17 nymph	Thuringia	+	+	+	+
40 nymph	Thuringia	+	-	-	-
43 nymph	Thuringia	+	+	-	+
5 female	Thuringia	+	+	+	-
II. Human skin					
61 BV1	Berlin	+	+	-	-
61 BV2	Berlin	+	+	-	-
61 BV3	Berlin	+	+	+	+

Table 4. Classification of strains into serogroups (Barbour, 1986)

	Reaction with		Positive Strains
	H 5332	H 3TS	
Type I	+	+	6
Type II	+	-	14
Type III	-	-	1

have shown that an antigenic determinant in OspA of 31 kD as well as in OspA of 32 kD to 33 kD must have reacted with the IgM-MAB 11 G1. Of six strains positive with MAB 11 G1, three belong to serogroup type I and three to type II.

For confirmation, it is suggested to use Western Blot on the 6 strains which are positive to MAB 11 G1. The demonstrated heterogeneity of proteins may have relevance for epidemiology and pathogenesis of Lyme borreliosis. According to Wilske et al. (1993) the OspA serotyping system for B. burgdorferi with monoclonal antibodies may provide a possibility in differentiating virulent from avirulent B. burgdorferi strains.

REFERENCES

Barbour, A.G., 1986, A proposal for a serotyping system based upon major outer surface proteins of Borrelia burgdorferi, Lyme Borreliosis Newsletter 2: 7-9.

Barbour, A.G., Hayes, S.F., Heiland, R.A., Schrumpf M.E., and Tessier, S.L., 1986, A Borrelia-specific monoclonal antibody binds to a flagellar epitope, Infect. Immun. 52: 549-554.

Barbour, A.G., and Schrumpf, M.E., 1986, Polymorphisms of major surface proteins of Borrelia burgdorferi, Zentralbl. Bakteriol. Hyg. A 263: 83-91.

Barbour, A.G., Tessier, S.L., and Todd, W.J., 1983, Lyme disease spirochetes and Ixodid tick share a common surface antigenic determinant defined by a monoclonal antibody Infect. Immun. 41: 795-804.

Bark, S., 1986, Zur Diagnose und Verbreitung der einheimischen Zecken-Borreliose beim Tier, Vet. med. Diss., München.

Benach, J.L., Coleman, J.L., and Golightly, M.G., 1988, A murine IgM monoclonal antibody binds to an antigenic determinant in outer surface protein A, immunodominant basic protein of the Lyme disease spirochete, J. Immunol. 140: 265-272.

Schmidt, K., and Schönberg, A., 1990, Serological identification of spirochetes isolated from Ixodes ricinus (Acari: Ixodidae) in Berlin (West), in: Abstracts IV. Conference on Lyme Borreliosis, Stockholm.

Wilske, B., Preac-Mursic, V., Schierz, G., and v. Busch, K., 1986, Immunochemical and immunological analysis of European Borrelia burgdorferi strains, Zentralbl. Bakteriol. Hyg A 263: 92-102.

Wilske, B., Preac-Mursic, V., Göbel, U.B., Graf, B., Jauris, S., Soutschek, E., Schwab, E., and Zumstein, G., 1993, An OspA serotyping system for Borrelia burgdorferi based on reactivity with monoclonal antibodies and OspA sequence analysis, J. Clin. Microbiol. 31: 340-350.

THE USE OF PCR IN THE DIRECT DETECTION OF *Borrelia burgdorferi* FROM *Ixodes dammini*

Rita Sun,[1] Scott L. Barmat,[1] Scott H. McQuilkin,[1] Lance S. Risley,[2] and Robert Diaco[1]

[1]Roche Molecular Systems, Inc., Branchburg, New Jersey 08876-1760, USA

[2]William Paterson College, Department of Biology, New Jersey 07470, USA

ABSTRACT

Detection of *Borrelia burgdorferi* spirochetes, the etiologic agent of Lyme disease, from *Ixodes dammini* ticks was accomplished by a polymerase chain reaction (PCR) based DNA probe assay. This assay involves a simple tick extraction procedure, amplification of *B. burgdorferi* 16s rDNA gene sequences, followed by detection of the amplified material (amplicon) using an enzyme based microwell plate hybridization assay.

One hundred and sixty-five field-collected ticks from northern New Jersey, USA, were screened for the presence of the spirochete with both darkfield microscopy (DFM) and PCR. Using PCR, we identified 29 infected ticks resulting in 17.6% infectivity. Twenty-seven of these PCR positive specimens were also positive by DFM. One sample was positive by DFM but PCR negative. A total of three discrepants were found resulting in 98.2% concordance between DFM and PCR.

Our PCR procedure was proven to be a useful, accurate, and time saving technique for the detection of *B. burgdorferi* in ticks. The assay takes about 4 h to evaluate up to 92 ticks, and does not require the use of live ticks, as with DFM and direct fluorescent antibody (DFA).

INTRODUCTION

Lyme borreliosis is presently the most common vector-borne illness in Europe and North America. The infection is caused by the spirochete *Borrelia burgdorferi* which is primarily transmitted by *Ixodes* ticks. Frequently, the incidence of Lyme disease parallels the degree of deer tick infestation. This physically debilitating disease continues to invade new areas as the geographical distribution of deer ticks expands. It is therefore important, for environmental and public health purposes, to have an accurate measure of the spirochete infected deer tick population, especially in potentially high risk areas.

Detection of *B. burgdorferi* from *I. dammini* is commonly performed by direct fluorescent antibody (DFA) or dark field microscopy (DFM). These techniques require live ticks and labor intensive microscopic visualization. In addition, DFM is used to detect spirochetes only; confirmation of *B. burgdorferi* requires further analysis. We have developed a tick extraction procedure that utilizes live or ethanol-preserved ticks and takes about 30 seconds for initial processing. The homogenate is then subjected to PCR, followed by a colorimetric microwell plate DNA-probe hybridization assay for the detection of specific targets. Also, in order to eliminate false positives, which can arise from carry-over contamination in PCR, uracil-N-glycosylase (UNG) has been incorporated into our system.

This procedure was compared directly to DFM performed by LSR. More than 500 deer ticks were collected by LSR in northern New Jersey during mid-June to mid-July of 1992, of which 165 ticks were examined individually by DFM and preserved with 75% ethanol. Through our collaborative studies, these ticks were then subjected to the Roche PCR assay.

MATERIAL AND METHODS

Tick Collection and DFM

Questing *I. dammini* nymphs were collected with a 1 m^2 white drag cloth in suitable habitats in northern New Jersey. Sampling took place in June and July, 1992. In the laboratory, living ticks were placed in phosphate buffered saline (pH 7.4) and dissected to remove a portion of the midgut. Midgut tissues from each tick were examined by DFM and the presence or absence of spirochetes was recorded. Tick cadavers containing the remainder of midgut tissues were individually preserved in 75% ethanol.

Tick Homogenization

Individual ticks preserved in 75% ethanol were placed in 1.5 ml Eppendorf microfuge tubes with 12.5 to 15.0 mg of pre-washed glass beads and 100 μl of homogenization solution. Each tick was homogenized for 10-20 sec with a disposable dowel attached to a hand held Bio-Vortexer (BioSpec Products, P.O. Box 722, Bartlesville, OK 74005, USA). Fifty μl of the supernatant was then subjected to PCR.

PCR Set-Up

Fifty μl of extracted sample was added to 50 μl of pre-made PCR master mix containing optimal concentrations of Tris-HCl, KCl, MgCl$_2$ dNTPs, Taq polymerase, UNG, Glycerol, and the biotinylated genus-specific primer pair DD02 and DD06. The mixture was then placed in a Perkin-Elmer 9600 thermalcycler. The thermalcycling profile included an initial sterilization step at 50°C for 2 min (1 cycle); denaturation of UNG at 95°C for 5 min (1 cycle); annealing at 55°C for 25 sec and denaturation at 95°C for 25 sec (50 cycles); the PCR products were held at 72°C until ready for analysis. Immediately after removing the amplified material (amplicon) from the thermalcycler, 100 μl of alkaline denaturation solution was added to each tube.

Detection (Microwell Plate Assay)

One hundred µl of hybridization solution was added to each well of a microwell plate pre-coated with BSA-conjugated capture oligonucleotide, followed by the addition of 25 µl of denatured amplicons. The plate was incubated at 37°C for 1 hour and washed 5 times with 350 µl of wash buffer per well. The captured products were then incubated with 100 µl of avidin-HRP at 37°C for 15 min and again washed 5 times with wash buffer as described above. One hundred ul of substrate/chromogen was then added to each well, incubated 10 min at RT, stopped with 100 µl of stop solution, and the absorbance at 450 nm was measured in a microwell plate reader.

Interpretation

Samples with absorbance values (O.D. 450 nm) greater than 0.25 were interpreted as positive, whereas samples with absorbance values (O.D. 450nm) less than 0.25 were interpreted as negative.

Validation

To validate our PCR procedure, 90 tick extracts (prepared using a classical organic extraction procedure by Dr. Ira Schwartz, New York Medical College, Valhalla, NY) were examined by both the Roche Molecular Systems (RMS) and NYMC PCR assays. Both systems amplify and detect 16s rDNA gene sequences of *B. burgdorferi*, but with different primer pairs and capture probes.

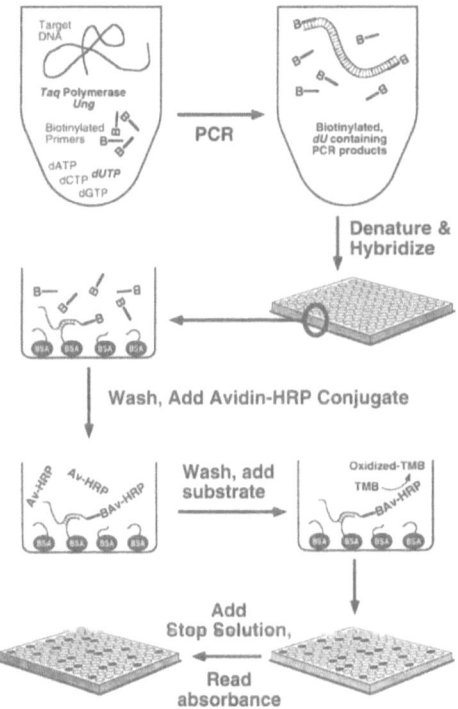

Figure 1. PCR with biotinylated primers, dUTP and uracil-N-glycosylase, and microwell plate detection of amplified products.

RESULTS AND DISCUSSION

In the validation studies, of 90 samples examined, only one discrepant was found between RMS and NYMC, leading to a 98.8% concordance. Subsequently, a direct comparison of PCR and DFM in determining the infectivity of field-collected ticks (Table 1) showed that of 165 samples tested, a total of 3 discrepants were found between PCR and DFM. Two samples were positive by PCR but were DFM negative. One sample was positive by DFM but negative by PCR. A chi-square analysis revealed that no significant difference was found between PCR and DFM (Table 2). These data indicate that the infectivity rate in certain areas of New Jersey can be as high as 50%. Although the sample size was not large enough to determine the actual rate of infection, these results certainly draw attention to public awareness of Lyme disease in northern New Jersey.

PCR methodology has wide applications for epidemiologists, researchers, veterinarians, and Public Health officials concerned about the incidence of *B. burgdorferi* infection in endemic tick populations. Our method of PCR testing is faster, and less labor intensive than DFM when testing large numbers of field-collected ticks, and is useful for testing preserved as well as live specimens (the ticks must be alive for DFM testing). Also, PCR provides for specific detection of *B. burgdorferi* infection whereas DFM can only non-specifically detect the presence of spirochetes. Additionally, by amplifying and detecting a highly conserved region of the genes encoding for *B. burgdorferi* 16s rDNA, our test is not subject to problems with regional strain variation often encountered when testing with serological-based assays. In conclusion, this study revealed the significance of *B. burgdorferi* infectivity in ticks around northern New Jersey, and that RMS PCR could be used as an easy and accurate tool for environmental studies that necessitate testing of large numbers of ticks that are observed during the peak season of the disease.

Table 1. Comparison of PCR and DFM in determining the infectivity rate of field collected ticks.

Tick Batch No.	RMS-PCR		DFM	
	+/n*	Infectivity rate (%)	+/n*	Infectivity rate (%)
LR920609	0/4	0	0/4	0
LR920618	0/6	0	0/6	0
LR920625	2/3	67	1/3	33.3
LR920629	0/10	0	0/10	0
LR920701	0/16	0	0/16	0
LR920702	3/26	11.5	3/26	11.5
LR920707	9/18	50	8/18	42.1
LR920708	0/9	0	0/9	0
LR920709	0/1	0	0/1	0
LR920710	1/15	6.7	1/15	6.7
LR920715	3/17	17.7	4/17	23.5
LR920716	0/3	0	0/3	0
LR920717	0/2	0	0/2	0
LR920721	11/35	31.4	11/35	31.4
Total	29/165	17.6	28/165	17

* number of positive for B. burgdorferi/total number of samples tested

Table 2. Chi-Square Analysis (PCR vs. DFM)

Hypothesis: PCR and DFM are independent

Rejection of hypothesis: PCR and DFM are not independent

Observed data:		PCR +	PCR -	Total
	DFM +	27	1	28
	DFM -	2	135	137
	Total	29	136	165
Expected data:		PCR +	PCR -	Total
	DFM +	4.92	23.08	28
	DFM -	24.08	112.92	137
	Total	29	136	165
Chi - Square value		99.06	21.12	**144.74**

Conclusion: PCR and DFM are significantly dependent (**144.74** $> X^2 = 3.841$, Chi-square table f=1, p=0.05)

ACKNOWLEDGMENTS

We thank Dr. Ira Schwartz for providing us with tick extracts and Dr. Steve Herman for assistance in preparing the manuscript.

BORRELIA BURGDORFERI DETECTED IN THE BLOOD, SYNOVIUM AND SKIN OF PATIENTS WITH LYME ARTHRITIS

Marika Valešová, Jana Hercogová,[1] and Dagmar Hulínská [2]

1st Department of Internal Medicine 3rd Medical School, Dermatological Department, 2nd Medical School,[1] Charles University, Prague, National Institute of Public Health[2], Prague, Czech Republic

INTRODUCTION

Originally, Lyme borreliosis (LB) was considered and described as an epidemic form of arthritis and classified among inflammatory rheumatic diseases (1). Identification of the causative agent occured in 1982 (2), when spirochetes were isolated and cultured from the midgut of ticks. At the same time, sera from patients with Lyme disease were found to contain antibodies to that organism. Subsequently, spirochetes were cultured from the blood, CSF and skin in patients with LB, confirming that the infection was, indeed, a spirochetosis (3,4).

The characterizing lesion during the acute stage of the disease is erythema chronicum migrans (ECM), which may be accompanied by fever, malaise and fatigue (5). Further signs of systemic involvement appear later:most frequently various neurological manifestations (6), cardiac disorders (7), and signs of other organ involvement (8). Often,months and even years after the initial infection, joints can be involved (1,9,10). Complaints may begin with simple arthralgia and intermittent arthritis. In some cases this may result in chronic erosive arthritis (10). Further study sugested that the involvement of the joints is similar in patients in the United States and Europe, but it seems to be a less frequent manifestation of this illness in Europe (9,11,12).

The difficulties in establishing a correct diagnosis of Lyme arthritis (LA) have inspired us to attempt to prove the presence of borrelia in synovium, blood and skin of patients suspected to be infected by these spirochetes.

The demonstration of spirochetes in the blood and synovium of Lyme arthritis patients has been successful by means of electron microscopy (13,14). In patients with ECM and ACA borreliae were cultivated also from the skin biopsy of these acute and chronic stages of LB (16,18).

PATIENTS AND METHODS

Patients

Between November 1986 and January 1993 we examined 163 patients with Lyme arthritis (LA). This diagnosis was established from patients, medical and epide- miological history, clinical findings, and results of laboratory tests. Investigated were: ESR, cell counts, rheumatoid factor test, immunoglobulins level, antinuclear antibodies and HLA-B 27 haplotype. In indicated cases rentgenograms of affected joints were done. In all patients there were repeatedly confirmed higher antiborreliar antibody titres, especially in the IgG class. The ELISA method for detection of borrelia antibodies in serum was used.

Methods

Of a selected group of 40 of 163 patients with Lyme arthritis (LA) 20 samples of synovium, 17 of blood and 5 of skin were examined for presence of B. burgdorferi (B.b.). The blood samples were high speed centrifuged and borrelie were harvested then by trapping on nickel grids. The spirochetes were partially negatively stained by immunocytochemistry (IEM). For immunological proving monoclonal antibodies followed by colloidal gold labelling were used. Bioptic tissue specimens of synovium and skin were fixed, dehydrated with alcohol, embedded in epoxy resin and sectioned. Ultra-thin sections for electron microscopy (70 um) were cut from tissue showing the presence of spiral-shaped structures in semi-thick sections. These sections were stained with uranyl acetate and lead citrate and observed in a transmission electron microscope JEOL 100 CXII.

All this examinations were provided in the National Health Institute, Prague, in the reference laboratory for Lyme Borreliosis headed by Dr. D. Hulínská.

RESULTS

These examinations were performed in 40 patients with Lyme arthritis. The presence of B. b. was electron microscopically visualized in 2 specimens of synovium, 6 of blood and 1 of skin. All attemps at the cultivation of these organism failled.

One of our positive blood samples was further specified by its reactivity with monoclonal antibodies. It was found that the antibody H 5332 reacted with some of the determinants associated with the outer membrane.

All 6 patients where borreliae were detected in blood and 2 where they were seen in synovium or 1 where it was visualised in the sample of skin in ACA, had chronic arthritis. (Table 1).

At least in this sample, chronic Lyme arthritis affected knees as monoarthritis in 2 patients. Another 2 had chronic involvement of a few joints (knees, shoulder, elbows, hips-oligoarthritis).

Three patients had polyarthritis involving various joints and showing clinicaly many similarities to rheumatoid arthritis. In one of them (NO.6 in the Table) primary diagnosis was Reiter s syndrome (as he developed arthritis, urethritis and conjunctivitis). Later on several joints were involved, especially both sternoclavicular joints, knees, talocalcaneal joints and some MCP and PIP joints of the hands. In this patients B.b. was found in synovium of the knee. At that time serological results were negative, but IgG positive titers developed soon after antibiotic therapy. In this patient arthritis was so severe that X-ray examinations showed destructive changes of the sternoclavicular joint at the one year follow-up.

In one patient who developed arthritis together with skin lession acrodermatitis chronica atrophicans (ACA) borrelia was found only in sample of the skin.

Table 1. Clinical signs and proof of Borrelia burgdorferi

PAT.	SEX	AGE	JOINTS AFFECTED	IgM	IgG	EL.MICR. Blood	EL.MICR. Synovium
1	f	60	MONOARTHRITIS knee	-	+	+	-
2	m	53	POLYARTHRITIS wrist, knee, MPC talocalcaneal	-	+	+	nd
3	m	56	OLIGOARTHRITIS knee, elbow, shoulder	-	+	+	nd
4	m	83	OLIGOARTHRITIS hips, knee	-	++	+	nd
5	f	65	MONOARTHRITIS knee	-	+	+	-
6	m	28	POLYARTHRITIS sternoclavicular wrists, MCP, MTP, talocalcaneal	-	-	-	+
7	f	62	POLYARTHRITIS knees, wrists, hip, MCP	-	+	-	+

DISCUSSION

Several years experience with Lyme borreliosis in Czech republic (ll,l3) confirms the assumed high incidence of this disease in our country. Observation from our patients indicate that involvement in the muskuloskeletal system is maybe more common that was previously recognized.

In patients who later developed Lyme arthritis, B.Burgorferi probably spreads to the joints early in the illness. However, the fact that only a small percentage of patients develop chronic arthritis suggests that host factors determine the severity and duration of the arthritis (5,12).

The proof of borreliaemia in chronic arthritis has great theoretical and practical importance. A positive microscopical result can establish the diagnosis of borreliosis without any doubt. We know little about the natural history of chronic L.A. It appears to fit in the pathogenetic framework suspected to be important in other rheumatic diseases. In rheumatoid arthritis, systemic lupus erythematosus and Reiter s syndrome, infectious agents or endogenous immune stimulation may trigger a disturbed or inappropriate immune response that leads to chronicity. Arthritis may become chronic, a proces that may involve pannus formation and cartilage erosion. This is probably the case in our patient No.6. In this patient, in whom the illness started as Reiter s syndrome, erosions could be seen in the sternoclavicular joint at one year follow-up. These findings emphasized the major role of B.b. in development of destructive arthritis.

Acrodermatits chronica atrophicans usually begins with inconspicioustly gradually developing skin inflammation and is localised mostly on the joints of the extremities, especially the knee, foot, elbow, wrist and along the extensors (l6,l7). At least in Europe ACA is probably the most common late and chronic manifestation of LB (l6,l7). ACA patients commonly complain of musculoskeletal pains. In about 2O % of ACA patients knee arthritis was found to have preceded or to occur simultaneoustly with the ACA (16) Radiographic changes with involvement of joints underneath skin lesion have been found in patients with long standing ACA (17).

In patients with ECM and ACA borreliae were cultivated also from the skin biopsy of these acute and chronic stages of LB (18).

CONCLUSIONS

Electron optical studies are believed to contribute to a better understanding of the possible pathogenesis of Lyme borreliosis. The direct visualization of spirochetes in blood, synovium and skin by IEM would certainly be an important step towards their morfologic characterisation and could help in understanding the ways in which borreliae act in infected organism. The demonstration of viable spirochetes in the blood of the patients with chronic arthritis provides evidence that this joint disorder is related to active infection, at least in some cases. We assume that also in late stages of Lyme borreliosis and arthritis spirochetemia can persist under special conditions.

REFERENCES

1. A.C. Steere, S.E. Malawista, D.R. Snydman, R.E. Shope, W.A. Andiman, M.R. Ross and F.M. Streele, Lyme arthritis. An epidemic of oligo-articular arthritis in children and adults in three Connecticut communities,Arthritis Rheum 20:7-17 (1977).
2. W. Burgdorfer, A.G. Barbour, S.F. Hayes, J.L. Benach, F. Grunwaldt and J.P. Davis, Lyme disease: A tick-born spirochetosis? Science 216:117-119 (1982).
3. Y.E.Johnston, P.H.Duray, A.C.Steere, M.Kashgarian, J.Bura, P.W.Askenase and S.E.Malawista, Lyme arthritis:Spirochetes found in synovial microangiopathic lessions.Am J Pathol 118:26-34 (1985).
4. J. Koning,A.A.Jacomina and M.Hoogkamp-Korstanje,Diagnosis of Lyme disease by demonstration of spirochetes in tissue biopsies.Zbl bakt Hyg A 263:179-188 (1986).
5. A.C.Steere, S.E.Malawista, J.A.Hardin, S.Ruddy, P.W.Askenase and W.A.Andiman,Erythema chronicum migrans and Lyme arthritis the enlarging clinical spectrum.Ann Intern Med 86:685-688 (1977).
6. A.R.Pachner and A.C.Steere,The triad of neurologic manifestation of Lyme disease:meningitis,cranial neuritis and radiculoneuritis.Neurology 35:47-53 (1985).
7. A.C.Steere, W.P.Batsford, M.A.Weinerg, J.Alexander, H.J.Berger, S.Wolfson and S.E. Malawista, Lyme carditis: cardiac abnormalities of Lyme disease. Ann Intern Med 93:8-16 (1980).
8. E.Alas, S.N.Novak, P.H.Duray and A.C.Steere, Mussle invasion by Borrelia burgorferi., Ann Intern Med 108:707-708 (1988).
9. J.P.Huax, G.Bigagnon, S.Stadtsbader, P.F.Zangerle and C.N.Deuxchaisnes. Pattern of Lyme arthritis in Europe:report of 14 cases.Ann Rheum Dis 47:164-167 (1988).
10. J.P.Lawson and A.C.Steere, Lyme arthritis: Radiological findings. Radiology 1954:39-43 (1985).
11. M.Valešová and K.Trnavsky, Joint manifestation of Lyme borreliosis in Czech patients, Z.Rheumatol 49: 192-196 (1990).
12. P.Herzer, B.Wilske, V.Preac-Mursic, G.Schierz, M.Stattenkirchner and N.Zollner, Lyme arthritis: Clinical features, seological and radiographic findings of cases in Germany, Wochenchr 64:206-215 (1986).
13. M.Valešová ,K.Trnavský, D.Hulínská, S.Alušík, J.Janoušek and J.Jirouš,Case report:Detection of Borrelia in the synovium tissue of a patient with Lyme borreliosis, detected by electron microscopy, J.Rheumatol 16:1502-1505 (1989).
14. D.Hulínská, J.Jirouš, M.Valešová and J.Hercogová, Ultrastructure of Borrelia burgorferi in tissues of patients with Lyme disease, J Basic Microbiol 29:2,73-83 (1989).
15. J.P.Lawson and D.W.Rahn, Lyme disease and radiological findings in Lyme arthritis,AJR 158:1065-1070 (1992).
16. E.Asbrink,Cutaneous manifestation of Lyme borreliosis,Scand J Infect Dis-Suppl.77:44-50 (1991).
17. A.Howmark, E.Asbrink, I.Olson, Joint and bone involvement in Swedish patients with Ixodes ricinus-borne Borrelia infection. Zbl Bakt Microbiol Hyg A 263: 275-284 (1986).
18. J.Hercogová, M.Tománková, J.Plch, J.Jirouš, D.Hulínská, D.Fresslová and P.Barták,Borrelia burgdorferi isolates from erythema migrans, Archives Dermatol Research- in press (1993).

INDEX

Adhesion molecules, 202
Animal hosts, 89, 114, 117, 128, 132, 139
    176, 187
Animal models
  cat, 187
  hamsters, 189
  AKR/N mouse, 202
  C3H mouse, 25
  C3H/HeJ mouse, 227
  SCID mouse, 25
  BALB/c mouse, 96
Antibody prevalence, 8, 13, 91, 113, 147, 287
    292
Antibodies to *Borrelia burgdorferi*
  in serum, 3, 8, 13, 23, 48, 50, 56, 62, 79, 89
    136, 158, 190, 230, 279, 285, 291,
    297, 303, 307, 318, 321
  in cerebrospinal fluid, 17
Asymptomatic infection, 3

Borrelia burgdorferi
  chemotaxonomy of, 212
  culture of, 46, 77, 95, 151, 176, 212, 287
    317
  genospecies, 26, 86, 97, 212, 217, 249, 253
    261, 270, 308
  in vitro attenuation of, 95
  lipoproteins, 261
  phenotypes, 1, 26, 182, 211, 217, 249, 261
    315
  plasmids, 95, 176, 211, 253

Cellular immune responses, 3, 169, 201
Clinical manifestations
  arthritis, 1, 9, 15, 19, 33, 49, 85, 136, 155
    327
  cardiac, 8, 21, 39, 55, 85, 136
  in children, 13,40
  dermatological, 9, 15, 18, 33, 78, 85, 136
    308
  fibromyalgia, 4, 78
  laryngeal, 33
  neurological, 8, 14, 35, 78, 85, 136, 279, 295
  ocular, 9, 14
Cytokines, 2, 171

Diagnostic criteria, 3, 8, 14, 34, 53, 280

Ecology
  in China, 217
  in France, 105
  in Sweden, 113
  in the UK, 125, 147
  in the USA, 139
Electron microscopy, 241, 327
Epidemiology
  in Australia, 75
  in Children, 13
  in Czechslovakia, 33
  in France, 105
  in Germany, 13,83
  in Italy, 7, 83, 135
  in North America, 70, 83, 90
  occupational risk of *B.burgdorferi* infection,
    61, 89, 147, 287
  in United Kingdom, 49, 55, 61, 117, 287, 291

Histocompatibility antigens, 3, 156, 171, 201,
    245
Histology, 3, 22, 46, 188, 193, 234

Idiotypes, 155

Macrophages, 169
Maternal antibody transfer, 13

Outer surface proteins, 3, 25, 78, 171, 175, 181
    263, 270

Pathogenesis, 2, 3, 22, 86, 155, 169, 201, 330
Phagocytosis, 171, 241
Polymerase chain reaction, 3, 47, 77, 96, 122
    127, 149, 176, 253, 269, 295, 303, 321
Prognosis, 1, 2, 3, 19, 42

Rheumatoid factors, 63, 64

Tick bite
  prevalence of, 8, 17, 21, 63, 72, 83, 147
  management of, 69
  transmission by, 71
Tick-borne encephalitis, 89
Tick cycle, 113, 117, 125, 131, 141
Transovarial transmission, 131

Treatment
  antibiotics, 4, 19, 23, 36, 42, 82, 299, 304
  cardiac pacing, 23, 42
  synovectomy, 4
  steroids, 4, 42
  prohylaxis, 70

Vaccine, 25
Vectors
  Amblyomma americanum, 188
  cat flea, 188
  Dermucantor Variabilis, 188
  fly, 45
  Haemaphysalis punctata, 113, 127
  I canisuga, 147

Vectors *(continued)*
  I Damini, 70, 176
  I Frontalis, 106
  I Hexagonus, 113, 127, 147
  I Pacificus, 70, 188
  I Persulcatus, 113, 188
  I Ricinus, 70, 106, 113, 117, 125, 131, 147
    265
  I Scapularis, 188
  I triangulceps, 147
  I Uriae, 113, 127, 147
  I Ventalloi, 106
  mosquito, 188